Worker Health Chartbook, 2000

U.S. Department of Health and Human Services

Public Health Service
Centers for Disease Control and Prevention
National Institute for Occupational Safety and Health

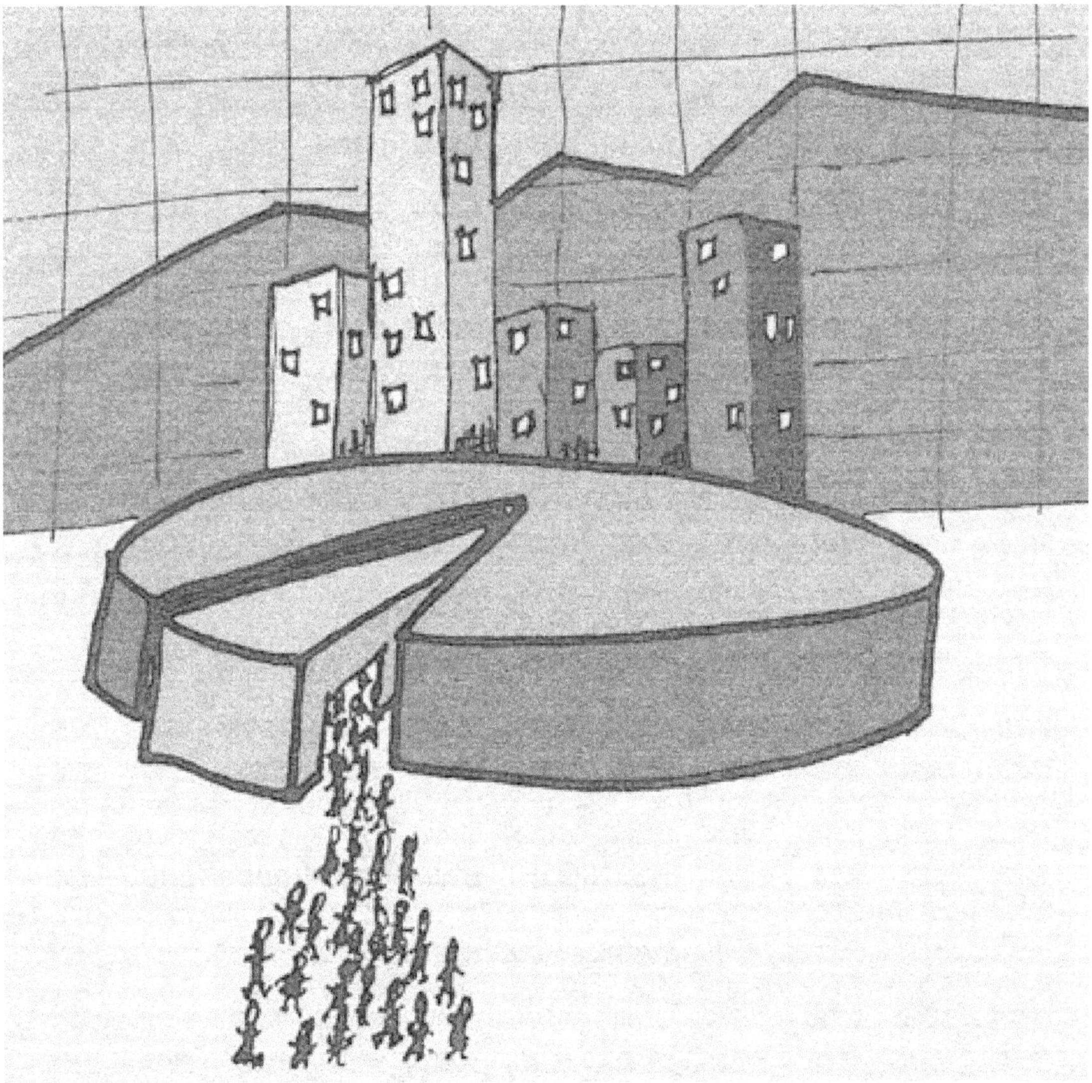

DISCLAIMER

ORDERING INFORMATION

To receive documents or more information about occupational safety and health topics, contact the National Institute for Occupational Safety and Health (NIOSH) at

**NIOSH—Publications Dissemination
4676 Columbia Parkway
Cincinnati, OH 45226-1998**

**Telephone: 1-800-35-NIOSH (1-800-356-4674)
Fax: 513-533-8573
E-mail: Pubstaff@cdc.gov**

or visit the NIOSH Web site at

www.cdc.gov/niosh

This document is in the public domain and may be freely copied or reprinted.

Disclaimer: Mention of any company or product does not constitute endorsement by NIOSH

DHHS (NIOSH) Publication Number 2000-127

September 2000

FOREWORD

Surveillance is the cornerstone of prevention: It helps us identify new and emerging problems, track and monitor issues over time, target and evaluate the effectiveness of intervention efforts, and anticipate future needs and concerns. Those who have long struggled with these issues in the occupational setting will share my enthusiasm for this first edition of the *Worker Health Chartbook*. I am grateful to the authors and contributors for accomplishing what has not been accomplished before—bringing together the patchwork of systems that monitor occupational illness and injury into one comprehensive and *comprehensible* guide.

One of the primary goals in compiling the chartbook was to create a resource that could be used by anyone interested in workplace safety and health, including occupational safety and health practitioners, legislators and policy makers, health care providers, educators, researchers, and workers and their employers. In an attempt to reach the widest possible audience, we have made the chartbook available in printed and electronic form.

Several Federal agencies worked together to organize the surveillance data sources required to produce this document. This is an important step toward identifying and filling significant gaps in occupational illness and injury information. The success of this initial effort has provided a framework for increased surveillance coordination between NIOSH and our partners in the future.

The *Worker Health Chartbook* serves NIOSH and the occupational safety and health community well by placing surveillance in the hands of those who work to prevent occupational injuries and illnesses. The forethought and collaborative spirit that made all of this possible are commendable and bode well for future efforts to integrate Federal, State, and private-sector surveillance information.

Linda Rosenstock

Linda Rosenstock, M.D., M.P.H.
Director, National Institute for
Occupational Safety and Health

Executive Summary

Understanding and preventing occupational injuries and illnesses require focused efforts to identify, quantify, and track both health outcomes and their associated workplace conditions. Occupational safety and health surveillance activities provide the ongoing and systematic collection, analysis, interpretation, and dissemination of data needed for prevention. Current occupational safety and health surveillance data reveal the staggering human and economic losses associated with occupational injuries and illnesses. Much work remains to be done to reduce those losses, despite overall decreases in occupational injuries and illnesses in recent years.

Our ability to survey and assess the state of occupational safety and health has improved over time. However, occupational safety and health surveillance data remain fragmented—collected for different purposes by different organizations using different definitions. We continue to have substantial gaps in surveillance information. Each surveillance system has limitations, particularly those that attempt to quantify occupational illness. Nonetheless, the data provide useful information for targeting and evaluating prevention efforts.

To make these data more accessible, the National Institute for Occupational Safety and Health (NIOSH) has assembled this chartbook, which provides occupational safety and health surveillance information from different sources in a single volume. This initial work focuses on injury and illness outcomes rather than on exposures or hazards. Included are contributions from several Federal agencies. Little information is included on public-sector employees or from State-based surveillance systems. Future editions of the chartbook will target additional data sources to provide a more comprehensive picture of occupational injury and illness for the U.S. workforce.

Trends Over Time

Recent overall decreases in occupational injuries and illnesses are apparent in the incidence rates for total recordable cases of injuries and illnesses in private industry reported by the U.S. Department of Labor in the Survey of Occupational Injuries and Illnesses (SOII). From 1973 to 1997, this rate declined from 11.0 to 7.1 cases per 100 full-time workers. The greatest

Executive Summary

change occurred among cases without lost workdays,* which decreased from 7.5 to 3.8 cases per 100 full-time workers over the same period. For 1988–1997, the rate of cases with days away from work declined 40%, but there was a 140% increase in the rate of cases with restricted work activity only.

Occupational injury fatality rates recorded by NIOSH in the National Traumatic Occupational Fatalities Surveillance System (NTOF) decreased substantially (43%) between 1980 and 1995, from 7.5 to 4.3 deaths per 100,000 workers. Injury fatality rates recorded by the U.S. Department of Labor in the Census of Fatal Occupational Injuries (CFOI) declined by 7% from 1992 to 1997.

Losses attributable to occupational illness over time are more difficult to describe. Although efforts have been made to estimate the burden of occupational disease in the United States, no surveillance system describes the magnitude of fatal occupational illnesses other than the pneumoconioses (dust diseases of the lung). These illnesses can be described because they are attributable entirely to occupation. Since 1968, more than 113,000 deaths have occurred with pneumoconiosis diagnosed as the underlying or contributing cause—mostly coal workers' pneumoconiosis (CWP). Deaths with CWP have decreased in recent years, whereas deaths with asbestosis increased from 1968 to 1996 (from fewer than 100 to nearly 1,200).

Recent Data

Fatal Occupational Injuries

About 17 workers were fatally injured on the job each day during 1997. Of the 6,238 fatal occupational injuries that year, 42% (2,605) were associated with transportation, excluding incidents that occurred while traveling to or from work. Most motor-vehicle-related fatalities (nearly 1,400) resulted from highway crashes. Homicides were the second leading cause of death, accounting for 14% of the total. The leading causes of death varied by sex, with motor vehicles being the leading cause for men and homicide the leading cause for women. Workers aged 65 and older had the highest rates of occupational injury death. Workplaces with 1 to 10 workers had the highest fatality rate (8.6 deaths per 100,000 workers), and workplaces with 100 or more workers had the lowest fatality rate (2 deaths per 100,000 workers). The highest numbers of fatalities occurred in construction, transportation

---

*Lost-workday cases include cases with days away from work and cases with restricted work activity only (i.e., cases in which workers report to their jobs for limited duty).

Executive Summary

and public utilities, and agriculture, forestry, and fishing industries. The highest fatality rates occurred in mining, construction, and agriculture, forestry, and fishing. The fatality rate in mining was more than five times the national average for all industries.

Fatal Occupational Illnesses

Deaths from diseases other than the pneumoconioses are difficult to attribute to the workplace for several reasons. For example, many diseases appear the same with or without occupational exposures; and some have latency periods of many years between exposure and disease development. Furthermore, health care professionals may not identify or consider occupational risk factors when making a diagnosis. Statistically elevated death rates for several diseases have been observed in a variety of occupations, but the degree to which these elevated rates can be directly associated with the workplace is not clear. However, these studies help set priorities for intervention and prevention as well as for future investigation. For example, death rates for persons with pneumoconiosis as an underlying or contributing cause varied by occupation and type of pneumoconiosis. Mining machine operators had high mortality rates from CWP and other/unspecified pneumoconiosis, and insulation workers and related occupations had high mortality rates from asbestosis. Various metalworking, plastic processing, and mining occupations had high mortality rates from silicosis, and textile machine operators and repairers had high mortality rates from byssinosis.

Nonfatal Injuries

Approximately 5.7 million injuries were reported in SOII in 1997. Those injuries represent 93% of the 6.1 million injuries and illnesses documented by employer records in the private sector. The nonfatal injury rate declined steadily in the 1990s. Agriculture, construction, manufacturing, and transportation reported rates above the average of 6.6 per 100 full-time workers for all industries. Sprains, strains, and tears accounted for a disproportionately large share of cases with days away from work (nearly 800,000 cases in 1997). Nearly half of those cases involved the back. Overexertion accounted for more than 60% of back injuries.

According to the National Electronic Injury Surveillance System (NEISS), occupational injuries treated in hospital emergency departments numbered 3.6 million in 1998. Rates for those injuries were highest among men and workers under age 25. Lacerations, punctures, sprains and strains, contusions, abrasions, and hematomas accounted for 70% of all injuries treated in emergency departments.

Executive Summary

Nonfatal Illnesses

Nearly 430,000 nonfatal occupational illnesses were recorded in SOII in 1997. About 60% of those illnesses occurred in the manufacturing sector. The illness incidence rate for 1997 was 49.8 cases per 10,000 full-time workers. Illness incidence rates varied by industry, with the highest rate occurring in manufacturing. The rates in private industry increased with establishment size, with the highest rate occurring in establishments employing 1,000 or more workers.

Disorders related to repeated trauma (including carpal tunnel syndrome [CTS], tendinitis, and noise-induced hearing loss) accounted for 64% of the occupational illnesses recorded in SOII in 1997. CTS accounted for more than 29,000 cases with days away from work in 1997. Half of the CTS cases required 25 or more days away from work. Most noise-induced hearing loss cases with days away from work occurred in manufacturing.

Skin diseases or disorders represented 13% (approximately 58,000 cases) of work-related illnesses recorded in SOII in 1997. Dermatitis, a subcategory of skin diseases or disorders, resulted in more than 6,500 cases with time away from work. Half of these cases required 3 or more days away from work.

SOII relies on employer records to identify work-related injuries and illnesses. Illnesses reported to SOII are those most easily and directly related to workplace activity (e.g., contact dermatitis). Diseases that develop over a long period (e.g., cancers) or that have workplace associations that are not immediately obvious are overwhelmingly underrecorded in SOII. Consequently, other approaches and data sources have been developed to track occupational illnesses in a more active way. For example, the Sentinel Event Notification System for Occupational Risks (SENSOR) establishes a variety of simultaneous data sources to increase the chances of identifying a work-related illness in State surveillance systems. The California SENSOR program has specifically targeted surveillance of occupational CTS. Of the CTS cases identified in that program through physician first reports filed with the State compensation system in 1998, 30% occurred in the services industry and 17% occurred in manufacturing. Currently, the Michigan SENSOR program monitors noise-induced hearing loss. Manufacturing accounted for 51% of the noise-induced hearing loss cases reported by clinicians in 1998. Seven States have had active SENSOR programs for silicosis surveillance. From 1993 to 1995, 75% of silicosis cases occurred in manufacturing. In addition, four States have had active SENSOR programs for occupational asthma surveillance. The industry divisions

Executive Summary

accounting for the most cases from 1993 to 1995 were manufacturing (42%) and services (31%).

Other public and private programs describe toxic exposures, pesticide poisonings, X-rays of working underground coal miners, infections in health care workers, and self-reported respiratory diseases among non-smokers by industry. For example, the Adult Blood Lead Epidemiology and Surveillance Program (ABLES) monitors elevated blood lead levels (BLLs) in persons aged 16 and older. In 1998, a total of 10,501 adults in 25 States had high BLLs (25 µg/dL or greater).

Conclusions

The data provided in this chartbook indicate encouraging decreases in the frequency of some occupational fatalities, injuries, and illnesses. Surveillance has helped identify new and emerging problems and trends such as occupational musculoskeletal disorders and asthma. Although our ability to monitor these outcomes has improved over time, this chartbook illustrates the continued fragmentation of occupational health surveillance systems as well as the paucity (or even total absence) of data for certain occupational disorders and groups. The data suggest a compelling need to improve, expand, and coordinate occupational safety and health surveillance activities to develop and augment the data needed to guide illness and injury prevention efforts. Working with government and nongovernment partners, NIOSH will continue efforts to enhance occupational health surveillance in the coming years.

CONTENTS

FOREWORD iii
EXECUTIVE SUMMARY iv
ABBREVIATIONS xii
ACKNOWLEDGMENTS xv

1 INTRODUCTION 3
Chartbook Organization and Data Systems 4
Demographics 6
Overview of Occupational Injuries and Illnesses 12
    The Burden of Occupational Injuries and Illnesses 12
        Fatal Injury 12
        Fatal Illness 14
        Nonfatal Injury and Illness Combined 15
    Characteristics of Workers and of Injuries and Illnesses Involving Days away from Work ...21
        Workers 21
        Injuries and Illnesses 23

2 FATAL INJURY 29
The Burden of Fatal Occupational Injuries 29
    Fatal Injuries by Age and Race 31
    Fatal Injuries by Leading Cause 32
    Fatal Injuries by Industry and Occupation 34
    Fatal Injuries by State 37
    Fatal Injuries by Establishment Size 40
Special Topics in Fatal Occupational Injury 41
    Fatal Injuries among Truck Drivers 41
    Homicides 43
    Fatal Falls 45
    Fire Fighter Fatalities 47

3 FATAL ILLNESS 53
Pneumoconiosis 53
    Pneumoconiosis Deaths by State 56
    Pneumoconiosis Deaths by Sex and Race 56
    Pneumoconiosis Deaths by Occupation 60
Malignant Pleural Neoplasm 63
Hypersensitivity Pneumonitis 66
PMRs for Selected Occupations and Causes of Death 69

CONTENTS

4 NONFATAL INJURY 89
    Nonfatal Occupational Injuries by Industry and Cases with Lost Workdays 93
    Characteristics of Injury Cases with Days away from Work 97
        Sprain, Strain, and Tear Cases with Days away from Work, 1997 98
        Back, Spine, or Spinal Cord Cases with Days away from Work, 1997 100
        Bruise and Contusion Cases with Days away from Work, 1997 102
        Cut and Laceration Cases with Days away from Work, 1997 104
        Fracture Cases with Days away from Work, 1997 106
        Heat Burn and Scald Cases with Days away from Work, 1997 108
        Amputation Cases with Days away from Work, 1997 110

5 NONFATAL ILLNESS 115
    Incidence of Occupational Illness in Private Industry 116
    Repeated Trauma Disorders 118
        Carpal Tunnel Syndrome 119
            Cases Recorded by SOII 119
            Cases Identified by SENSOR 121
        Tendinitis 123
    Noise-Induced Hearing Loss 124
    Skin Diseases or Disorders 127
    Respiratory Disorders 129
        Dust Diseases of the Lungs 129
            Coal Workers' Pneumoconiosis 130
            Silicosis 132
        Respiratory Disorders Attributable to Toxic Agents 134
        Asthma and Chronic Obstructive Pulmonary Disease 135
            NHANES III 135
            SENSOR 138
    Poisoning and Toxicity 141
        Poisoning 141
        Lead Toxicity 142
        Pesticide and Insecticide Toxicity 144
    Infections in Health Care Workers 148
        Consequences of Bloodborne Exposures 151
            Hepatitis B Virus 151
            Hepatitis C Virus 152
            Human Immunodeficiency Virus 153
        Tuberculosis (TB) 154
    Physical Agents 155
    Anxiety, Stress, and Neurotic Disorders 156
    All Other Nonfatal Occupational Illnesses 158

6 FOCUS ON MINING 161
    Fatal Injuries 161
        Historical Perspective 161
        Fatal Injuries during 1988–1997 162
    Lost-Workday Injuries 172

Contents

REFERENCES 183

APPENDIX A: SURVEILLANCE SYSTEM DESCRIPTIONS..191
   Overview..191
   Bureau of Labor Statistics (BLS) of the U.S. Department of Labor 191
      Current Population Survey (CPS) 191
      Survey of Occupational Injuries and Illnesses (SOII) 197
      Census of Fatal Occupational Injuries (CFOI)..199
   Centers For Disease Control and Prevention (CDC), U.S. Department of
   Health and Human Services 201
      National Institute for Occupational Safety and Health (NIOSH) 201
         National Electronic Injury Surveillance System (NEISS)...202
         National Occupational Mortality Surveillance System (NOMS) 203
         National Surveillance System for Pneumoconiosis Mortality (NSSPM).....................205
         Coal Workers' X-Ray Surveillance Program (CWXSP) 205
         National Traumatic Occupational Fatalities Surveillance System (NTOF)................207
         Sentinel Event Notification System for Occupational Risk (SENSOR) 208
         Adult Blood Lead Epidemiology and Surveillance Program (ABLES) 210
         Mining Injury and Employment Statistics...211
      National Center for Health Statistics (NCHS) 215
         National Hospital Ambulatory Medical Care Survey (NHAMCS)............................215
         National Health and Nutrition Examination Survey (NHANES) 216
         Multiple-Cause-of-Death Data..217
      National Center for Infectious Diseases (NCID) 217
         National Surveillance System for Hospital Health Care Workers (NaSH)..................218
         Sentinel Counties Study of Acute Viral Hepatitis 219
         Viral Hepatitis Surveillance Program (VHSP) 220
      National Center for HIV, STD, and TB Prevention (NCHSTP) 220
         Surveillance of Health Care Workers with HIV/AIDS...221
         Surveillance for Tuberculosis Infection in Health Care Workers (staffTRAK–TB) 222
   References Cited..223
   Bibliography..227
      Bureau of Labor Statistics 227
      National Electronic Injury Surveillance System (NEISS)..228
      National Occupational Mortality Surveillance System (NOMS) 229
      Adult Blood Lead Epidemiology and Surveillance Program..230
      Multiple-Cause-of-Death Data 230

**APPENDIX B: DESCRIPTION OF INDUSTRY AND
OCCUPATION CODING SYSTEMS**...233
   Overview..233
   North American Industry Classification System/Standard Industrial Classification (NAICS/SIC) 233
   Standard Occupational Classification (SOC) 237
   Bureau of the Census...240
   References Cited..242

GLOSSARY 245

Abbreviations

AAPCC	American Association of Poison Control Centers
ABLES	Adult Blood Lead Epidemiology and Surveillance Program
AIDS	acquired immune deficiency syndrome
BLL	blood lead level
BLS	Bureau of Labor Statistics
CDC	Centers for Disease Control and Prevention
CDPR	California Department of Pesticide Regulation
CFOI	Census of Fatal Occupational Injuries
CFR	*Code of Federal Regulations*
CI	confidence interval
COPD	chronic obstructive pulmonary disease
CPS	Current Population Survey
CPSC	Consumer Product Safety Commission
CSTE	Council of State and Territorial Epidemiologists
CTS	carpal tunnel syndrome
CWP	coal workers' pneumoconiosis
CWXSP	Coal Workers' X-Ray Surveillance Program
dL	deciliter(s)
DHHS	U.S. Department of Health and Human Services
DNA	deoxyribonucleic acid
ECPC	U.S. Economic Classification Policy Committee
EPA	U.S. Environmental Protection Agency
FACE	Fatality Assessment and Control Evaluation
HARS	HIV/AIDS Reporting System
HIV	human immunodeficiency virus
ICD–8	International Classification of Diseases, Eighth Revision (World Health Organization)
ICD–9	International Classification of Diseases, Ninth Revision (World Health Organization)
IHD	ischemic heart disease
ILO	International Labour Organization or Office
ITCIC	Interagency Technical Committee on Industrial Classification
MSHA	Mine Safety and Health Administration

ABBREVIATIONS

NAICS .... North American Industry Classification System
NAMCS .... National Ambulatory Medical Care Survey
NaSH .... National Surveillance System for Hospital Health Care Workers
NCHS .... National Center for Health Statistics
NCHSTP .... National Center for HIV, STD, and TB Prevention
NCI .... National Cancer Institute
NCID .... National Center for Infectious Diseases
n.e.c. .... not elsewhere classified
NEISS .... National Electronic Injury Surveillance System
NFPA .... National Fire Protection Association
NHAMCS .... National Hospital Ambulatory Medical Care Survey
NHANES .... National Health and Nutrition Examination Survey
NHANES III .... Third National Health and Nutrition Examination Survey
NOICC .... National Occupational Information Coordinating Committee
NIOSH .... National Institute for Occupational Safety and Health
NNIS .... National Nosocomial Infections Surveillance System
NOMS .... National Occupational Mortality Surveillance System
n.o.s. .... not otherwise specified
NTOF .... National Traumatic Occupational Fatalities Surveillance System
NSSPM .... National Surveillance System for Pneumoconiosis Mortality
OMB .... Office of Management and Budget
OSHA .... Occupational Safety and Health Administration
PEST .... Pesticide Exposure Surveillance in Texas Program
PMR .... proportionate mortality ratio
RADS .... reactive airways dysfunction syndrome
SENSOR .... Sentinel Event Notification System for Occupational Risk
SIC .... standard industrial classification
SOC .... standard occupational classification
SMR .... standardized mortality ratio
SOII .... Survey of Occupational Injuries and Illnesses
staffTRAK–TB ... Surveillance for Tuberculosis Infection in Health Care Workers
STD .... sexually transmitted disease
SUDAAN .... Survey Data Analysis

ABBREVIATIONS

TB .... tuberculosis
TESS .... Toxic Exposure Surveillance System
VHSP .... Viral Hepatitis Surveillance Program
WHO .... World Health Organization
WoRLD .... *Work-Related Lung Disease Surveillance Report 1999*
µg .... microgram(s)

Acknowledgments

This document was prepared by the staff of the National Institute for Occupational Safety and Health (NIOSH). All contributors are affiliated with NIOSH unless otherwise indicated. We extend special thanks to our technical reviewers for their constructive comments and suggestions.

Editors
   Roger R. Rosa, Ph.D.
   Michael J. Hodgson, M.D.
   R. Alan Lunsford, Ph.D.
   E. Lynn Jenkins, M.A.
   Kathleen Rest, Ph.D.

Document Design
   David Peabody, Synectics for Management Decisions, Inc.
   Suzanne Meadows Hogan, M.A.
   Chris Cromwell, Synectics for Management Decisions, Inc.
   Toni Garrison, Synectics for Management Decisions, Inc.

Contributors
   Toni Alterman, Ph.D.
   Ricki Althouse, M.S.
   Ki Moon Bang, Ph.D.
   Margot Barnett, M.S., Strategic Options Consulting
   Jerome M. Blondell, Ph.D., M.P.H., U.S. Environmental
      Protection Agency
   Winifred L. Boal, Ph.D.
   Richard Braddee, M.S.
   Carol Burnett, M.S.
   Geoffery Calvert, M.D., M.P.H.
   Scott Campbell, National Center for Infectious Diseases
   Denise M. Cardo, M.D., National Center for Infectious Diseases
   Robert Castellan, M.D., M.P.H.
   Virgil Casini
   Janice Devine, M.S., Bureau of Labor Statistics
   Ann N. Do, M.D., National Center for HIV, STD, and
      TB Prevention
   Barbara Fotta, M.S.
   Jennifer Flattery, M.P.H., California Department of
      Health Services
   Janie L. Gittleman, Ph.D., M.R.P.
   Robert Harrison, M.D., M.P.H., California Department of
      Health Services
   Dan Hecker, M.S., Bureau of Labor Statistics

ACKNOWLEDGMENTS

Steven Hipple, M.S., Bureau of Labor Statistics
Janice Huy, M.S.
Larry Jackson, Ph.D.
Angela Booth Jones, M.S.
Larry Layne, M.A.
Suzanne Marsh
Elizabeth Marshall, M.S., M.P.H., New York State Department of Health
Linda McCaig, M.P.H., National Center for Health Statistics
Louise N. Mehler, M.D., California Environmental Protection Agency
Teri Palermo
Adelisa L. Panlilio, M.D., M.P.H., National Center for Infectious Diseases
Audrey Podlesny
Mary Jo Reilly, M.S., Michigan Department of Public Health, Michigan State University
Robert Roscoe, M.S.
Kenneth Sacks, Ph.D., M.B.A.
Lee Sanderson, Ph.D.
John Sestito, J.D.
Jackilen Shannon, Ph.D., Texas Department of Health
Rosemary Sokas, M.D., M.O.H.
Lisa Thomas
Catherine Thomsen, M.P.H., Oregon Health Division
James Walker, Ph.D.
John M. Wood, M.S.
William Weber, M.S., Bureau of Labor Statistics
Ian T. Williams, Ph.D., M.S., National Center for Infectious Diseases

Technical Reviewers
Heinz Ahlers, J.D.
Letitia Davis, Sc.D., Massachusetts Department of Health
Rick Ehrenberg, M.D., M.P.H.
William Eschenbacher, M.D.
Larry Grayson, Ph.D.
William Halperin, M.D., M.P.H.
Joseph Hurrell, Ph.D.
Jeff Kohler, Ph.D.
Gail McConnell, V.M.D.
Kenneth Rosenman, M.D., Michigan State University
Mitchell Singal, M.D., M.P.H.
Nancy Stout, Ed.D.
Gregory Wagner, M.D.
Carol Wilkinson, M.D., IBM Corporation

Acknowledgments

Editorial and Production Support
- Vanessa Becks
- Shirley Carr
- Susan Feldmann
- Lawrence Foster
- Anne C. Hamilton
- Marie Haring-Sweeney, Ph.D.
- Susan Kaelin
- Barbara Landreth
- Charlene Maloney
- Lucy Schoolfield
- Michelle Thompson
- Kristina Wasmund
- Jane Weber, M.Ed.
- Wendy Wippel, M.S.

1 Introduction

1 Introduction

More than 131 million people are employed in the United States. As we enter the new millennium, the U.S. workforce will be older and more diverse and will continue to shift from traditional heavy industry to services. Alternative work arrangements such as job sharing, part-time scheduling, and temporary or contingent work will become more common in response to rapid technological and economic changes. These changes will present new challenges to assuring the safety and health of Americans in the workplace.

Preventing occupational injuries and illnesses depends on our ability to quantify and track them. Through occupational safety and health surveillance, we can provide ongoing and systematic collection, analysis, interpretation, and dissemination of data for the purposes of prevention. Surveillance increases the effectiveness of prevention activities by targeting them to industries, workplaces, and occupations that have the greatest needs. Surveillance also expands knowledge about which prevention programs are effective.

Current occupational safety and health surveillance efforts indicate that 6.1 million injuries and illnesses were recorded in 1997 in private-sector establishments in the United States. During the same year, 6,238 workers died of occupational injuries. Since 1968, more than 113,000 worker deaths have been attributed to pneumoconioses (dust diseases of the lung). This number represents only a small portion of the total deaths attributable to occupational lung disease. Workplace injuries and illnesses also take a toll on workers' ability to earn a living and on economic productivity overall. From 1973 to 1997, the number of lost-workday cases* rose from 1.9 million to 2.9 million per year. During this period, there was a decrease in the number and rate of cases with actual days away from work and an increase in the number and rate of cases with restricted work activity only. Taken together, the surveillance data indicate that the human and economic losses associated with occupational injuries and illnesses are staggering. Much work remains to reduce those losses, but some improvements have been observed in recent years.

---

*Lost-workday cases include cases with days away from work and cases with restricted work activity only (i.e., cases in which workers report to their jobs for limited duty).

INTRODUCTION

Our ability to survey and assess the state of occupational safety and health has improved over time. Publications such as the *Work-Related Lung Disease Surveillance Report, 1999* (WoRLD) [NIOSH 1999], fatality summaries from the National Traumatic Occupational Fatalities Surveillance System (NTOF) [NTOF 1999], and the annual series of occupational safety and health data publications from the Bureau of Labor Statistics (BLS) provide periodic updates on occupational injuries and illnesses. Despite these efforts, occupational safety and health surveillance data are fragmented and have substantial gaps, making it difficult to characterize the overall health of working America. To make existing data more accessible, the National Institute for Occupational Safety and Health (NIOSH) has assembled this chartbook to provide a variety of occupational safety and health surveillance information in a single volume. The book includes contributions from several Federal agencies that collect data relevant to occupational safety and health. These agencies include BLS; the National Center for Health Statistics (NCHS); the National Center for Infectious Diseases (NCID); the National Center for HIV, STD, and TB Prevention (NCHSTP); the National Cancer Institute (NCI); the U.S. Environmental Protection Agency (EPA); the Mine Safety and Health Administration (MSHA); and the Consumer Product Safety Commission (CPSC). We hope this chartbook will be useful to anyone interested in workplace safety and health, including researchers, legislators and policy makers, health care professionals, educators, and occupational safety and health practitioners in labor, management, and consulting environments. NIOSH and the contributing agencies invite everyone to use the information provided here to see where we have been, where we are, and where we might go toward our common goal of protecting the safety and health of American workers.

Chartbook Organization and Data Systems

This is the first edition of the *Worker Health Chartbook*, which will be an ongoing effort to assemble and integrate occupational safety and health surveillance information. The document is organized into sections on occupational fatal injuries, fatal illnesses, nonfatal injuries, and nonfatal illnesses. The *Focus on Mining* section describes safety and health in the mining industry. Topics in this section will change in future editions to summarize available data about other high-risk industries (e.g., agriculture and construction) or special populations (e.g., women and working adolescents).

INTRODUCTION

The information presented here was obtained from several data systems maintained by a variety of Federal agencies. Each system has strengths and limitations. The systems may use different definitions, recording approaches, and sample populations, so they may produce different results for the same topic. For example, some of the reported values represent estimates based on statistical samples of populations, whereas others represent actual counts of cases. Some systems concentrate on workers only, and others report on all U.S. residents aged 15 and older. In addition, most data are restricted to private-sector workers. Public-sector workers (i.e., Federal, State, and municipal workers) may be included in some data systems such as fatality surveys and case-based surveillance systems, but the coding may not permit exact numbers to be determined. Consequently, public workers, a large segment of the U.S. workforce, are not described adequately. Appendix A and the Glossary describe the surveillance systems and terms used throughout this book.

Data on fatal injuries were obtained from NTOF and from the Census of Fatal Occupational Injuries (CFOI). These systems record rates and numbers of fatal injuries by industry and occupation, changes over time, and rates for high-risk industries and occupations. Information about fatal occupational illness was taken from the National Occupational Mortality Surveillance System (NOMS), the National Surveillance System for Pneumoconiosis Mortality (NSSPM), and the Vital Statistics, Mortality, and Multiple-Cause-of-Death data files from the National Center for Health Statistics (NCHS). NOMS presents an overview of the risk of death from several chronic diseases, whereas NSSPM is restricted to the pneumoconioses. The NCHS multiple-cause-of-death data provide information about mortality due to malignant neoplasms of the pleura and hypersensitivity pneumonitis. Data on nonfatal occupational injuries were obtained from the BLS annual Survey of Occupational Injuries and Illnesses (SOII), the National Electronic Injury Surveillance System (NEISS), and the National Hospital Ambulatory Medical Care Survey (NHAMCS). Data on nonfatal occupational illnesses were taken from SOII, the Sentinel Event Notification System for Occupational Risk (SENSOR), the California Department of Pesticide Regulation (CDPR), the Adult Blood Lead Epidemiology Surveillance System (ABLES), the Coal Workers' X-Ray Surveillance Program (CWXSP), the third National Health and Nutrition Examination Survey (NHANES III), the National Surveillance System for Hospital Health Care Workers (NaSH), the Toxic Exposure Surveillance System (TESS), the Viral Hepatitis Surveillance Program (VHSP), the Sentinel Counties Study of Acute Viral Hepatitis, Surveillance for

Introduction

Tuberculosis Infection in Health Care Workers *(staffTRAK–TB)*, and the Centers for Disease Control and Prevention (CDC) national HIV/AIDS Reporting System (HARS).

Demographics

In 1998, approximately 131 million people were employed in the United States; 54% were male. By race/ethnicity, 84% of these workers were white, 11% were black, and 10% were Hispanic (of any race). These distributions vary by industry division (Table 1–1) and occupation (Table 1–2). In the 15-year period from 1983 to 1998, the largest increases in employment occurred in the services industry division (Figure 1–1) and in the executive, administrative, and managerial occupations and the professional specialty occupations (Figure 1–2). Future employment trends are projected from 1983 to 1998 changes. By 2008, 20 occupations (of 500 listed by the BLS) are projected to gain the largest number of jobs—about 8 million, or 39% of growth (Figure 1–3). Four occupations from this list also appear on the list of fastest growing occupations (Figure 1–4): computer engineers, computer support specialists, computer systems analysts, and personal care and home health aides.

The distribution of the labor force is projected to change by age, with workers aged 45 and older increasing from 33% to 40% of the workforce, and those aged 25 to 44 decreasing from 51% to 44% (Figure 1–5). From 1998 to 2008, the number of women in the labor force will increase by 15% compared with 10% for men and 12% overall. Women's total share of the workforce is projected to increase from 46% in 1998 to 48% in 2008. The share of labor force by race/ethnicity also is projected to shift, with decreases for whites, little or no change for blacks, and increases for Hispanics (of any race), Asians, and other races (Figure 1–6).

INTRODUCTION

Table 1–1. Persons aged 16 and older employed in the United States in 1998, by major industry, sex, and race/ethnicity

Industry division	Number employed (thousands)	% male	% female	% white	% black	% Hispanic*
All industries	131,464	53.8	46.2	84.4	11.1	10.1
Agriculture	3,378	75.6	24.4	93.5	4.1	22.0
Mining	620	86.3	13.7	93.4	4.5	8.9
Construction	8,518	90.6	9.4	91.1	6.4	12.7
Manufacturing—durable goods	12,566	72.7	27.3	85.9	9.1	9.6
Manufacturing—nondurable goods	8,168	61.2	38.8	82.1	12.8	13.7
Transportation and public utilities	9,307	70.9	29.1	80.8	15.1	9.5
Wholesale and retail trade	27,203	52.8	47.2	85.2	9.6	11.4
Finance, insurance, and real estate	8,605	41.3	58.7	85.4	10.4	7.1
Services	47,212	37.9	62.1	83.0	12.2	8.7
Public administration	5,887	56.4	43.6	79.5	16.5	6.6

Source: BLS [1999].
*Of any race.

Table 1–2. Persons aged 16 and older employed in the United States in 1998, by major occupation, sex, and race/ethnicity

Occupation	Number employed (thousands)	% male	% female	% white	% black	% Hispanic*
All occupations	131,463	53.8	46.2	84.4	11.1	10.1
Executive, administrative, and managerial	19,054	55.6	44.4	88.7	7.2	5.4
Professional specialty	19,883	46.7	53.3	86.3	7.9	4.6
Technicians and related support	4,261	46.4	53.6	83.5	10.4	6.6
Sales occupations	15,850	49.7	50.3	86.5	8.9	7.9
Administrative support, including clerical	18,410	21.4	78.6	82.7	13.1	9.0
Service occupations	17,836	40.5	59.5	77.4	17.6	15.0
Precision production, craft, and repair	14,411	91.7	8.3	88.3	8.0	12.4
Operators, fabricators, and laborers	18,256	75.4	24.6	80.0	15.7	16.0
Farming, forestry, and fishing	3,502	81.0	19.1	92.3	4.9	22.6

Source: BLS [1999].
*Of any race.

INTRODUCTION

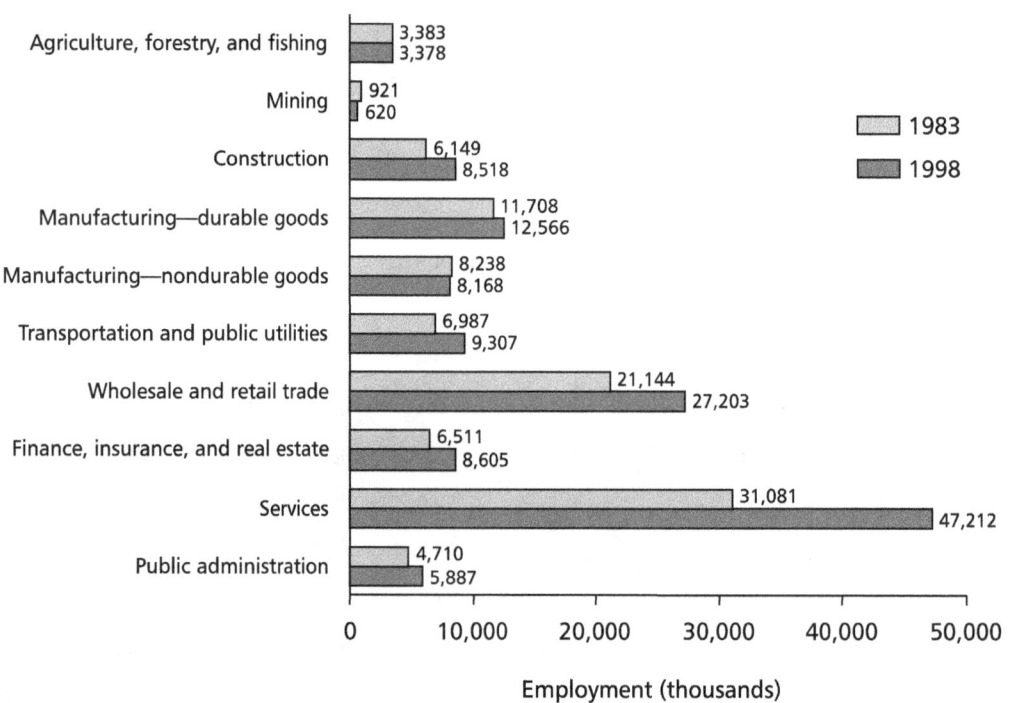

Figure 1–1. Employment by major industry division, 1983 and 1998. (Source: BLS [1999].)

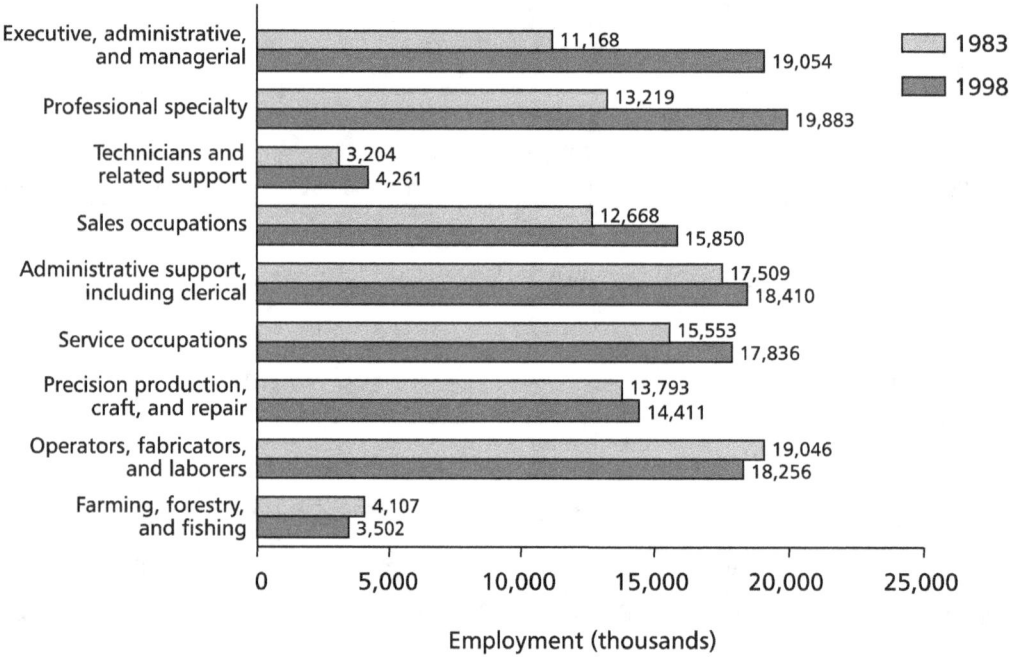

Figure 1–2. Employment by major occupational category, 1983 and 1998. (Source: BLS [1999].)

INTRODUCTION

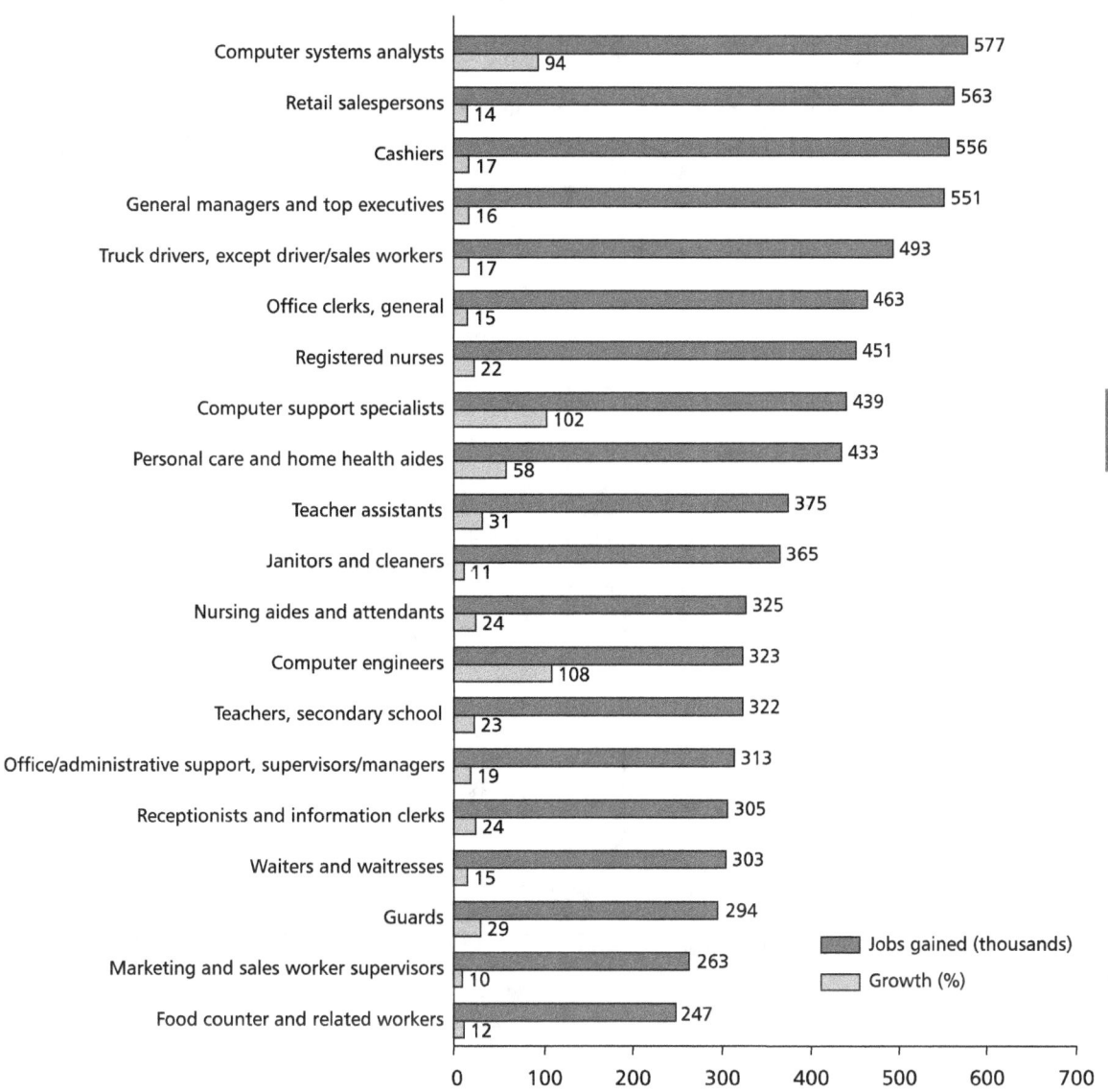

Figure 1–3. Employment growth in occupations gaining the largest number of jobs, projected for 1998–2008. (Source: BLS [2000].)

INTRODUCTION

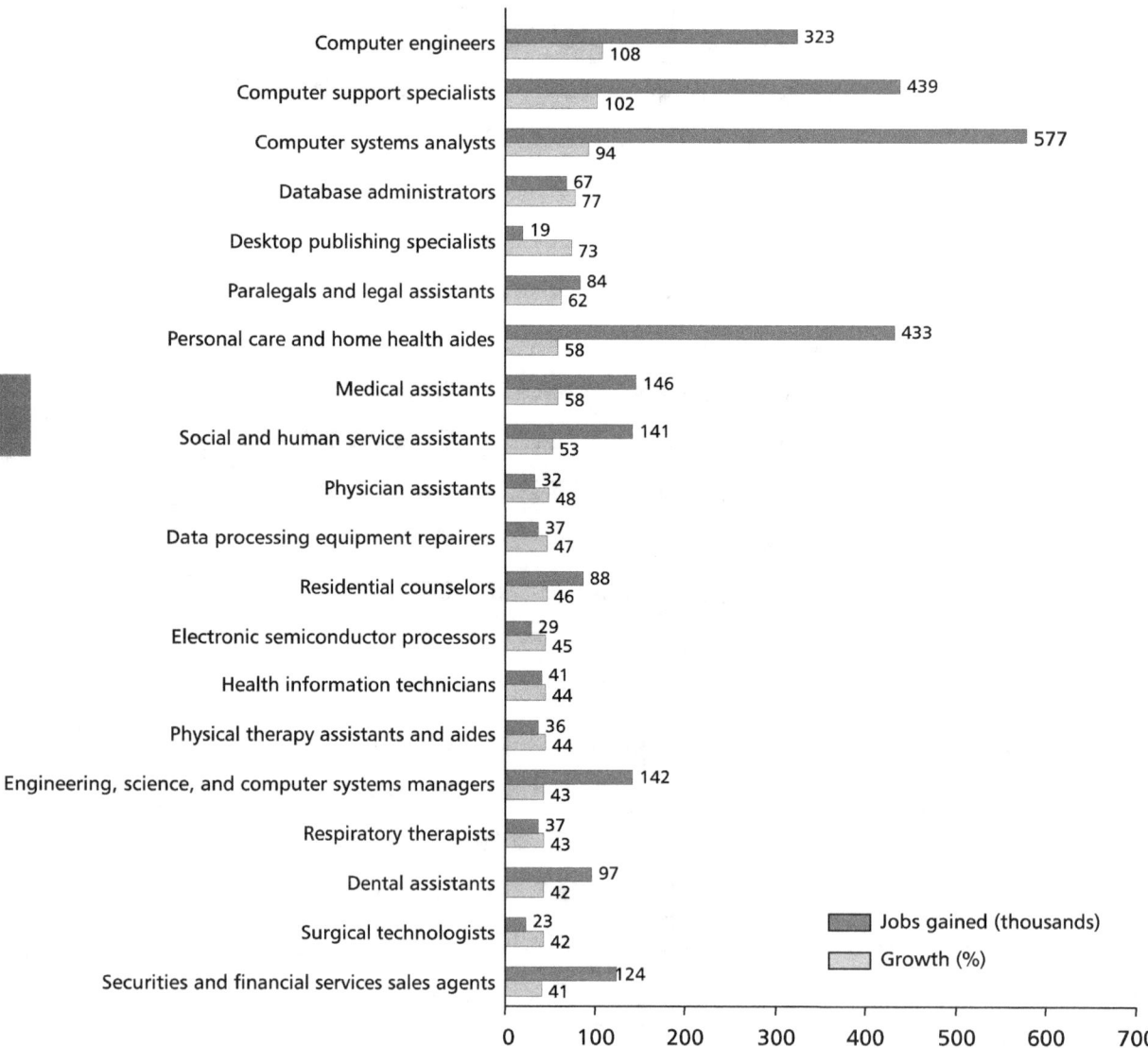

Figure 1–4. Employment growth in the fastest growing occupations, projected for 1998–2008. (Source: BLS [2000].)

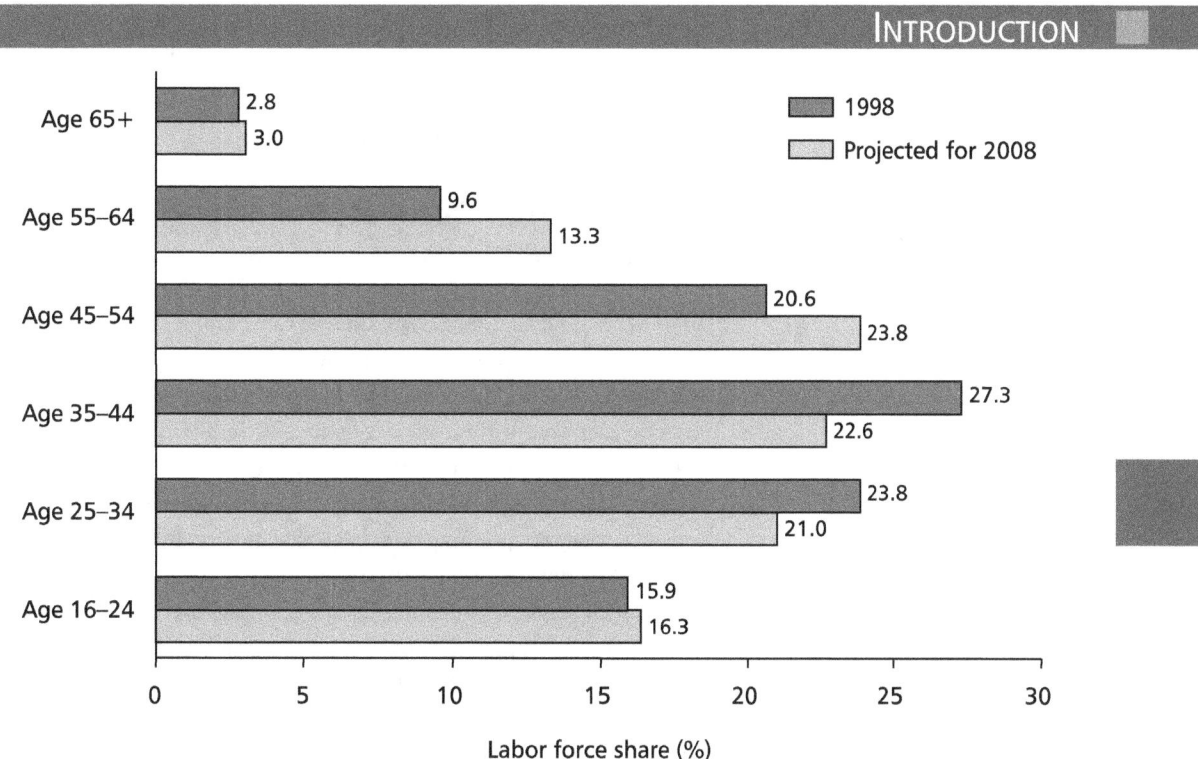

Figure 1–5. Distribution of the civilian labor force by age group, 1998 and projected for 2008. (Source: Fullerton [1999].)

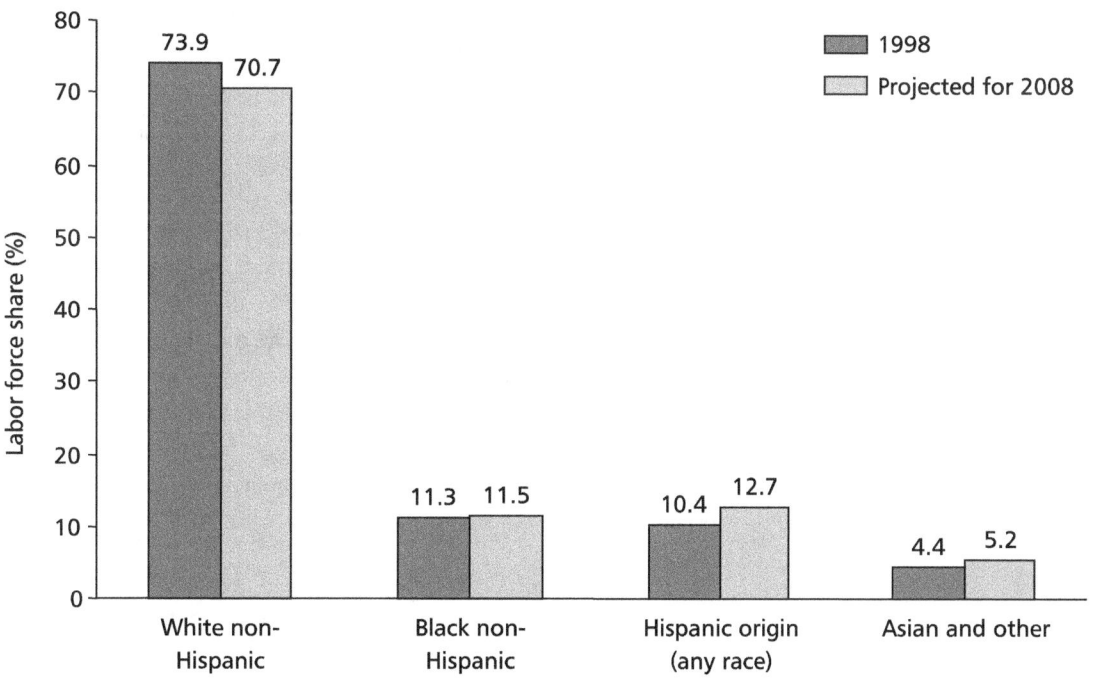

Figure 1–6. Civilian labor force share by race/ethnicity, 1998 and projected for 2008. (Source: Fullerton [1999].)

INTRODUCTION

Overview of Occupational Injuries and Illnesses

The Burden of Occupational Injuries and Illnesses

Injuries are generally easier than illnesses to categorize as occupationally related because their occurrence at the workplace or during work activities is usually obvious. Designating illnesses as occupational in origin is not as straightforward because illnesses often take a long time to develop and may be influenced by nonoccupational factors such as age, family history, or lifestyle habits such as tobacco use or avocational noise exposure. For example, a cancer appearing in old age may be very difficult to associate with work performed many years earlier. No single data system describes deaths from all occupational illnesses, but several data systems describe deaths from all occupational injuries. Therefore, the burden of occupational injuries is more apparent than the burden of occupational illnesses.

Fatal Injury

About 17 workers were fatally injured each day in 1997, yielding a total of 6,238 deaths that year; this total is about the same as that for 1992 (Figure 1–7) according to CFOI. Data from NTOF suggest that the overall rate of traumatic occupational fatalities declined during the 1980s and was stable in the early 1990s (Figure 1–8). CFOI fatality estimates exceeded those of NTOF by 1,000 or more for years reported in both surveillance systems (1992–1995).

Fatal transportation incidents accounted for 42% of all occupational injuries in 1997 (Figure 1–9), with highway crashes being the most frequent cause of death. Other frequent transportation incidents included crashes on the side of the road, jackknifings, and overturns. Assaults and other violent acts, including suicide, were the second most common fatal occupational events in 1997, accounting for 18% of total cases (Figure 1–9). Most violent acts were homicides, the second single leading type of fatality. Eighty percent of the homicides resulted from shootings, and most (85%) occurred during a robbery or another crime.

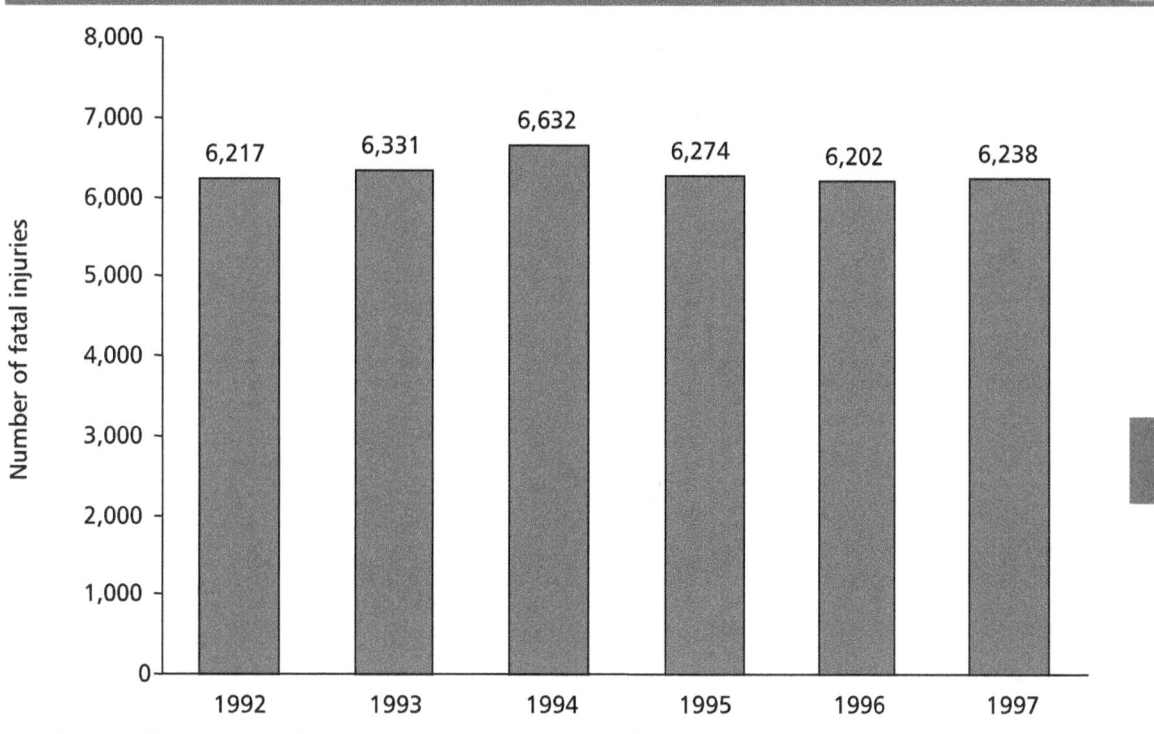

Figure 1–7. Number of fatal work injuries, 1992–1997. (Source: CFOI [1999].)

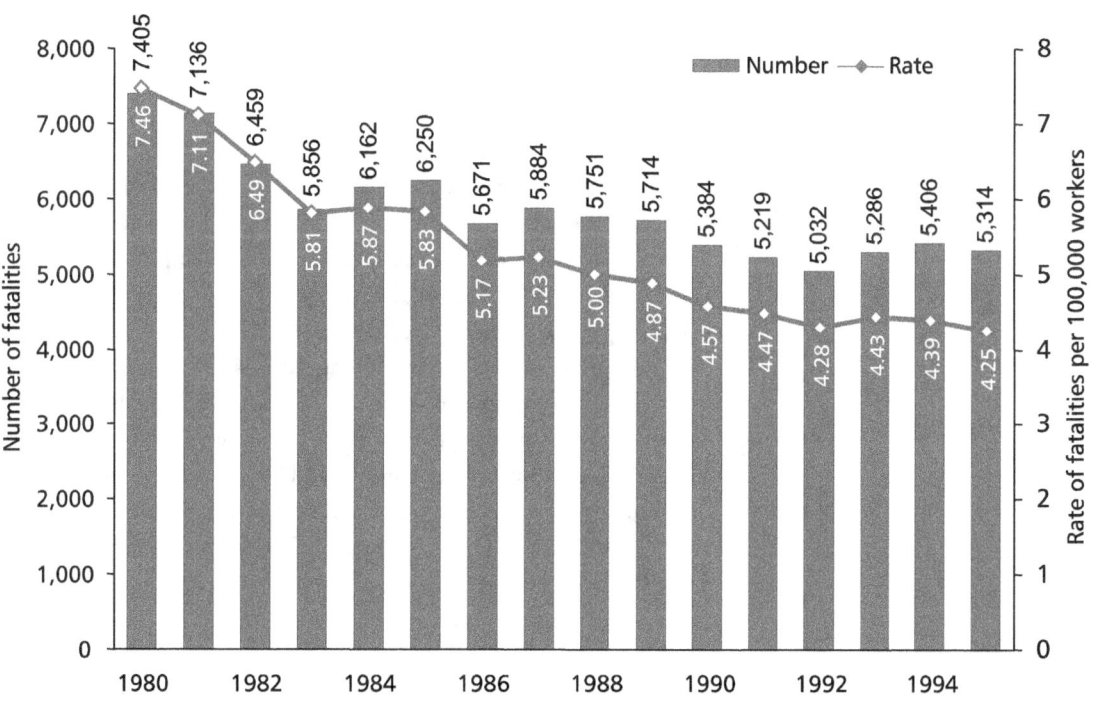

Figure 1–8. Number and annual rate of traumatic occupational fatalities, 1980–1995. (Source: NTOF [1999].)

INTRODUCTION

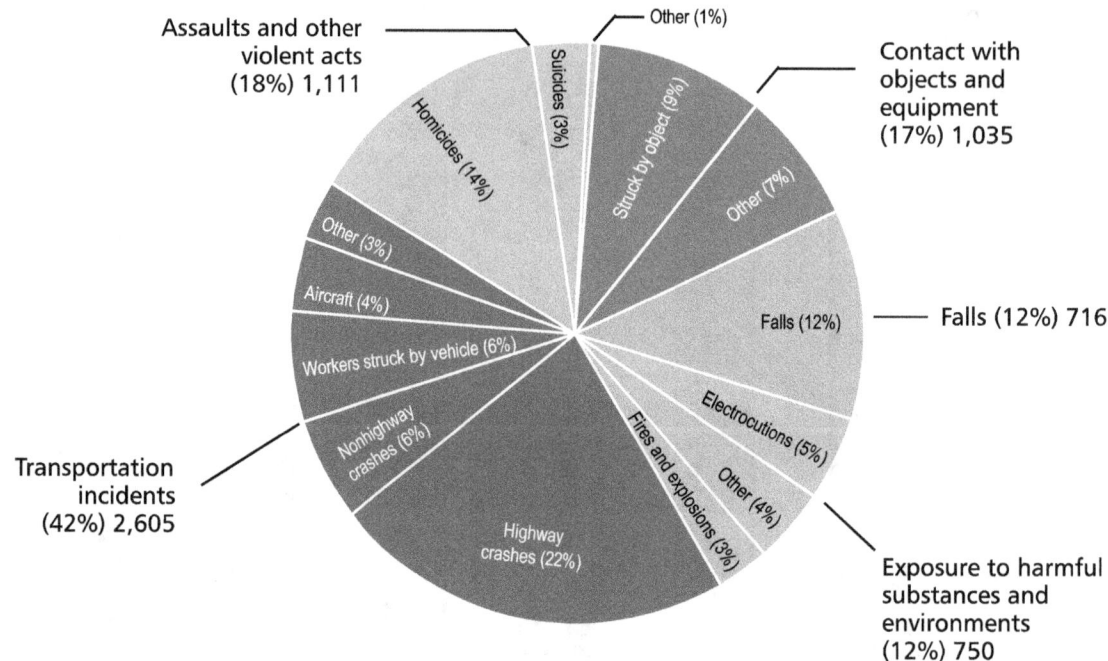

Figure 1–9. Number and distribution of fatal occupational injuries in 1997, by event and exposure. An additional 21 fatalities were attributed to other events and exposures, including *bodily reaction and exertion*. (Source: CFOI [1999].)

Fatal Illness

No surveillance data exist for most fatal occupational illnesses. One reason for this lack of data is that most occupational illnesses can be caused by factors other than workplace exposures. Lung diseases such as asthma, tuberculosis (TB), respiratory cancers, and chronic obstructive pulmonary disease (COPD) are examples of these diseases. However, the pneumoconioses, a small subset of lung diseases, are among the few illnesses attributable entirely to occupation. Since 1968, more than 113,000 fatalities have occurred with pneumoconiosis listed as an underlying or contributing cause of death (Figure 1–10). The number of deaths declined from a maximum of more than 5,400 in 1972 to slightly more than 3,100 in 1996. Coal workers' pneumoconiosis (CWP) deaths accounted for more than 50% of those deaths. Among the pneumoconioses, only asbestosis deaths have continued to increase.

Introduction

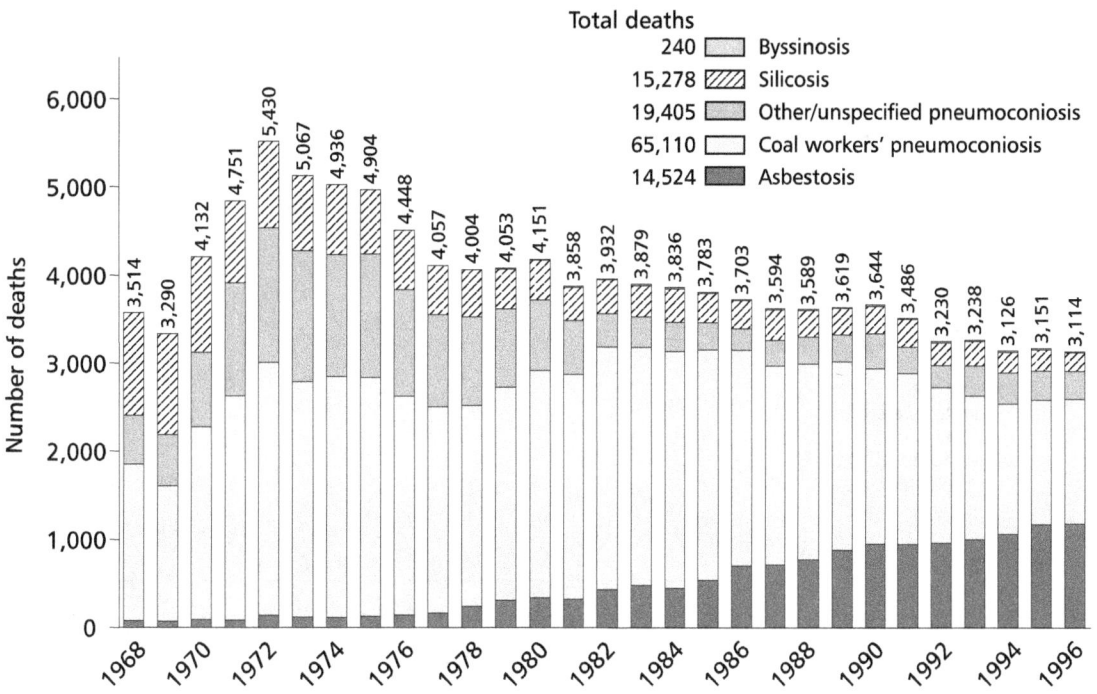

Figure 10. Number of deaths with pneumoconiosis listed as an underlying or contributing cause, U.S. residents aged 15 and older, 1968–1996. The stacked bars slightly overstate the numbers because some deaths are associated with more than one type of pneumoconiosis. The actual numbers are shown above the bars. (Source: NSSPM [1999].)

Nonfatal Injury and Illness Combined

Injuries accounted for 5.7 million (93%) of the 6.1 million injuries and illnesses reported by SOII for private-sector employers in 1997. The percentage of injuries in the combined count of illness and injury cases varied by industry division according to SOII. In manufacturing, 87% of all cases were injuries; in construction, almost 99% of the cases were injuries (Figure 1–11).

Incidence rates for total recordable cases of injuries and illnesses decreased from 11.0 to 7.1 cases per 100 full-time workers between 1973 and 1997 (Figure 1–12). The greatest change occurred among cases without lost workdays, which decreased from 7.5 in 1973 to 3.8 in 1997. In cases with lost workdays, the incidence rate in 1997 (3.3) was similar to that in 1973 (3.4) despite the fact that the total number of lost-workday cases rose from 1.9 million in 1973 to 2.9 million in 1997.

INTRODUCTION

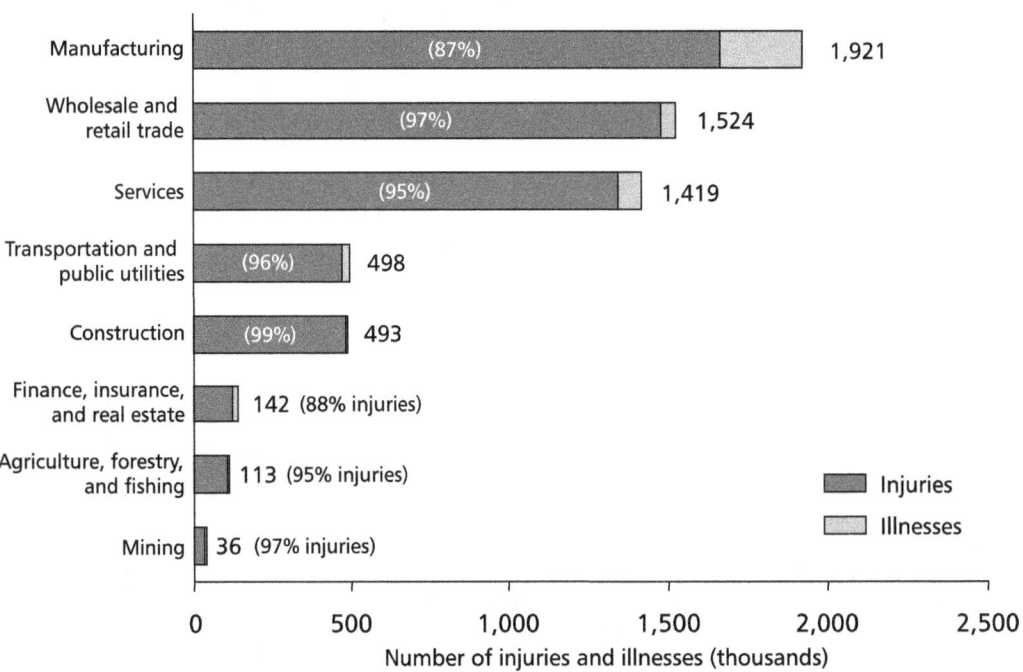

Figure 1–11. Number of nonfatal occupational injury and illness cases in private industry, by industry division, 1997. Injuries as the % of total cases for each industry division are shown in parentheses. (Source: SOII [1999].)

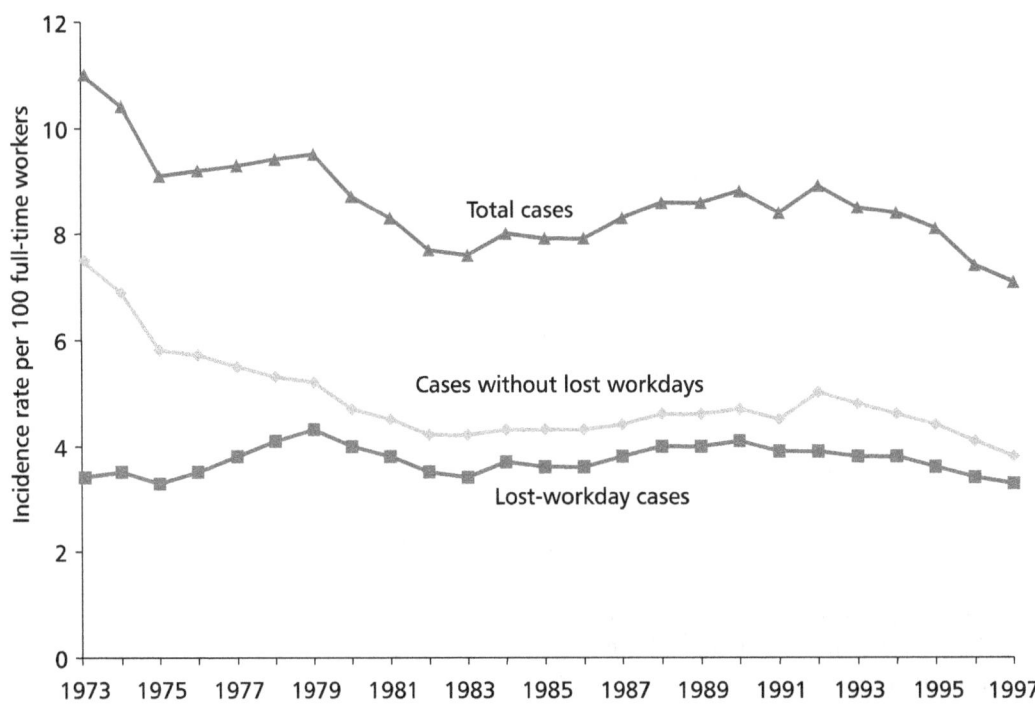

Figure 1–12. Incidence rates for occupational injury and illness cases in private industry, 1973–1997. (Cases without lost workdays and lost-workday cases are subsets of total cases.) (Source: SOII [1999].)

INTRODUCTION

As noted earlier, lost-workday cases include cases with days away from work and cases with restricted work activity only (i.e., cases in which workers report to their jobs for limited duty). From 1988 to 1997, there was a decrease in the rate of cases with days away from work and an increase in the rate of cases with restricted work activity only (Figure 1–13).

One factor contributing to the decline in overall injury and illness incidence rates is the shift in hours worked from a sector with a high rate of injuries (manufacturing) to other sectors with lower rates of injury. Manufacturing hours decreased from 35% of all hours worked in 1973 to 17% of all hours worked in 1997. Hours worked in the service industries increased from 18% to 23% during that period. Actual injury and illness incidence rates from 1973 to 1997 are compared with incidence rates based on the 1973 industry distribution of hours worked (i.e., adjusted rates) in Figure 1–14. In all years, the rates would be higher if the number of manufacturing hours worked was as high as in 1973. However, the decrease over time is still apparent, suggesting that the shift away from work in manufacturing does not account completely for the decrease in injury and illness incidence rates. The results of a similar analysis performed on incidence rates for lost-workday cases are shown in Figure 1–15. Again, the rates would be higher if the number of manufacturing hours worked was as high as in 1973. However, no decrease over time is apparent in Figure 1–15 in either the actual or the adjusted rates.

Incidence rates in 1997 by State for total nonfatal occupational injuries and illnesses in private industry (not available for some States) ranged from a low of 4.4 cases per 100 full-time workers in New York to a high of 10.0 cases per 100 full-time workers in Wisconsin (Figure 1–16). The national rate was 7.1 cases per 100 full-time workers. Rates of nonfatal occupational injury and illness cases with days away from work ranged from 1.4 cases per 100 full-time workers in Georgia to 3.5 cases per 100 full-time workers in Alaska (Figure 1–17). The national rate for lost workdays was 2.1 cases per 100 full-time workers. For nonfatal occupational injuries and illnesses with restricted work activity only, rates ranged from 0.3 cases per 100 full-time workers in New York to 2.3 cases per 100 full-time workers in Maine (Figure 1–18). The national rate of cases with restricted work activity only was 1.2 per 100 full-time workers.

INTRODUCTION

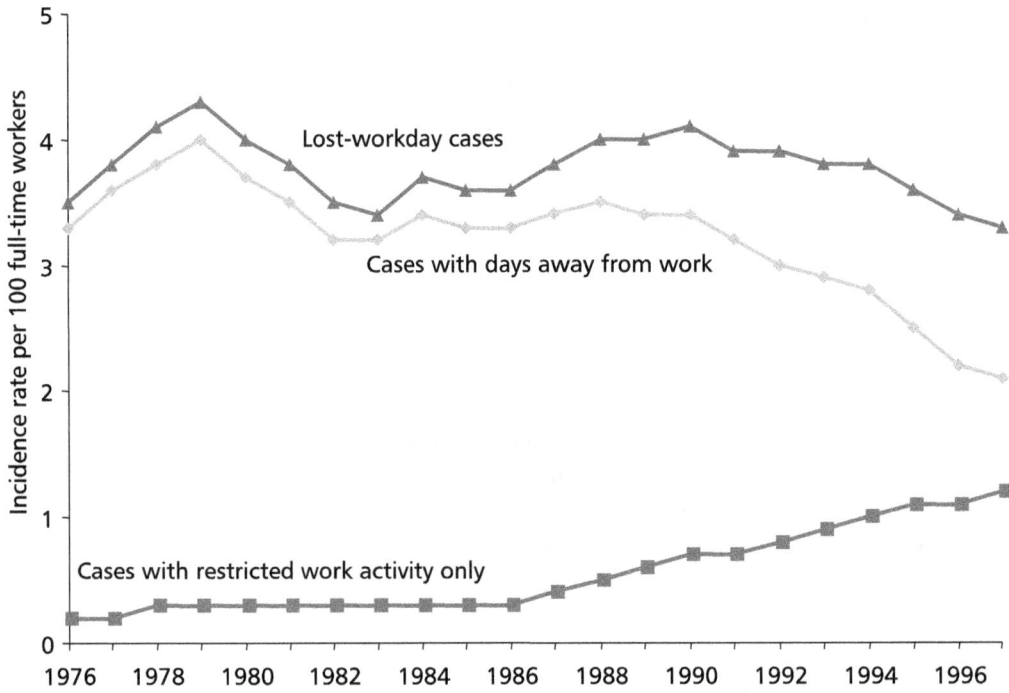

Figure 1–13. Incidence rates of lost-workday cases associated with nonfatal occupational injuries and illnesses in private industry, 1976–1997. (Cases with days away from work and cases with restricted work activity only are subsets of lost-workday cases.) (Source: SOII [1999].)

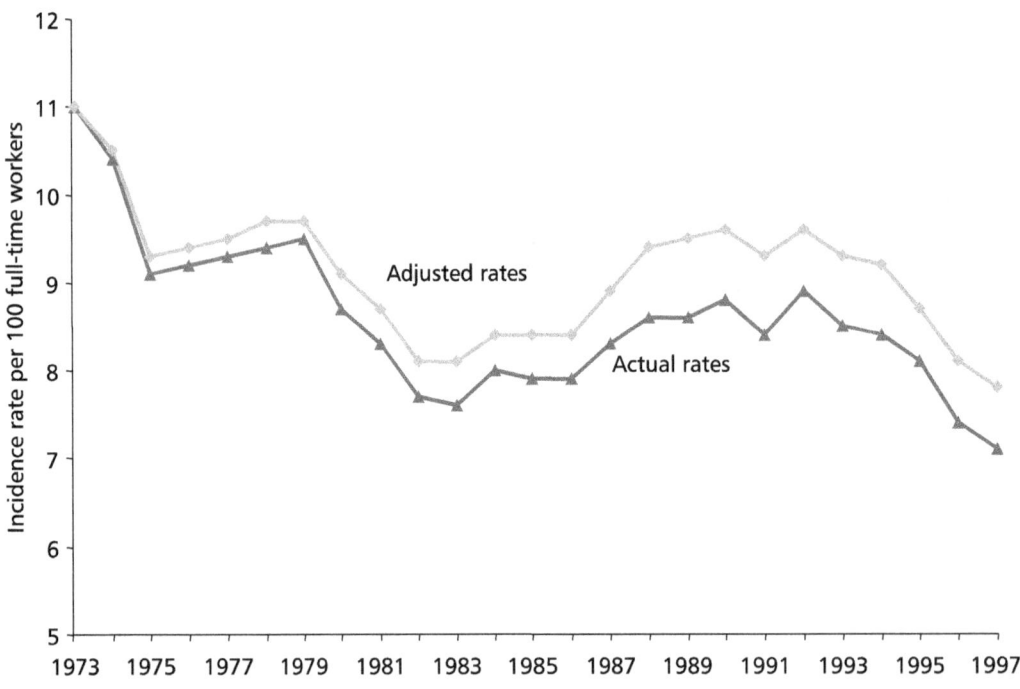

Figure 1–14. Total injury and illness incidence rates in private industry: actual rates compared with rates adjusted to 1973 hours series, 1973–1997. (Source: SOII [1999].)

INTRODUCTION

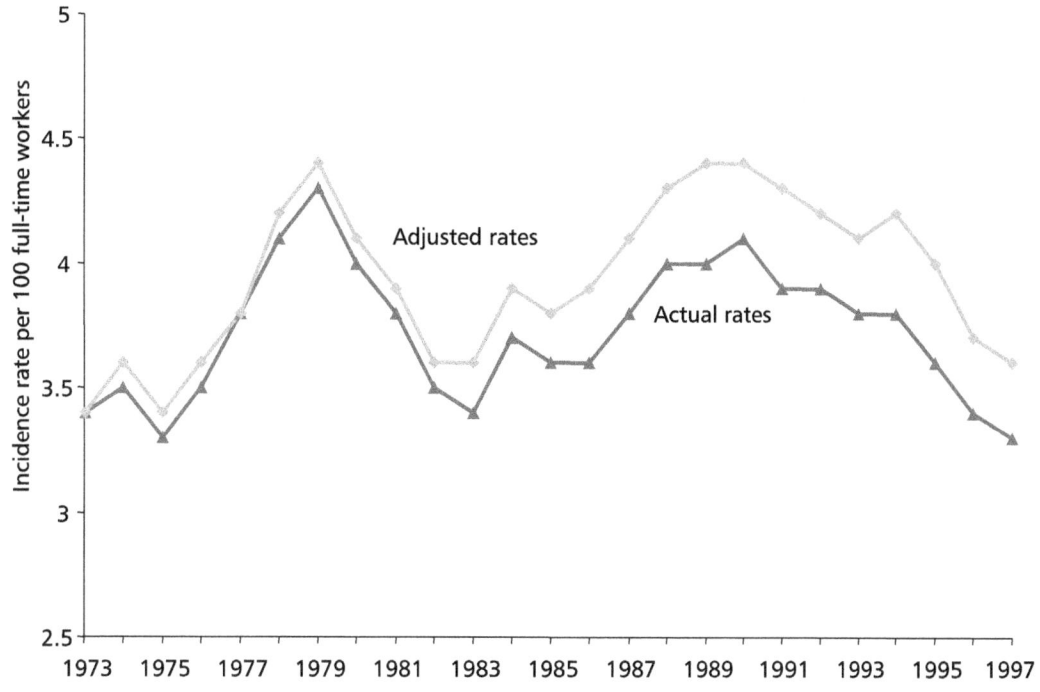

Figure 1–15. Incidence rates of lost-workday injury and illness cases in private industry: actual rates compared with rates adjusted to 1973 hours series, 1973–1997. (Source: SOII [1999].)

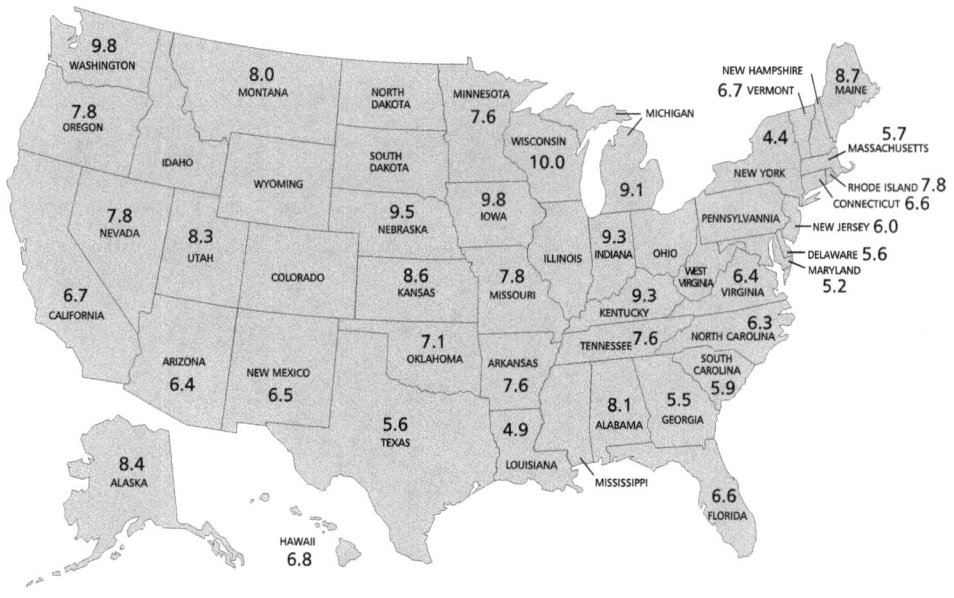

Figure 1–16. Incidence rates of nonfatal occupational injury and illness cases per 100 full-time workers in private industry, by State, 1997. National rate was 7.1. (Source: SOII [1999].)

INTRODUCTION

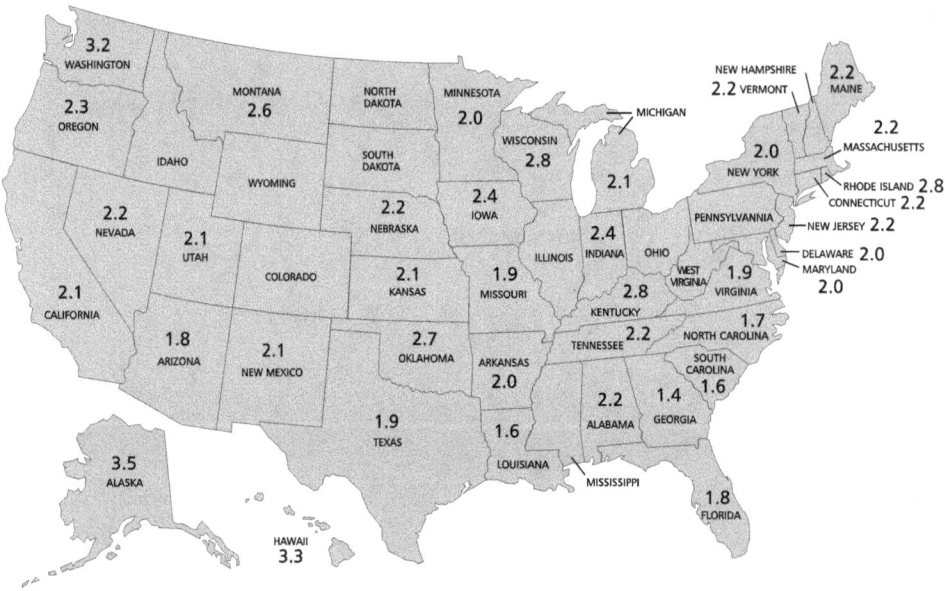

Figure 1–17. Incidence rates of nonfatal occupational injury and illness cases with days away from work per 100 full-time workers in private industry, by State, 1997. National rate was 2.1. (Source: SOII [1999].)

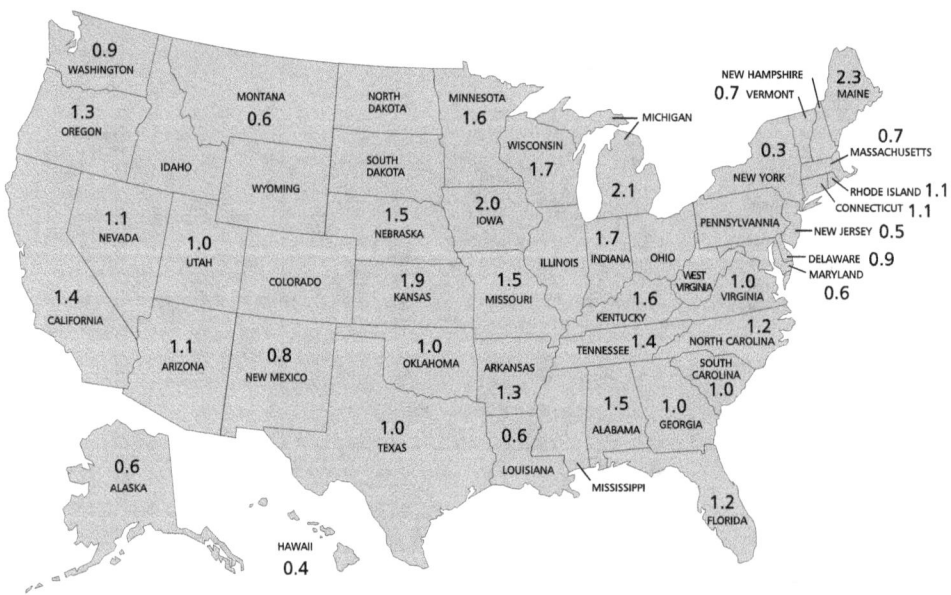

Figure 1–18. Incidence rates of restricted-workday cases of nonfatal occupational injury and illness cases per 100 full-time workers in private industry, by State, 1997. National rate was 1.2. (Source: SOII [1999].)

Introduction

Characteristics of Workers and of Injuries and Illnesses Involving Days away from Work

Workers

Men constituted 55% of the employed workers covered by SOII in 1997, but they accounted for 67% of the 1.8 million occupational injury and illness cases with days away from work (Figure 1–19). Workers aged 25 to 44 constituted 53% of the employed workers covered by SOII in 1997 and accounted for 59% of injuries and illnesses involving days away from work (Figure 1–20).

Ten occupations accounted for nearly one-third of the 1.8 million injuries and illnesses involving days away from work in 1997 (Figure 1–21). Truck drivers, nonconstruction laborers, and nursing aides and orderlies each accounted for more than 90,000 job-related injuries and illnesses involving days away from work. Injuries and illnesses in these three groups represent almost 19% of the total cases with days away from work in 1997. The five occupational groups with the largest numbers of injuries involving days away from work during 1993–1997 are shown in Figure 1–22. Truck drivers accounted for the largest number of lost-time injuries each year.

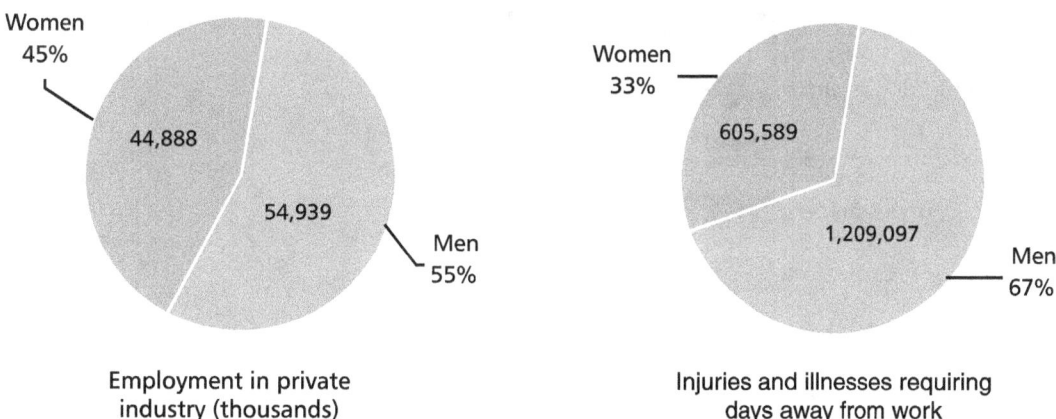

Figure 1–19. Employment in private industry and distribution of nonfatal occupational injury and illness cases with days away from work, by sex of worker aged 16 and older, 1997. Excludes cases in which sex of worker was not reported. Total number of injury and illness cases with days away from work was 1,833,380. (Source: BLS [1999]; SOII [1999].)

INTRODUCTION

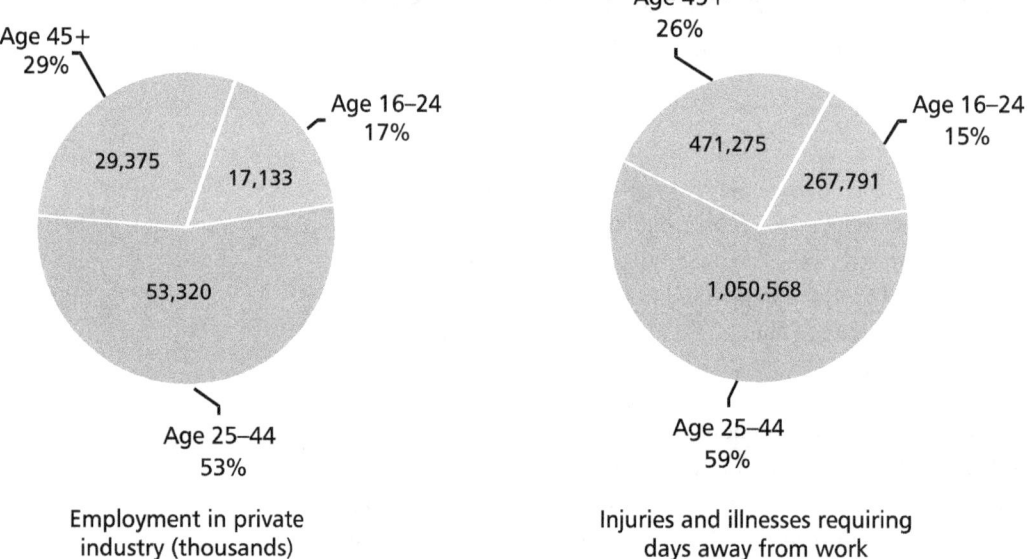

Employment in private industry (thousands)

Injuries and illnesses requiring days away from work

Figure 1–20. Employment in private industry and distribution of nonfatal injury and illness cases with days away from work, by age of worker, 1997. Excludes cases in which age of worker was not reported. Total number of injury and illness cases with days away from work was 1,833,380. (Source: BLS [1999]; SOII [1999].)

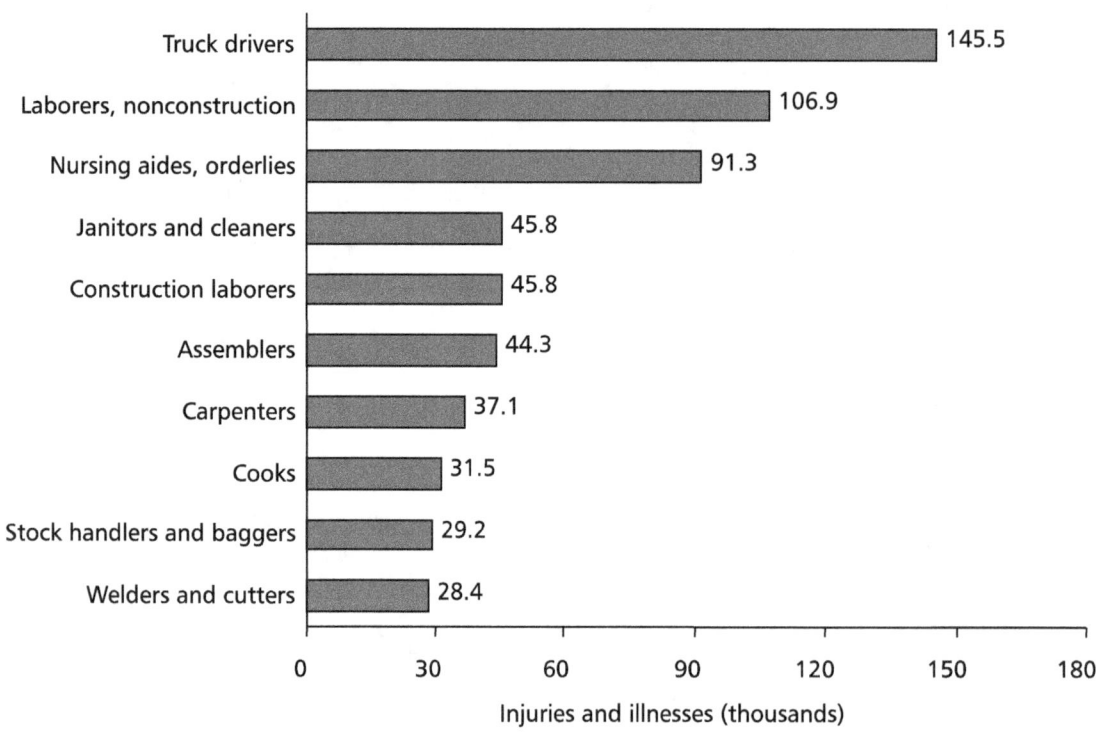

Figure 1–21. Ten occupations with the most injuries and illnesses involving days away from work, 1997. Total number of injuries and illnesses involving days away from work was 1,833,380. (Source: SOII [1999].)

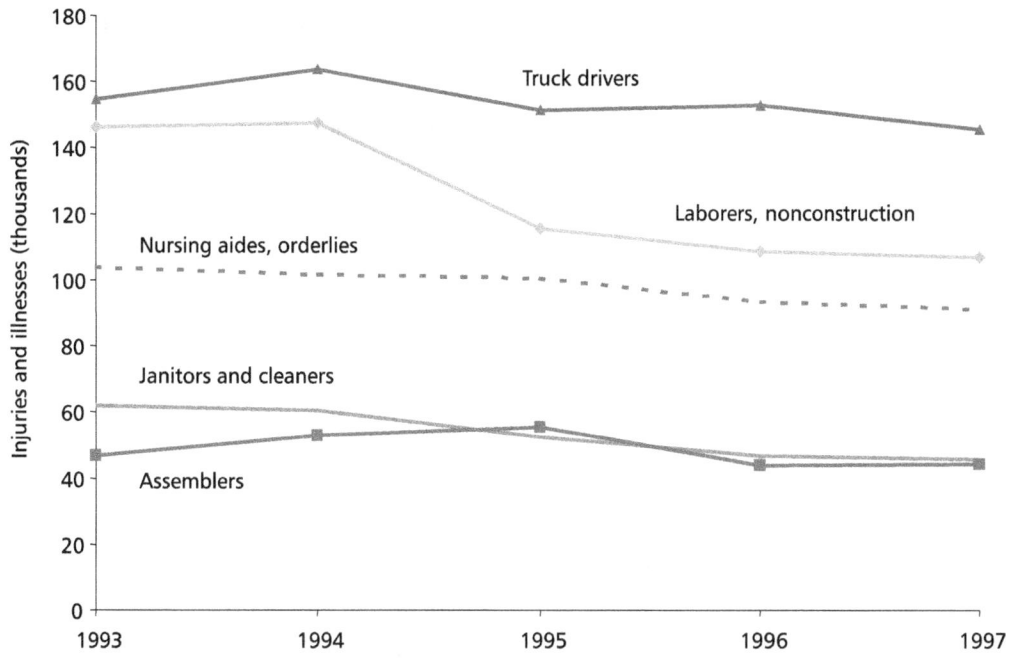

Figure 1-22. Number of occupational injuries and illnesses involving time away from work, for selected occupations, 1993–1997. (Source: SOII [1999].)

Injuries and Illnesses

Thirty-one percent of nonfatal injuries and illnesses involving days away from work in 1997 occurred among new workers (i.e., workers having less than 1 year of service with their employer). The percentages for new workers were even higher in mining (44%), agriculture, forestry, and fishing (43%), construction (41%), and wholesale and retail trade (34%) (Figure 1–23). Nearly two-thirds of injury and illness cases with days away from work occurred among workers with 5 or fewer years of service with their employer.

Sprains and strains were by far the most frequent disabling conditions, accounting for 799,012 cases (43.6%) with days away from work. Bruises accounted for 165,800 cases (9.0%), and cuts and punctures accounted for another 156,700 cases (8.5%) (Figure 1–24). The back was the body part most often affected by disabling work incidents (Figure 1–25). Bodily reaction and exertion, contact with objects and equipment, and falls were the most frequent events or exposures leading to work injury or illness that involved days away from work (Figure 1–26).

INTRODUCTION

Severity of illness or injury can be estimated from the number of days away from work. Five days was the median number of days away from work for all types of injury and illness. Carpal tunnel syndrome (CTS), fractures, amputations, tendinitis, multiple injuries, and sprains and strains had median days away from work greater than the 5-day median for all injuries and illnesses combined (Figure 1–27).

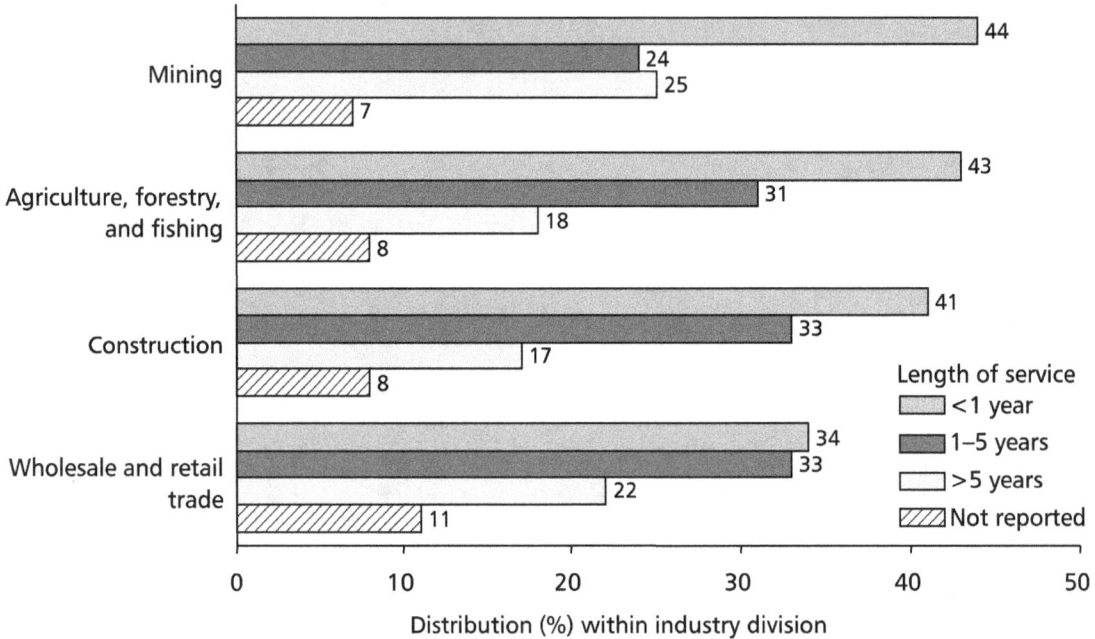

Figure 1–23. Distribution of nonfatal injuries and illnesses involving days away from work within selected private industry divisions, by length of service with employer, 1997. (Source: SOII [1999].)

INTRODUCTION

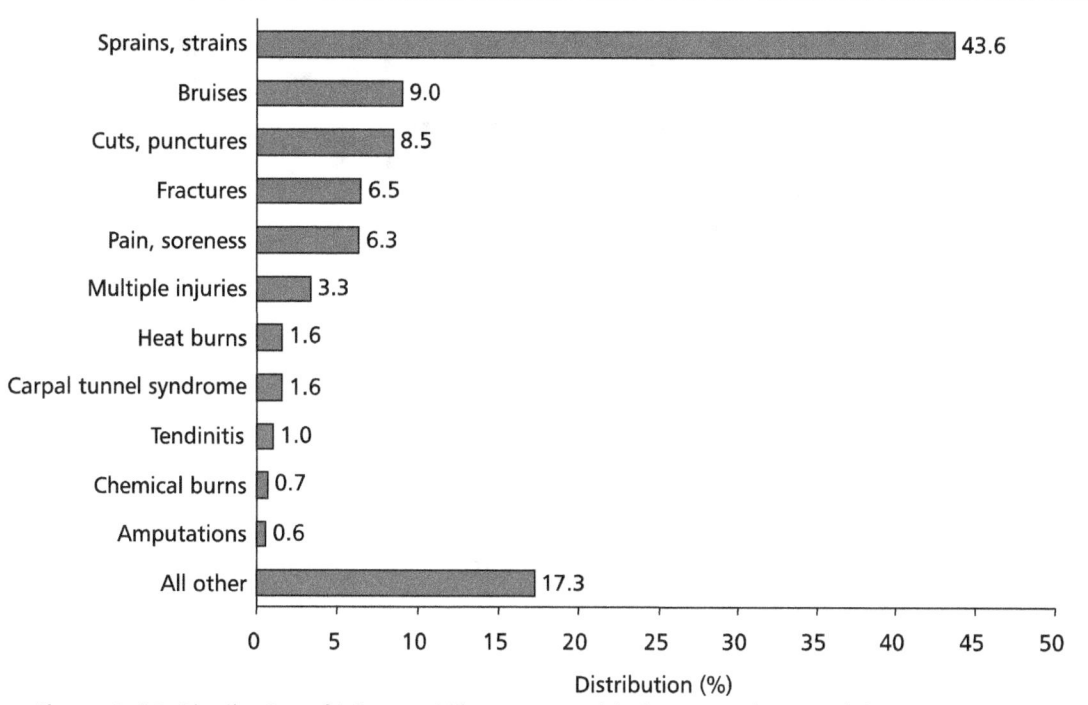

Figure 1–24. Distribution of injury and illness cases with days away from work in private industry, by nature of injury or illness, 1997. Total number of injury and illness cases with days away from work was 1,833,380. (Source: SOII [1999].)

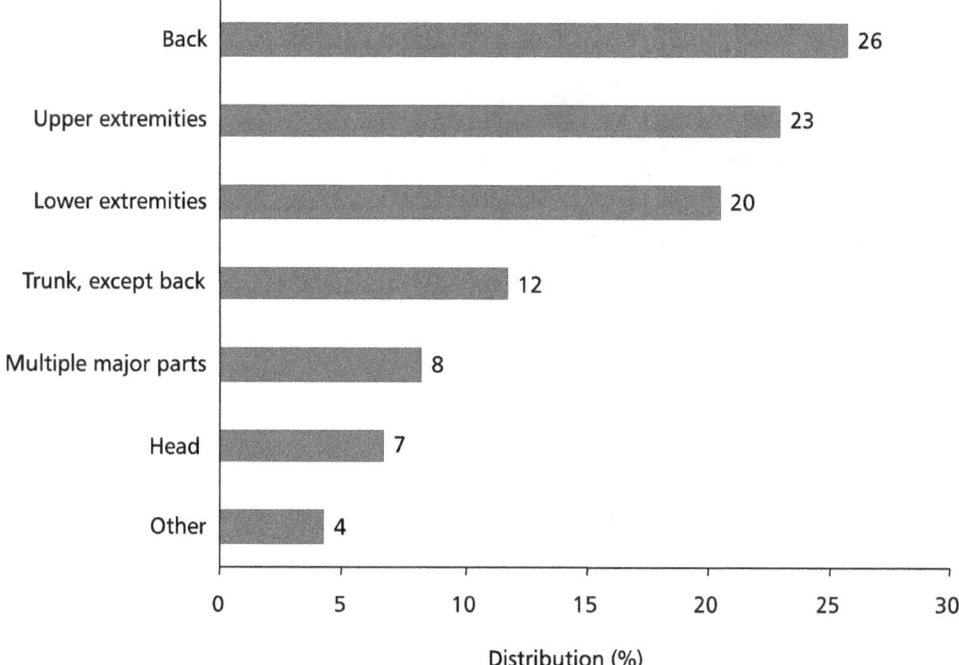

Figure 1–25. Distribution of injury and illness cases with days away from work in private industry, by part of body affected, 1997. Total number of injuries and illnesses involving days away from work was 1,833,380. (Source: SOII [1999].)

INTRODUCTION

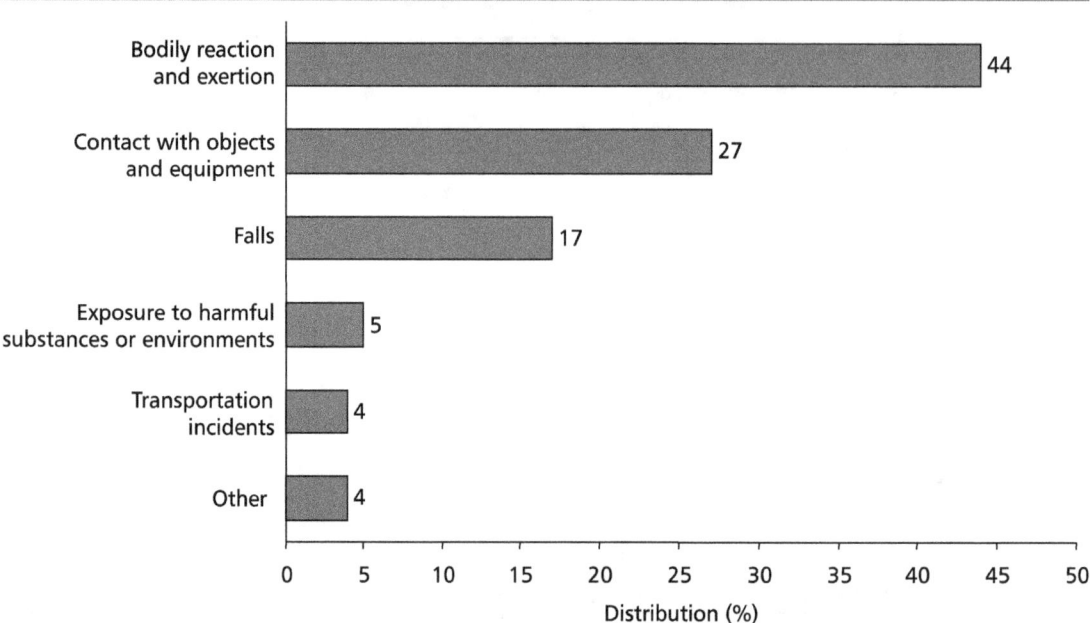

Figure 1–26. Distribution of occupational injuries and illnesses involving days away from work in private industry, by type of event or exposure, 1997. Total number of injuries and illnesses involving days away from work was 1,833,380. (Source: SOII [1999].)

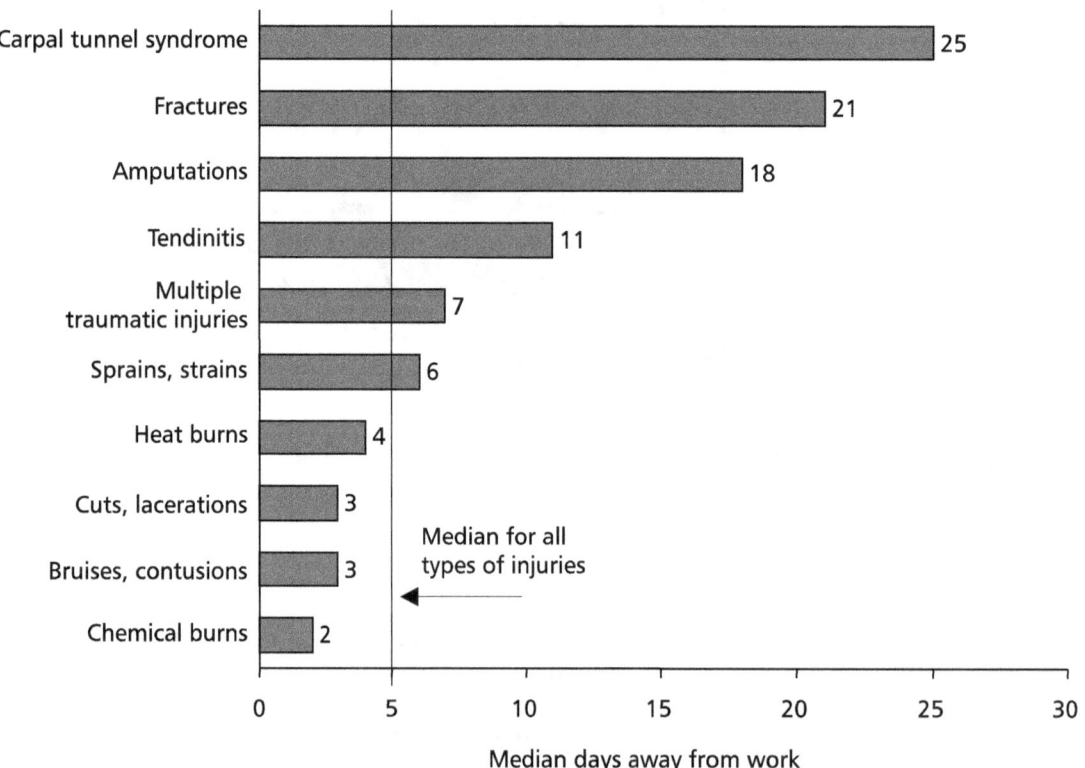

Figure 1–27. Median days away from work due to selected types of nonfatal occupational injury or illness in private industry, 1997. (Source: SOII [1999].)

2 Fatal Injury

2 Fatal Injury

The Burden of Fatal Occupational Injuries

The National Institute for Occupational Safety and Health (NIOSH), the Bureau of Labor Statistics (BLS), and the States share responsibility for the surveillance of fatal occupational injuries. NIOSH conducts surveillance of these injuries through the National Traumatic Occupational Fatalities Surveillance System (NTOF), which contains information from death certificates managed by the 52 U.S. vital statistics reporting units and has fatality data from 1980 onward. In response to a National Academy of Sciences recommendation, BLS began compiling fatal occupational injury data in 1992 through its Census of Fatal Occupational Injuries (CFOI). Data for CFOI are obtained from various Federal, State, and local administrative sources, including death certificates, workers' compensation reports and claims, reports to regulatory agencies, medical examiner reports, police reports, and news items. Differences in NTOF and CFOI definitions and data collection and recording procedures may result in different fatality counts. The two programs are complementary, each having unique features that contribute to the surveillance of fatal occupational injuries. Appendix A details the methodological differences between the surveillance systems.

Data from NTOF indicate that 93,929 civilians in the United States were killed on the job from 1980 through 1995. The average annual fatality rate for this period was 5.3 per 100,000 workers. From 1980 through 1995, the number of deaths recorded by NTOF decreased by 28% (from 7,405 to 5,314), and the rate of death decreased by 43% (from 7.46 to 4.25 cases per 100,000 workers) (Figure 2–1). CFOI fatality counts exceeded those of NTOF by about 1,000 in the years reported in both surveillance systems (1992–1995) (Figure 2–2). Based on CFOI data, the rate of fatal occupational injuries declined by 7% between 1992 and 1997.

Fatal Injury

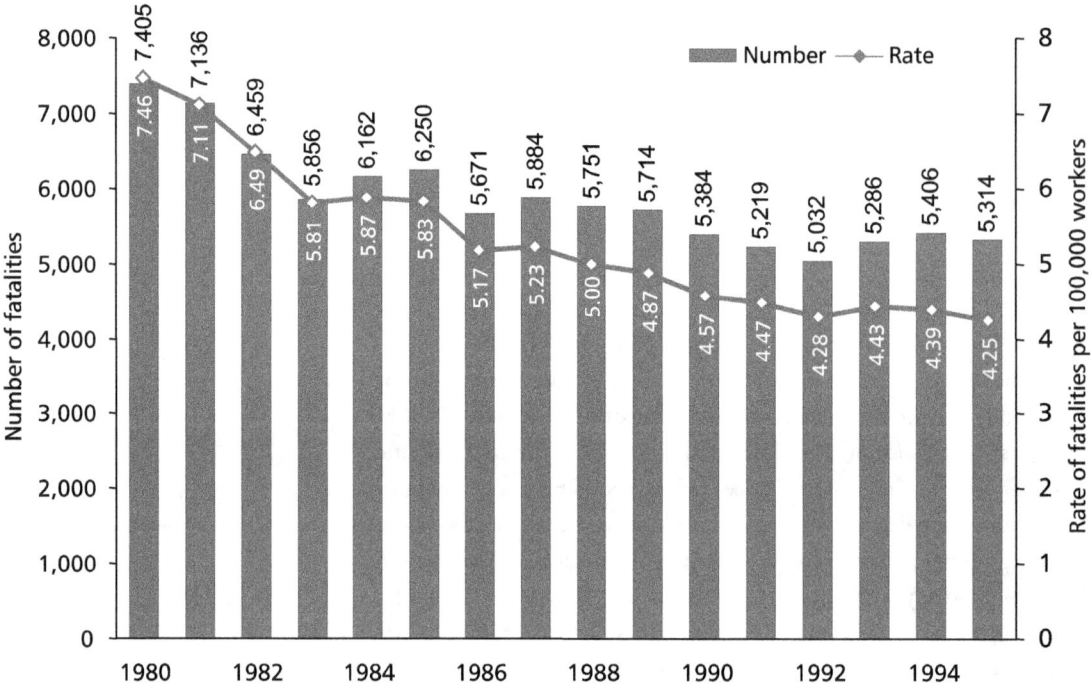

Figure 2–1. Number and annual rate of fatal occupational injuries, 1980–1995. (Source: NTOF [1999].)

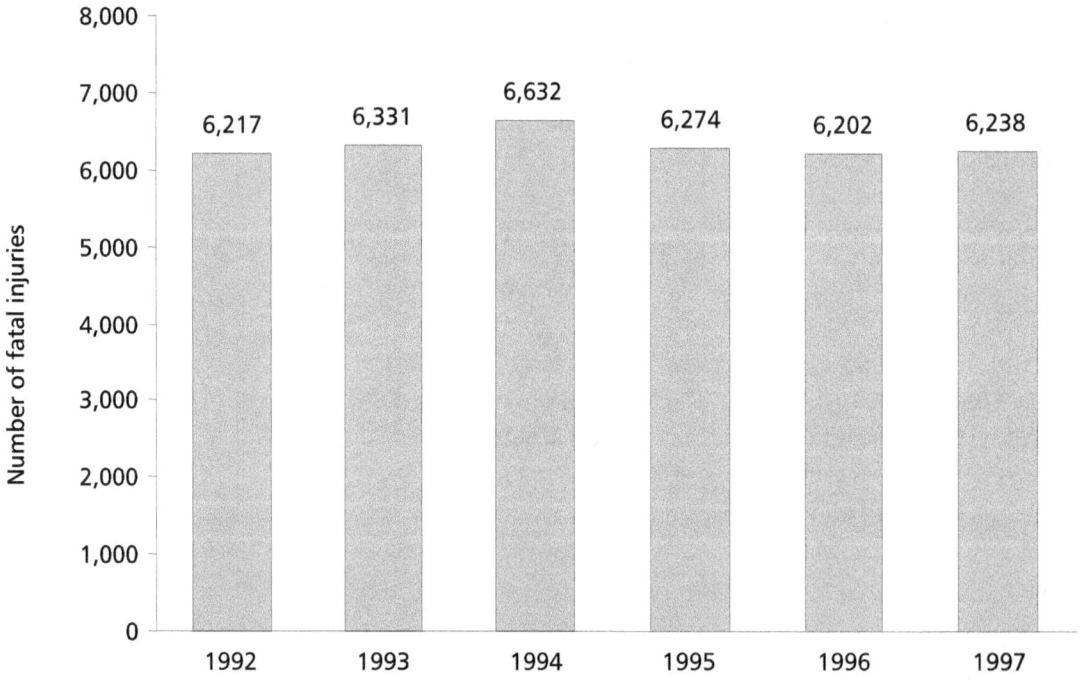

Figure 2–2. Number of fatal occupational injuries, 1992–1997. (Source: CFOI [1999].)

Fatal Injuries by Age and Race

The highest number of deaths recorded in NTOF from 1980 to 1995 occurred among workers aged 25 to 34 (Figure 2–3). CFOI data from 1992 to 1997 indicate that workers aged 35 to 44 had the highest number of fatal occupational injuries, similar to the share of employment for that age group. Rates of death recorded in NTOF were similar for the younger age groups, increased slightly in workers aged 55 to 64, and increased dramatically among workers aged 65 years and older (Figure 2–3). Death rates recorded in NTOF fell gradually from 1980 through 1995 for workers of all races (Figure 2–4).

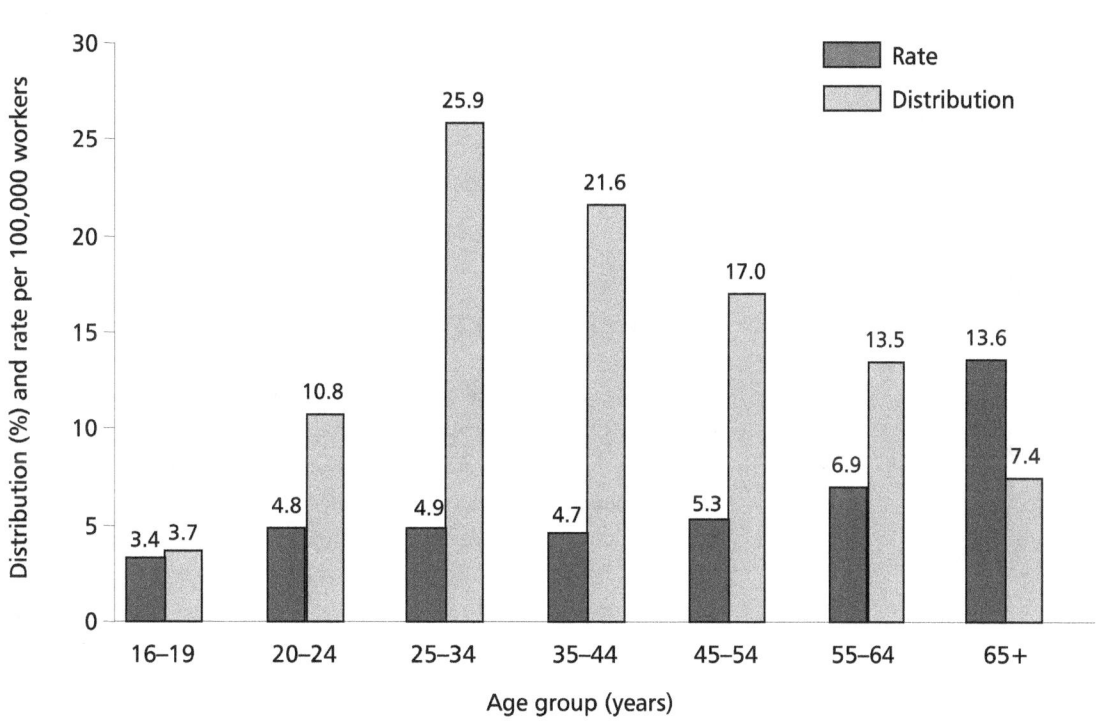

Figure 2–3. Distribution and average annual rate of fatal occupational injuries by age group, 1980–1995. (Source: NTOF [1999].)

Fatal Injury

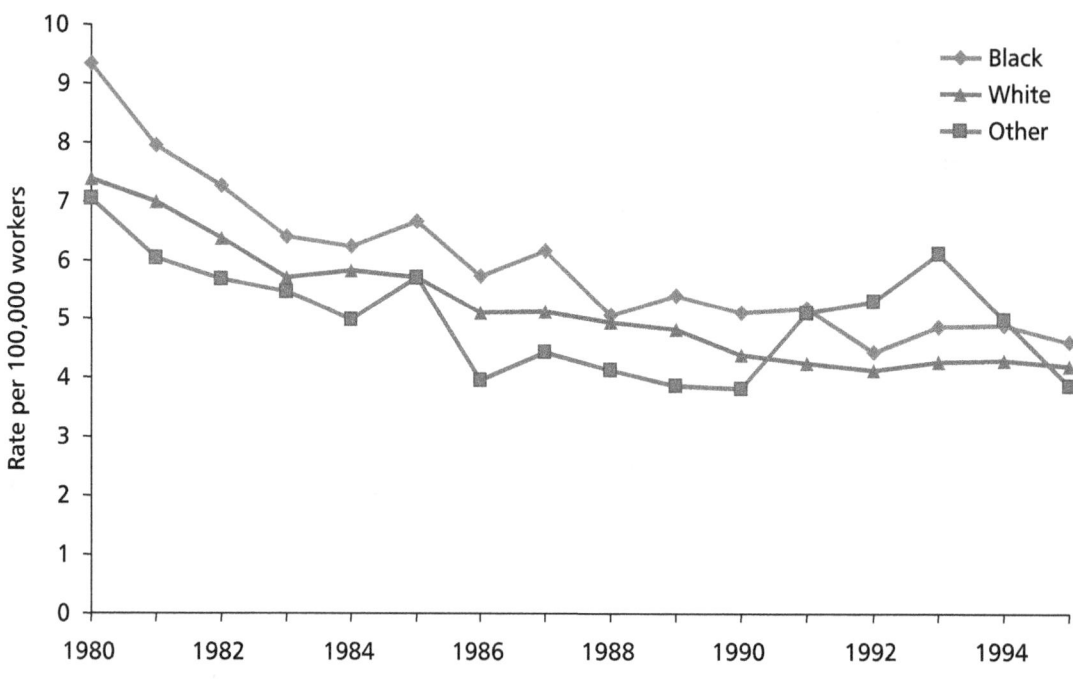

Figure 2–4. Rate of fatal occupational injuries by race, 1980–1995. (Source: NTOF [1999].)

Fatal Injuries by Leading Cause

The leading causes of fatal occupational injuries recorded in NTOF from 1980 to 1995 were motor vehicle incidents, machine-related injuries, homicides, falls, and electrocutions (Figure 2–5). During that period, rates for deaths from all causes declined, although not always consistently. Male workers died most frequently from motor vehicle incidents, machine-related injuries, homicides, and falls; female workers died most frequently from homicides and motor vehicle incidents, followed by falls and machine-related injuries (Figure 2–6). CFOI data, which are classified differently from NTOF data, indicate that transportation incidents accounted for 42% of all fatal occupational injuries in 1997 (Figure 1–9). Highway-related motor vehicle crashes and homicides accounted for about one-third of the fatalities recorded in CFOI.

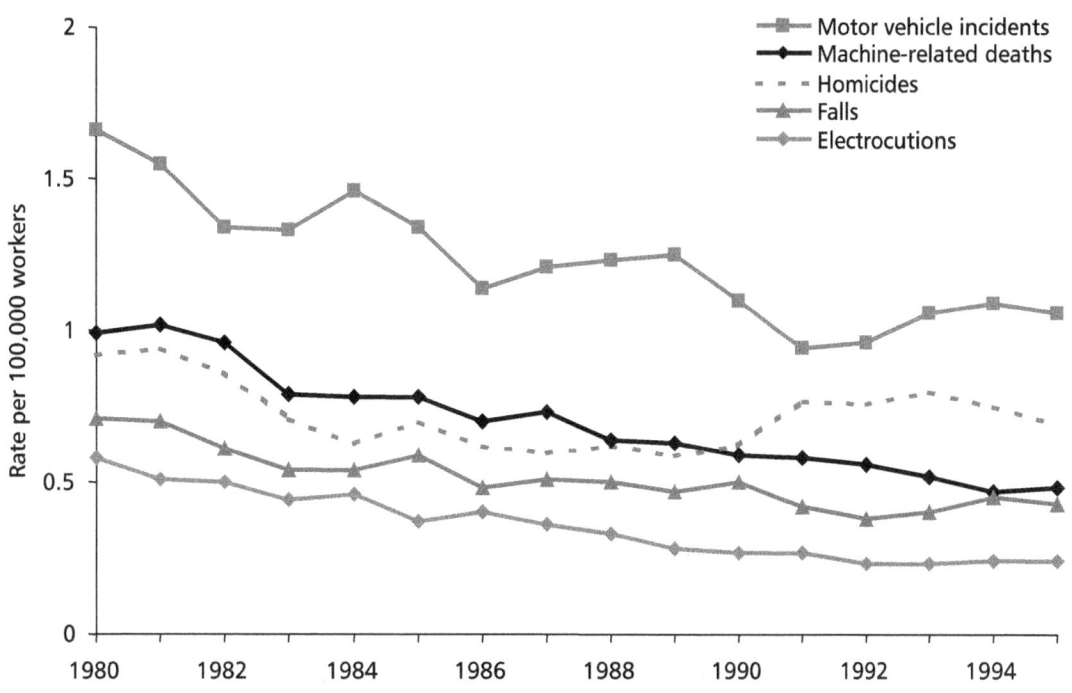

Figure 2–5. Rates of fatal occupational injuries by leading causes, 1980–1995. (Source: NTOF [1999].)

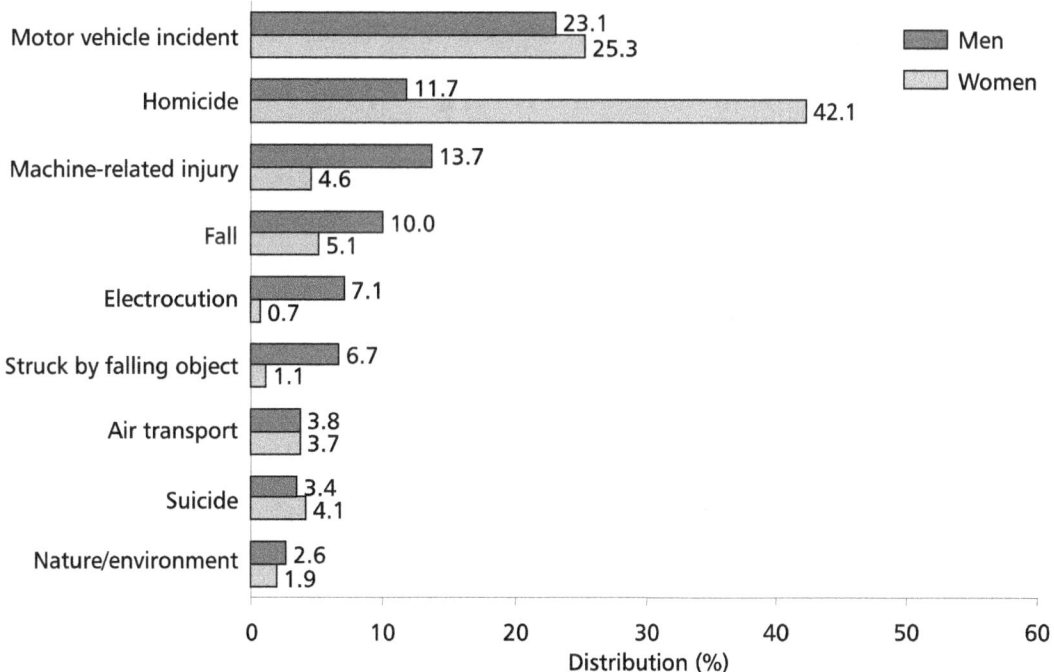

Figure 2–6. Distribution of fatal occupational injuries for male and female workers by selected causes of death, 1980–1995. Total deaths were 87,835 for male workers and 6,088 for female workers. (Source: NTOF [1999].)

Fatal Injury

Fatal Injuries by Industry and Occupation

NTOF classifies a fatality by the industry and occupation in which the worker was "usually" employed. By industry division, mining and agriculture, forestry, and fishing (followed by construction and transportation and public utilities), had the highest fatal occupational injury rates recorded in NTOF from 1980 to 1995. The most deaths occurred in construction, transportation and public utilities, and manufacturing (Figure 2–7). By occupational group, the highest rates of fatal injury occurred among transportation and agriculture, forestry, and fishing workers. Precision production, craft, and repair occupations (11% of the workforce) along with transportation workers (4% of the workforce) accounted for nearly 40% of the fatal occupational injuries from 1980 to 1995 (Figure 2–8).

Figure 2–7. Average annual rate and distribution (%) of fatal occupational injuries by industry division, 1980–1995. Total deaths were 93,929; 5.7% were not classified by industry. (Source: NTOF [1999].)

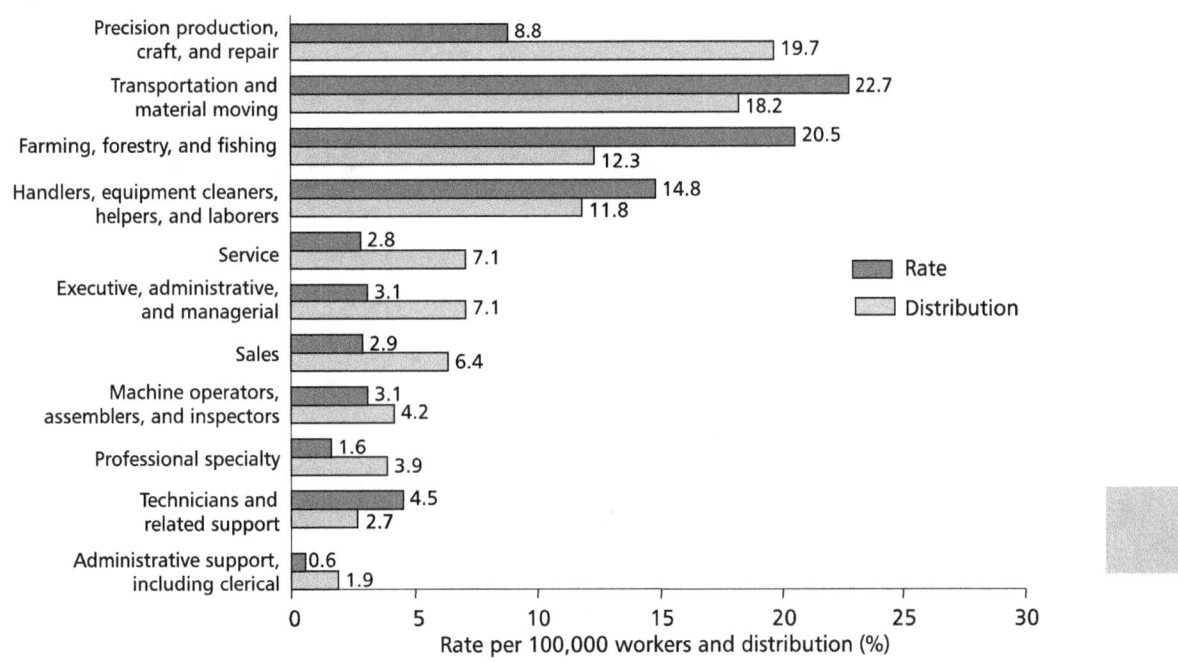

Figure 2–8. Average annual rate and distribution (%) of fatal occupational injuries by occupational group, 1980–1995. Total deaths were 93,929; 4.8% were not classified by occupation. (Source: NTOF [1999].)

CFOI classifies a fatality by the industry and occupation in which the worker was employed at the time of death. By industry division, construction accounted for the largest number of deaths recorded in CFOI in 1997, and mining had the highest fatality rate per 100,000 workers. Agriculture, forestry, and fishing ranked second in rate and third in number of fatal occupational injuries (Figure 2–9). By occupation, the largest number of fatalities occurred among truck drivers, farm occupations, sales occupations, and construction laborers (Figure 2–10). The leading causes of death for these groups were highway crashes and jackknifing for truck drivers, tractor-related injuries for farmers, homicides for sales occupations, and falls for construction laborers. The occupations with fatal occupational injury rates at least 10 times the national average of 4.8 per 100,000 workers include timber cutters, fishers, water transportation occupations, aircraft pilots, and extractive occupations (Figure 2–11).

FATAL INJURY

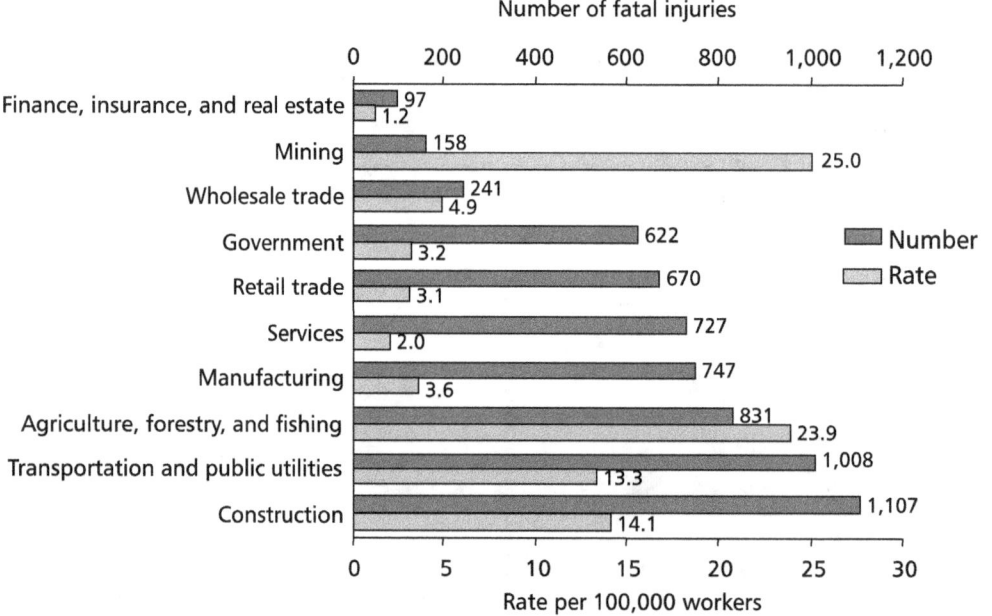

Figure 2–9. Number and rate of fatal occupational injuries by industry division, 1997. The total number of fatal occupational injuries was 6,238; the national rate was 4.8 per 100,000 workers. (Source: CFOI [1999].)

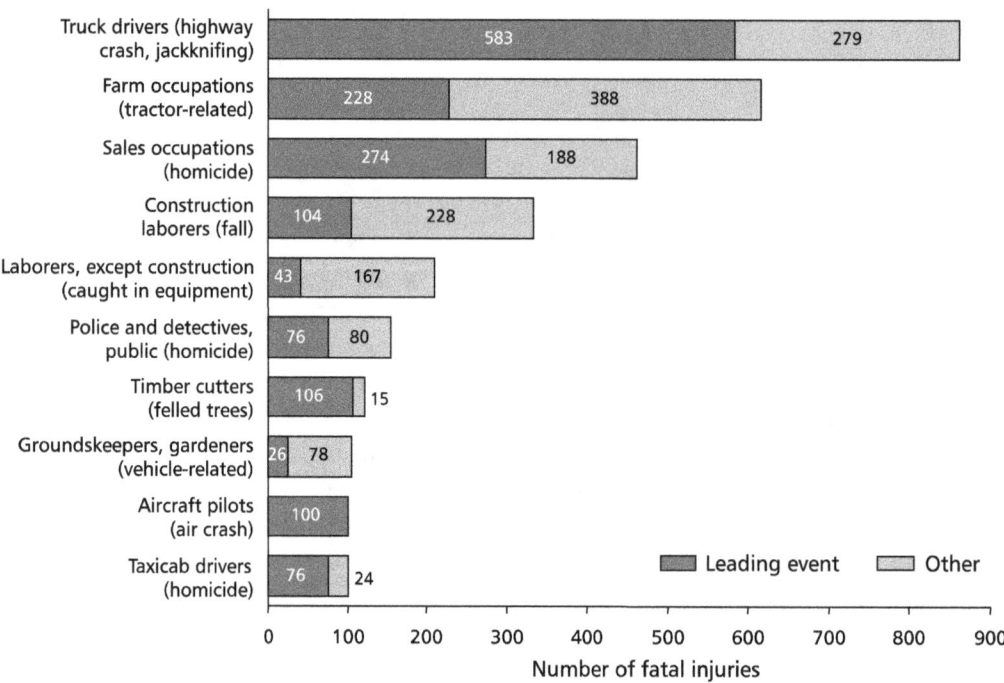

Figure 2–10. Number of fatal occupational injuries by selected high-risk occupations and leading event, 1997. The total number of fatal occupational injuries in 1997 for all occupations was 6,238. (Source: CFOI [1999].)

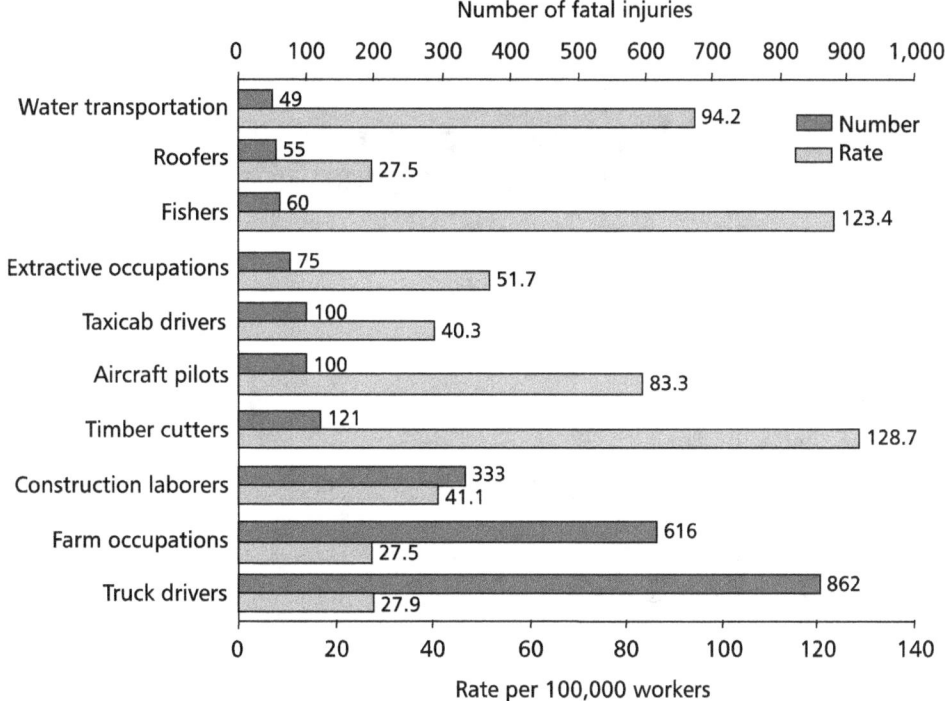

Figure 2–11. Number and rate of fatal occupational injuries per 100,000 workers in high-risk occupations, 1997. The national rate was 4.8 per 100,000 workers. (Source: CFOI [1999].)

Annual rates of fatal occupational injury by industry division for selected causes of death are shown in Figure 2–12 using NTOF data. Workers in mining and agriculture, forestry, and fishing had the highest rates of machine-related deaths, and workers in transportation and public utilities, mining, and agriculture, forestry, and fishing had the highest rates of work-related motor vehicle deaths. Workers in retail trade and public administration had the highest rates of workplace homicide.

Fatal Injuries by State

NTOF data for 1980–1995 (based on the State listed on the death certificate) indicate that Alaska, Wyoming, and Montana had the highest fatal occupational injury rates (Table 2–1). California, Texas, Florida, and Illinois had the greatest number of fatal occupational injuries. CFOI data (based on the State in which the fatal incident occurred) indicate that California, Texas, Florida, and New York had the greatest number of fatal occupational injuries in 1997.

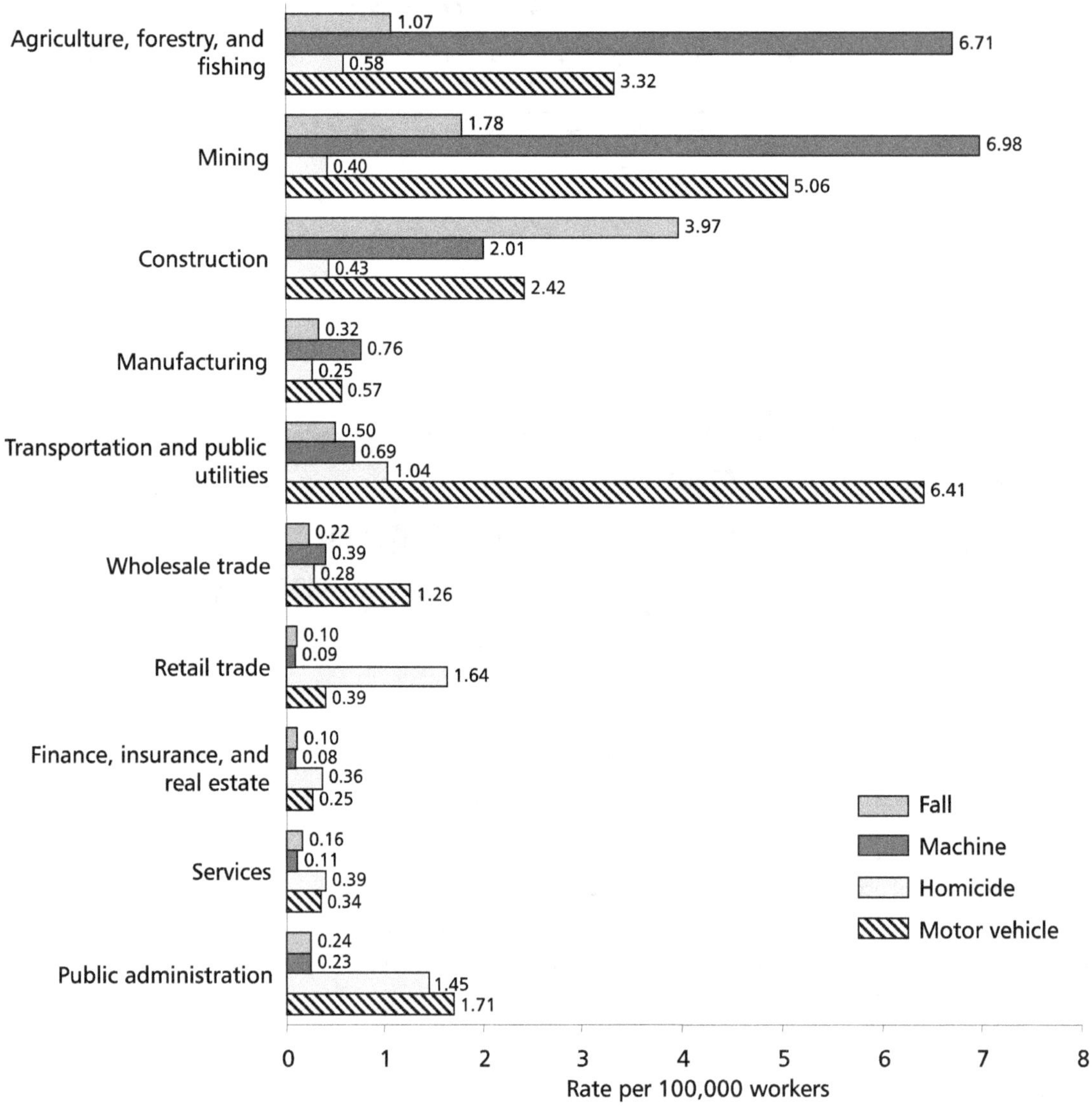

Fatal Injury

Industry	Fall	Machine	Homicide	Motor vehicle
Agriculture, forestry, and fishing	1.07	6.71	0.58	3.32
Mining	1.78	6.98	0.40	5.06
Construction	3.97	2.01	0.43	2.42
Manufacturing	0.32	0.76	0.25	0.57
Transportation and public utilities	0.50	0.69	1.04	6.41
Wholesale trade	0.22	0.39	0.28	1.26
Retail trade	0.10	0.09	1.64	0.39
Finance, insurance, and real estate	0.10	0.08	0.36	0.25
Services	0.16	0.11	0.39	0.34
Public administration	0.24	0.23	1.45	1.71

Rate per 100,000 workers

Figure 2–12. Average annual rate of fatal occupational injuries by industry division and selected causes of death, 1980–1995. (Source: NTOF [1999].)

Table 2–1. Distribution and average annual rate of fatal occupational injuries by State listed on death certificate, 1980–1995

State	Deaths	Rate*	Rank	State	Deaths	Rate	Rank
Alabama	1,875	6.9	19	Montana	750	12.4	3
Alaska	916	25.2	1	Nebraska	1,005	8.0	13
Arizona	596	2.5	49	Nevada	732	8.4	9
Arkansas	1,340	8.3	10	New Hampshire	257	3.0	45
California	9,821	4.8	33.5	New Jersey	1,523	2.6	47
Colorado	1,642	6.3	22	New Mexico	769	7.7	15.5
Connecticut	445	1.7	51	New York	3,567	2.9	46
Delaware	205	4.0	40	North Carolina	2,691	5.5	29.5
District of Columbia	298	6.5	21	North Dakota	442	9.0	8
Florida	5,643	6.6	20	Ohio	2,659	3.4	42.5
Georgia	3,284	7.2	18	Oklahoma	1,329	5.8	25.5
Hawaii	353	4.5	37	Oregon	1,552	7.3	17
Idaho	787	10.8	4	Pennsylvania	3,927	4.7	35
Illinois	4,171	4.9	32	Rhode Island	193	2.6	48
Indiana	2,284	5.5	29.5	South Carolina	1,416	5.8	25.5
Iowa	1,374	6.2	23	South Dakota	496	9.2	7
Kansas	1,178	6.1	24	Tennessee	1,980	5.7	27
Kentucky	2,040	8.0	13	Texas	9,449	7.7	15.5
Louisiana	2,282	8.2	11	Utah	938	8.0	13
Maine	403	4.6	36	Vermont	195	4.4	39
Maryland	1,227	3.4	42.5	Virginia	2,541	5.5	29.5
Massachusetts	987	2.1	50	Washington	1,833	5.4	31
Michigan	2,437	3.7	41	West Virginia	1,145	10.5	5
Minnesota	1,092	3.1	44	Wisconsin	1,730	4.5	38
Mississippi	1,692	10.1	6	Wyoming	615	16.7	2
Missouri	1,823	4.8	33.5				

Source: NTOF [1999].
*Per 100,000 workers.

Fatal Injuries by Establishment Size

CFOI collects information about the number of workers in establishments where fatally injured workers were employed. In 1997, this information was available for 79% of all records for private sector wage and salary workers. Based on the available data, the highest rate of fatal occupational injury (8.6 per 100,000 workers) occurred in establishments with 1 to 10 workers, whereas the lowest rate (2 per 100,000 workers) occurred in establishments with 100 or more workers (Figure 2–13). Self-employed workers accounted for 20% of the fatal occupational injuries in 1997. The fatality rate of 11.7 cases per 100,000 workers for the self-employed was nearly two and a half times the rate of 4.8 per 100,000 for all wage and salary workers (public and private sector combined).

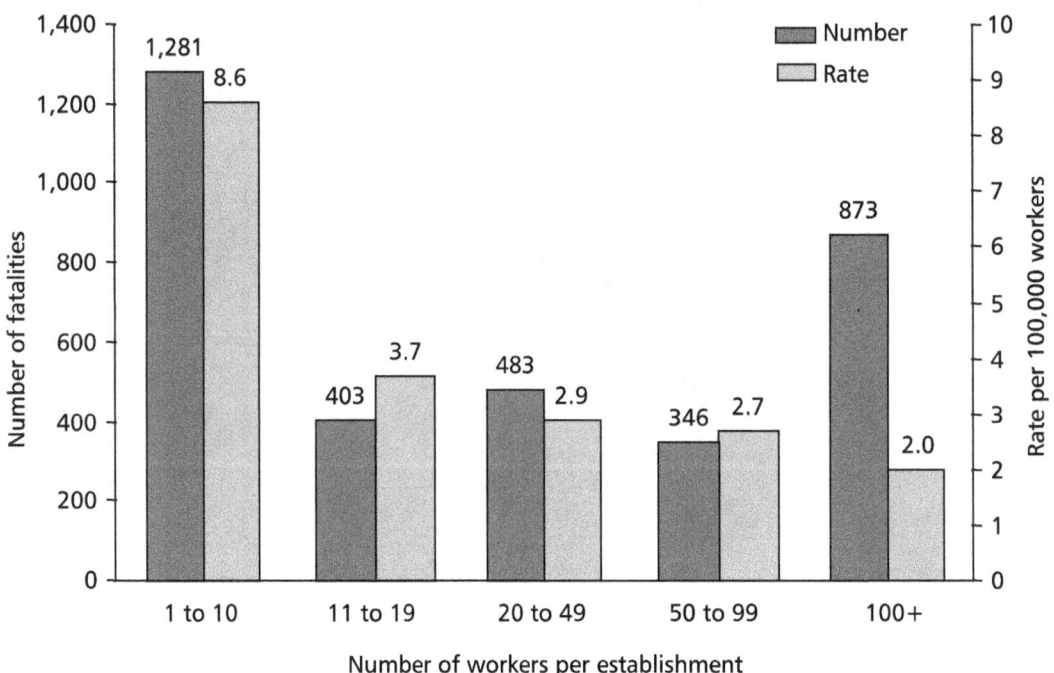

Figure 2–13. Number and rate of fatal occupational injuries in private sector wage and salary workers by employment size of establishment, 1997. The total number of fatal occupational injuries was 4,305. Employment size was not reported for 919 fatalities; these data could significantly change the above rates. (Source: CFOI [1999]. Employment data are from the *Employment and Wages Annual Averages, 1997* [BLS 1998].)

Special Topics in Fatal Occupational Injury

Fatal Injuries among Truck Drivers

Truck drivers suffered nearly 14% of the fatal occupational injuries during 1997 according to CFOI data. The number of fatalities among truck drivers has increased fairly steadily, from 699 in 1992 to 862 in 1997. Over the same period, the fatality rate increased from 26 to 28 per 100,000 workers. In 1997, more than 50% of the fatalities occurred in trucks with trailers or semitrailers (Figure 2–14), and more than 80% occurred in transportation-related incidents (Figure 2–15). Fatalities from jackknifing and from collisions increased by 16% and 9%, respectively, between 1996 and 1997. More than half of the fatal occupational injuries among truck drivers occurred on interstate highways, freeways, expressways, or other State or U.S. highways (Figure 2–16).

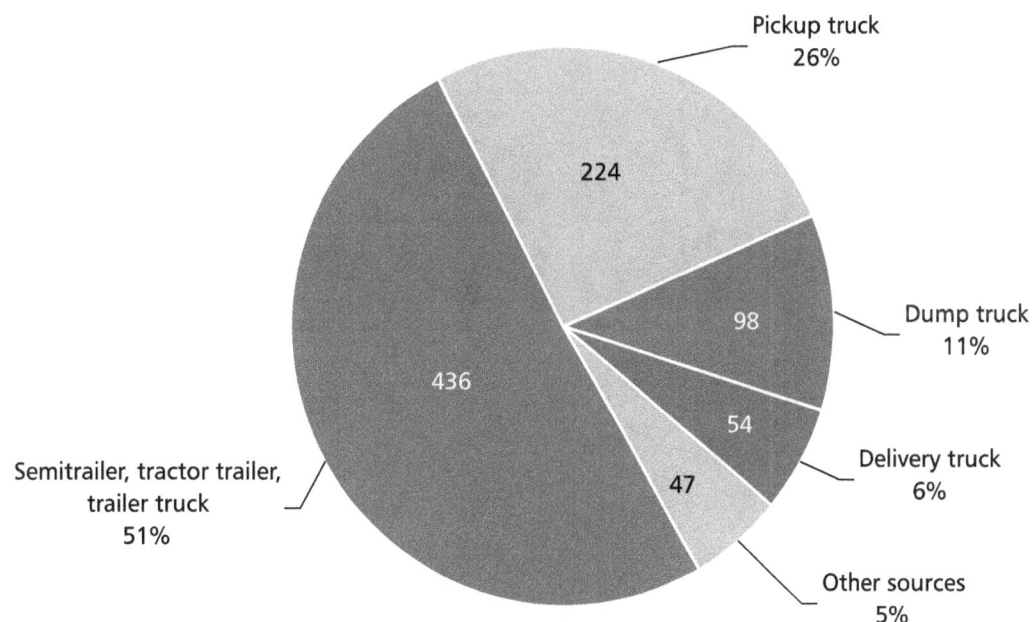

Figure 2–14. Number and distribution of fatal occupational injuries to truck drivers by source of fatal injury, 1997. (Source: CFOI [1999].)

FATAL INJURY

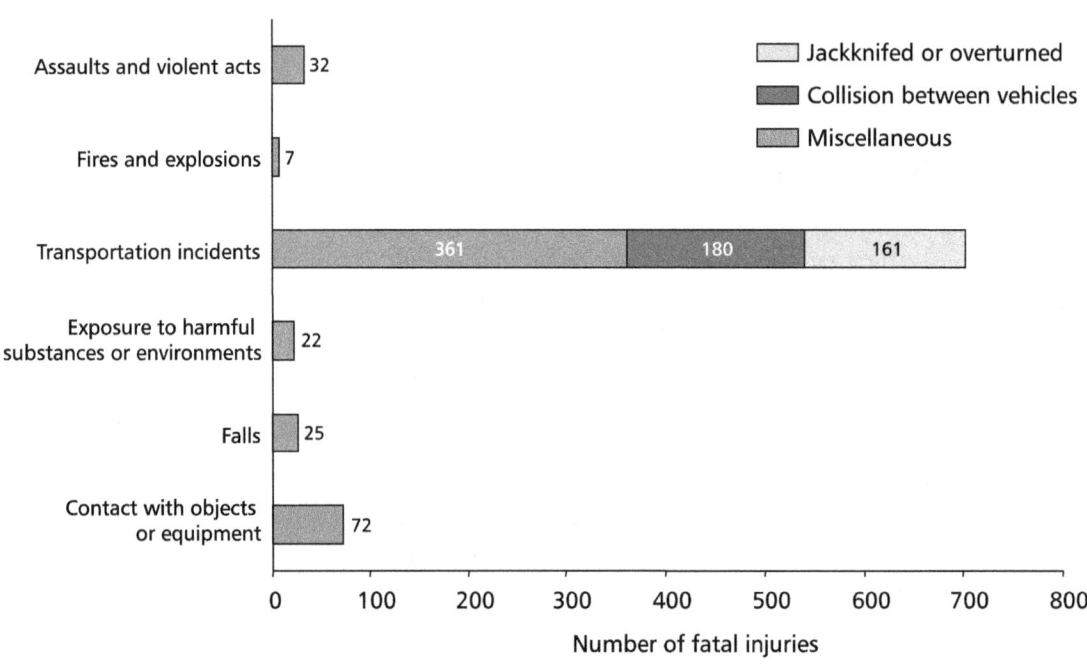

Figure 2–15. Number of fatal occupational injuries to truck drivers by event or exposure, 1997. (Source: CFOI [1999].)

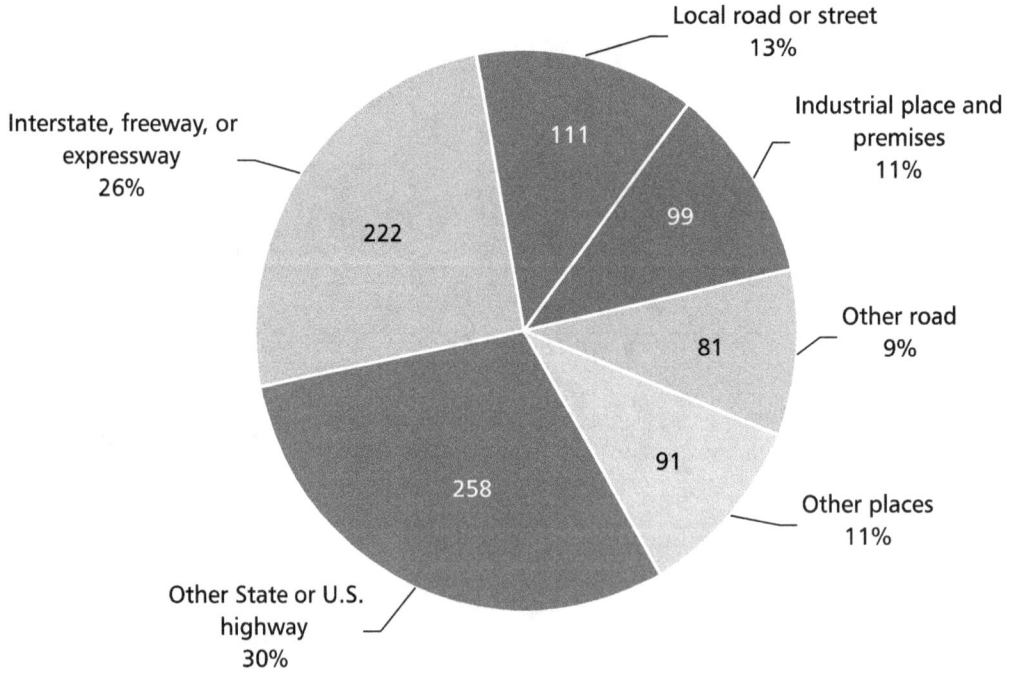

Figure 2–16. Number and distribution of fatal occupational injuries to truck drivers by location of fatal injury, 1997. (Source: CFOI [1999].)

Homicides

Homicides, the second leading cause of fatal occupational injuries, declined by 7% from 1996 to 1997. Taxi drivers had the highest rate of homicide (Figure 2–17); the highest number of homicides occurred in retail trade in grocery stores and eating and drinking establishments (Figure 2–18). Eighty percent of workplace homicides resulted from shootings [CFOI 1999]. Robbery was the primary motive for occupational homicide when a motive could be ascertained from the source documents (Figure 2–19).

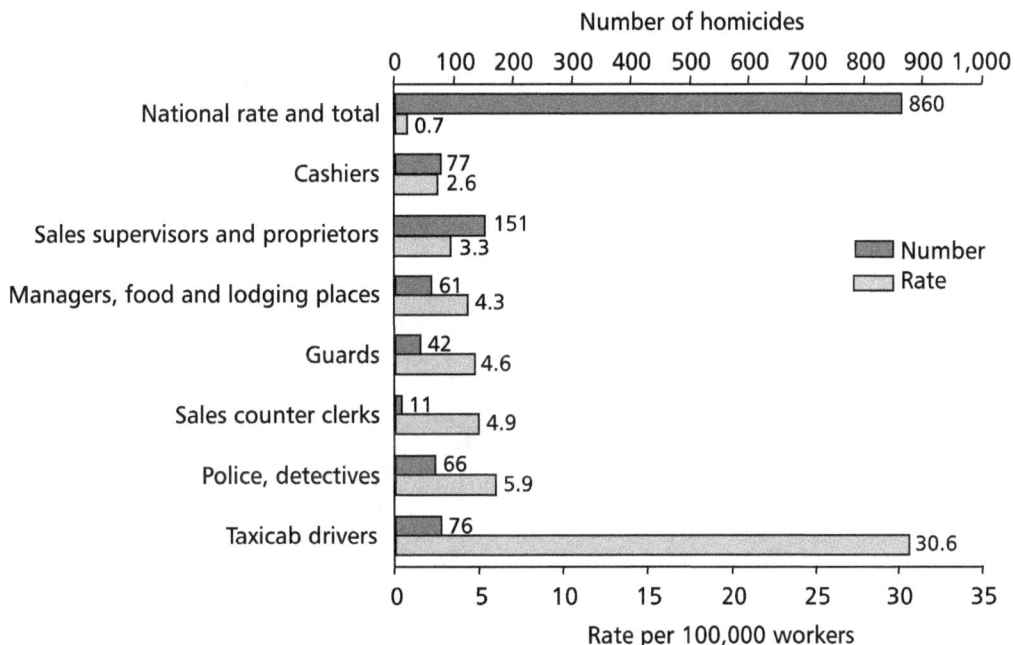

Figure 2–17. Number and incidence rate of homicides for high-risk occupations, 1997. (Source: CFOI [1999].)

FATAL INJURY

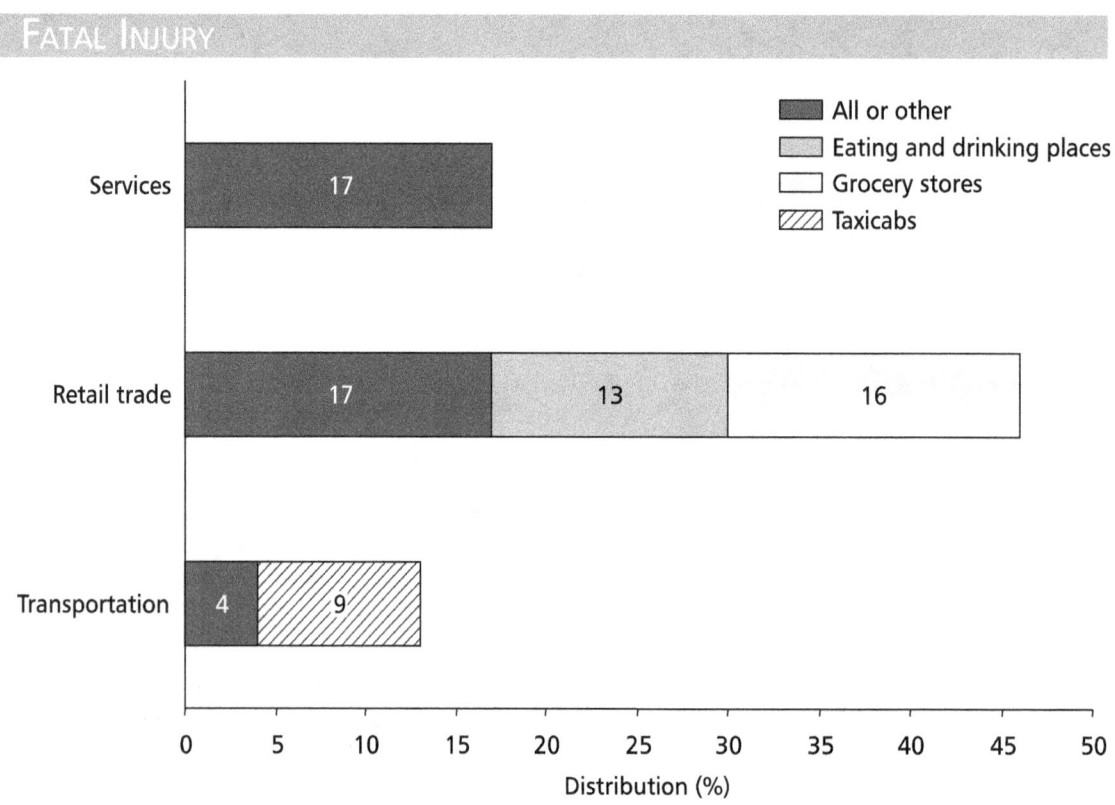

Figure 2–18. Distribution of homicides in high-risk industries, 1997. (Source: CFOI [1999].)

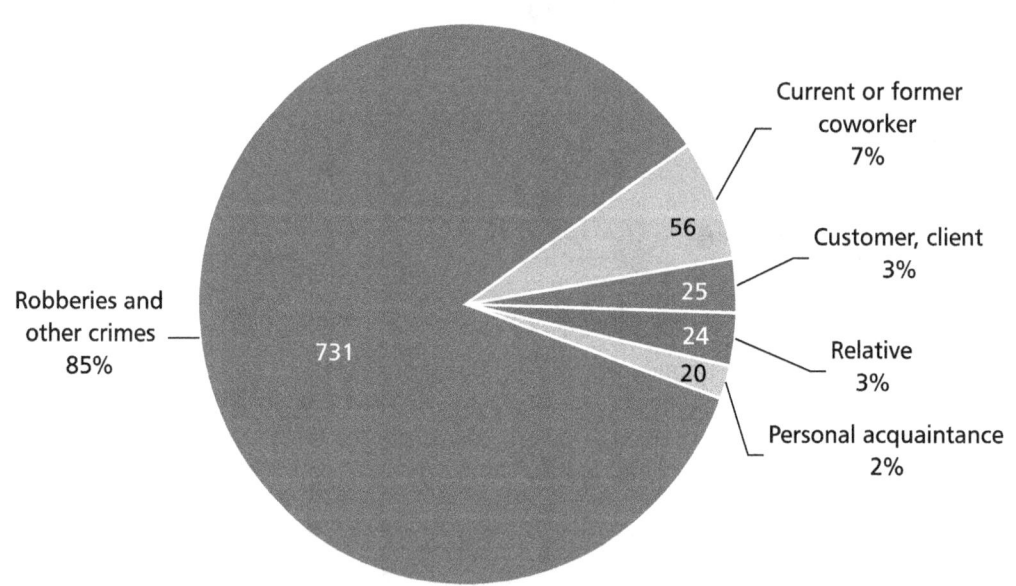

Figure 2–19. Number and distribution of work-related homicides, by circumstance or alleged perpetrator, 1997. (Source: CFOI [1999].)

Fatal Falls

Falls were the fifth leading type of fatal occupational event in 1997, accounting for more than 700 deaths, or 12% of all fatal occupational injuries (Figure 1–9). Fatalities from falls recorded in the CFOI increased by more than 19% from 1992 to 1997. Falls to a lower level, including falls from roofs, were the major contributors (Figure 2–20). Approximately half of the falls occurred in the construction industry (Figures 2–21 and 2–22).

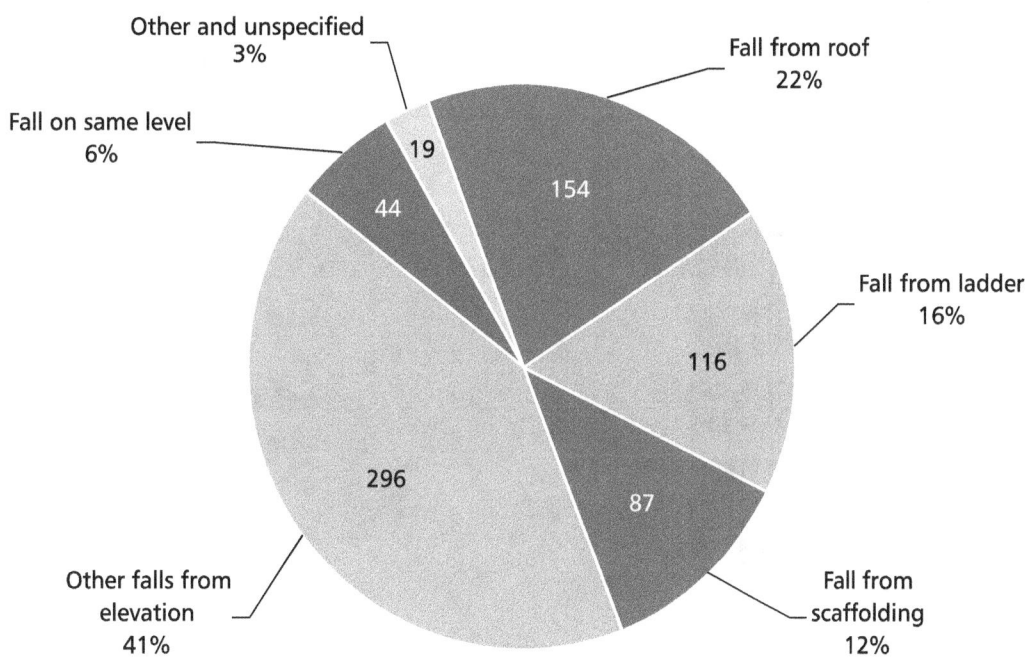

Figure 2-20. Number and distribution of fatal occupational falls by type of fall, 1997. (Source: CFOI [1999].)

FATAL INJURY

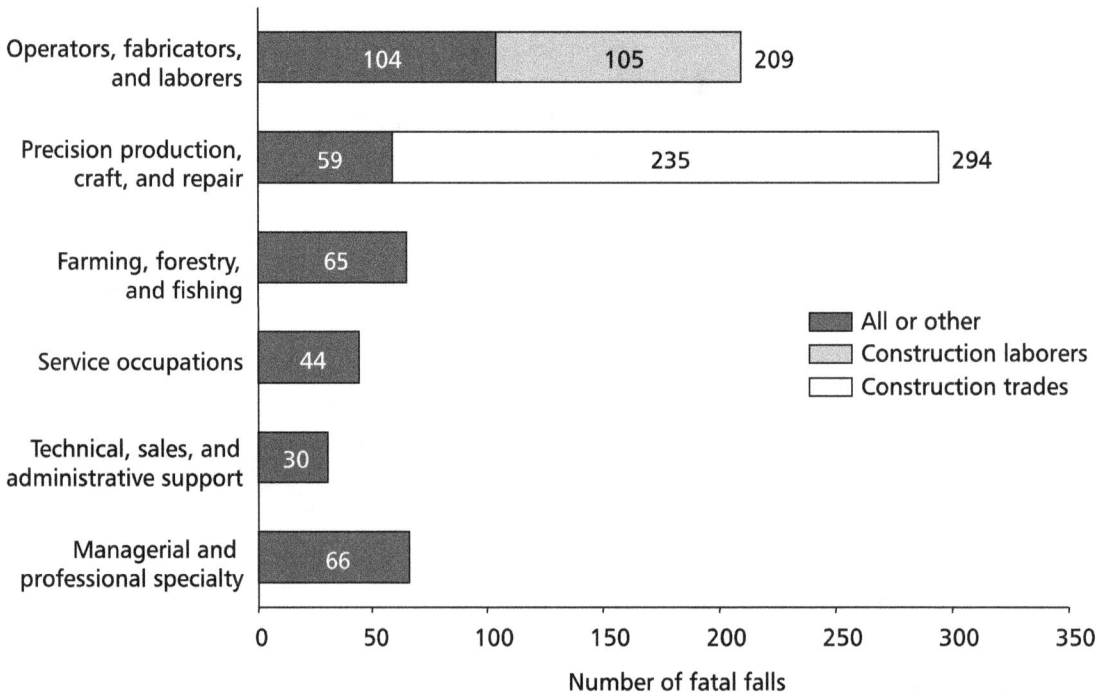

Figure 2–21. Number of fatal occupational falls by occupational group, 1997. (Source: CFOI [1999].)

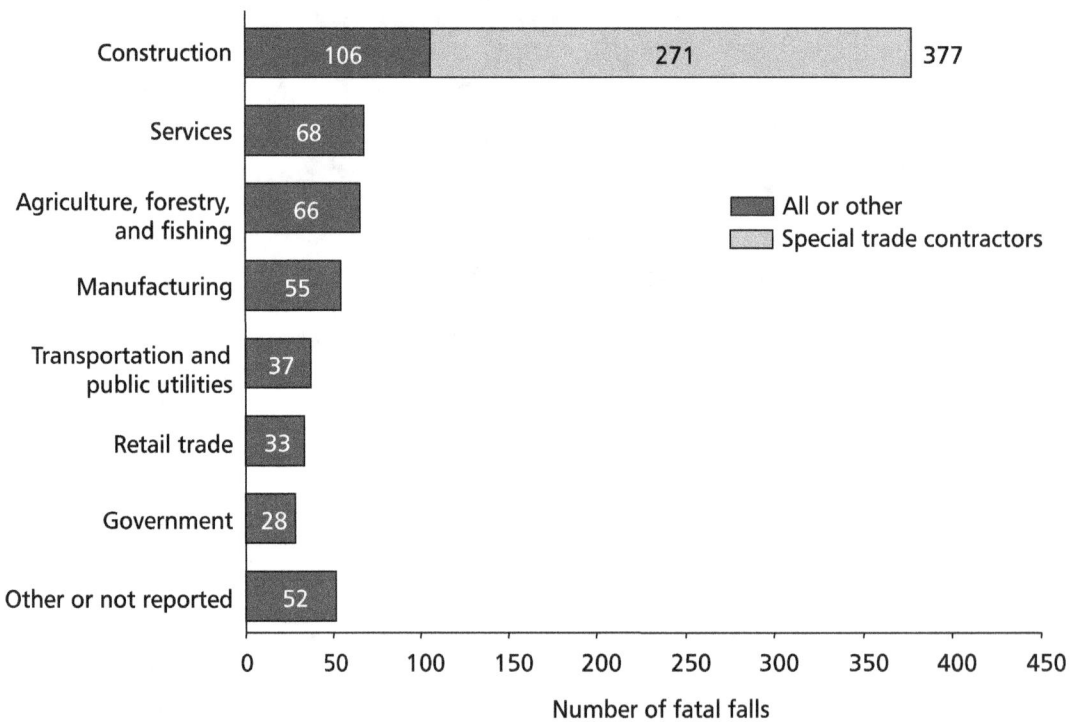

Figure 2–22. Number of fatal occupational falls by industry division, 1997. (Source: CFOI [1999].)

Fire Fighter Fatalities

NIOSH began investigating all fire fighter fatalities in October 1998. The goal of this initiative is to examine the magnitude and characteristics of occupational deaths and severe injuries among fire fighters and to develop recommendations for injury prevention. The investigations are being conducted through the fatality assessment and control evaluation (FACE) model developed by NIOSH. For each case investigated, information is collected on factors associated with the fire fighter who died, the physical agents contributing to the death, and the environment. These factors are identified during three phases: pre-event, event, and post-event. The contributing factors are investigated in detail for each incident and are summarized in the investigation report along with recommendations for preventing future incidents. Additional information about the NIOSH fire fighter program and individual investigation case reports are available on the NIOSH Web site at www.cdc.gov/niosh/firehome.html.

The National Fire Protection Association (NFPA) and the U.S. Fire Administration estimate that an average of 112 fire fighters died on the job each year between 1979 and 1998 (Figure 2–23). In 1998, 44% of the fire fighter deaths occurred at the fireground. Another 35% occurred while responding to or returning from alarms or performing other nonfire emergency duties (Figure 2–24). Heart attacks (43%), internal trauma (23%), and asphyxiation (10%) were the most frequent causes of death in 1998 (Figure 2–25).

FATAL INJURY

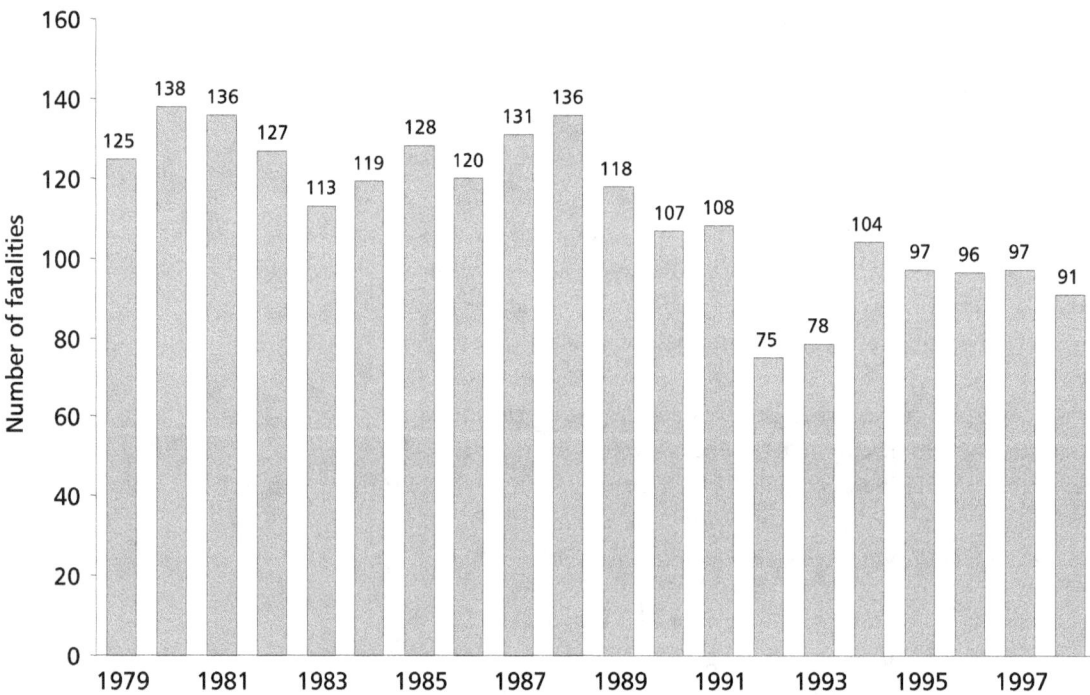

Figure 2–23. Number of fire fighter deaths, 1979–1998. Total number of deaths was 2,244. (Source: NFPA [Washburn et al. 1999].)

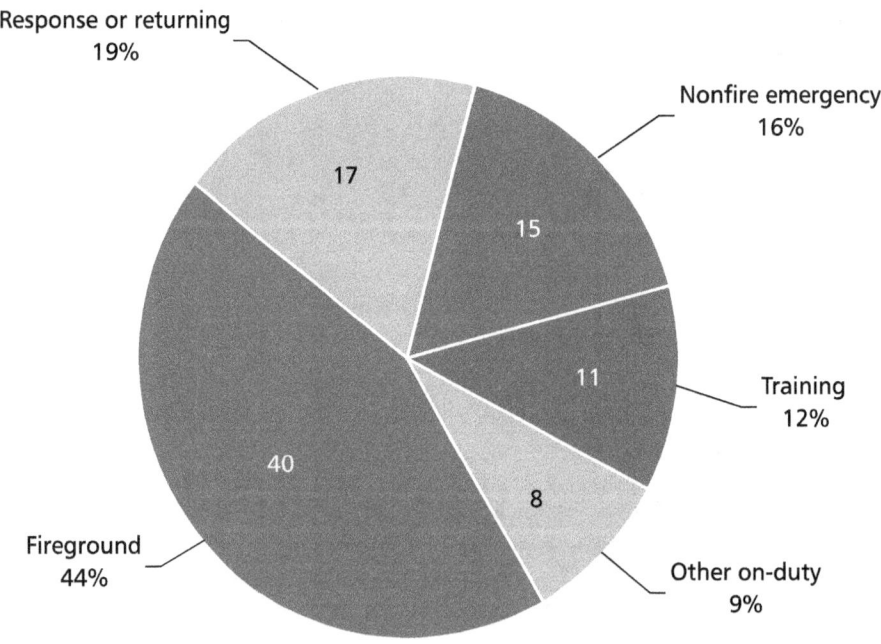

Figure 2–24. Number and distribution of fire fighter deaths by type of duty, 1998. Total number of deaths was 91. (Source: NFPA [Washburn et al. 1999].)

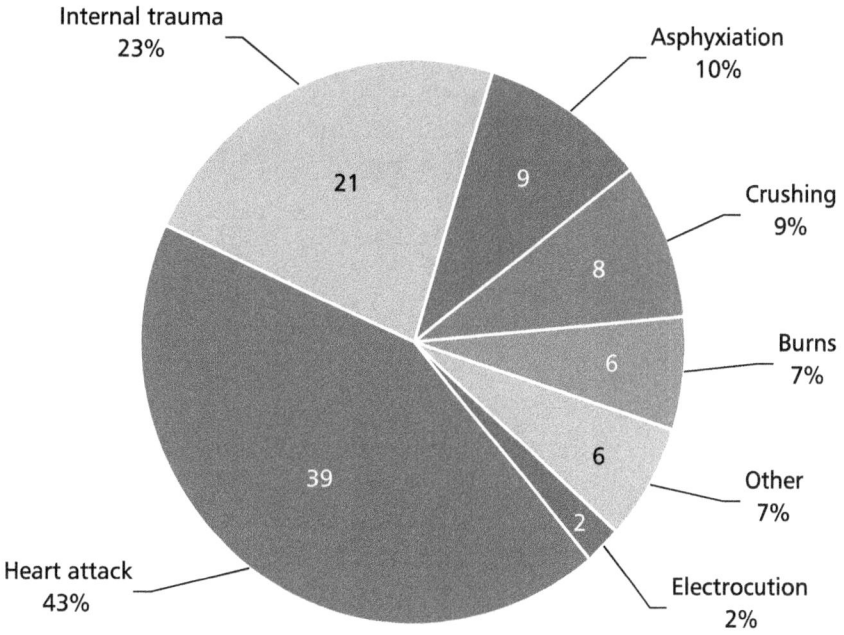

Figure 2–25. Number and distribution of fire fighter deaths by nature of injury, 1998. Total number of deaths was 91. (Source: NFPA [Washburn et al. 1999].)

3 Fatal Illness

3 Fatal Illness

Fatal illness in the workplace has been of interest to the public health community since at least the 18th century, when Bernardino Ramazzini compiled the first systematic description of the diseases of workers [Ramazzini 1713]. Diseases are generally more difficult to link with work than injuries. Many diseases related to occupational exposures (e.g., tuberculosis [TB], cancers, central nervous system disorders, and asthma) are no different when encountered in the absence of occupational exposures. Work-related aspects of illness may go unrecognized for many reasons, including long latency periods between the exposure and development of some diseases and the failure of health care professionals to recognize work-related illnesses or to obtain information about work history. This chapter covers conditions generally accepted to be solely or predominantly related to work. Excluded, for example, is lung cancer, even though 16% to 17% of cases in men and 2% of cases in women are considered to be work-related.

Pneumoconiosis

The pneumoconioses are a class of respiratory diseases attributed solely to workplace factors. From 1968 through 1996, pneumoconiosis was an underlying or contributing cause of 113,519 deaths in the United States (see Figure 1–10). The largest number of pneumoconiosis deaths were attributed to coal workers' pneumoconiosis (CWP), but deaths from this disease have declined over the years (Figure 3–1). By contrast, asbestosis deaths increased from fewer than 100 in 1968 to nearly 1,200 in 1996 (Figure 3–2). Over the same period, silicosis deaths decreased (Figure 3–3), byssinosis deaths varied substantially each year from 1979 to 1996 (Figure 3–4), and unspecified and other types of pneumoconiosis decreased (Figure 3–5).

Fatal Illness

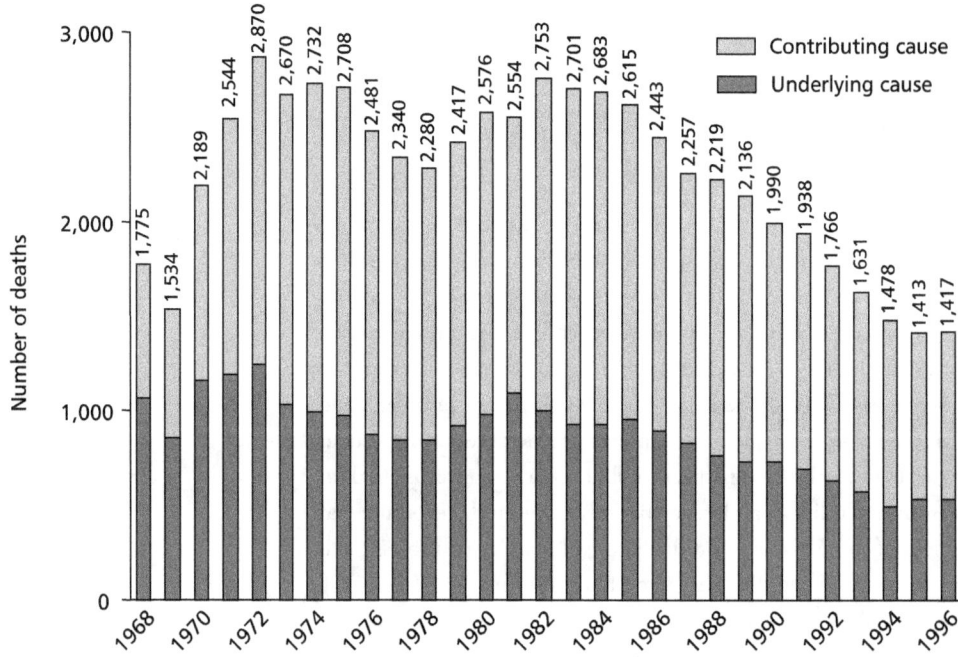

Figure 3–1. Number of deaths recorded with CWP as an underlying or contributing cause on the death certificate—U.S. residents aged 15 and older, 1968–1996. (Source: NSSPM [1999].)

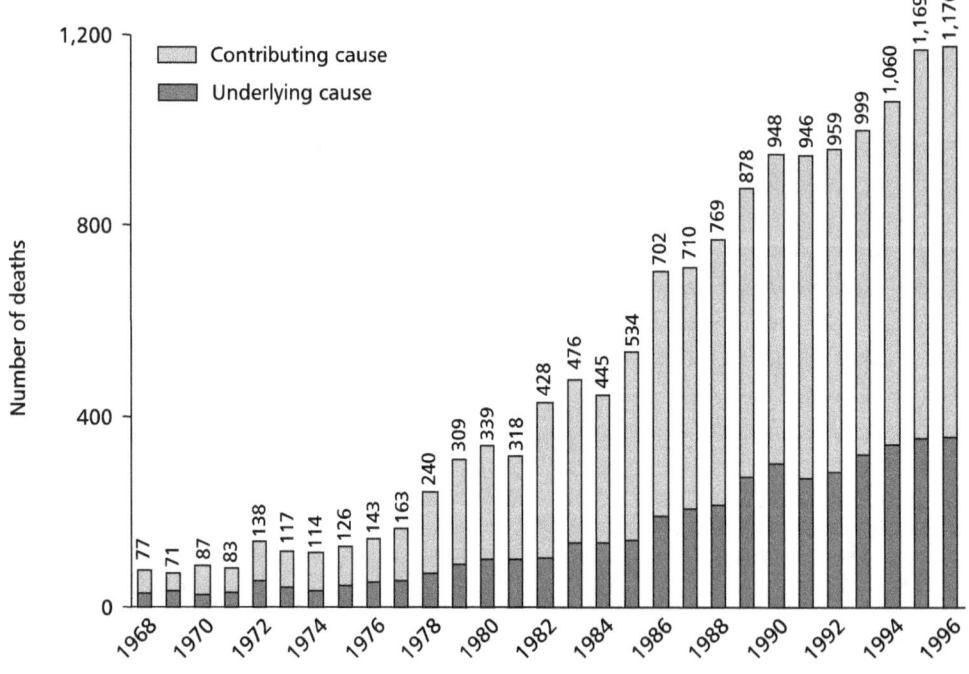

Figure 3–2. Number of deaths recorded with asbestosis as an underlying or contributing cause on the death certificate—U.S. residents aged 15 and older, 1968–1996. (Source: NSSPM [1999].)

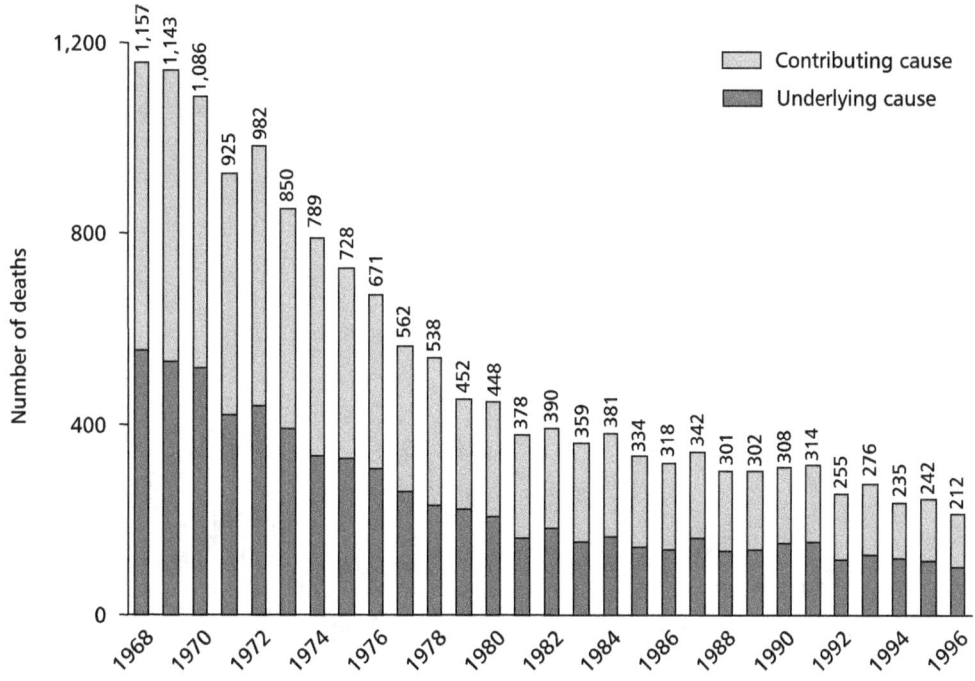

Figure 3-3. Number of deaths recorded with silicosis as an underlying or contributing cause on the death certificate—U.S. residents aged 15 and older, 1968–1996. (Source: NSSPM [1999].)

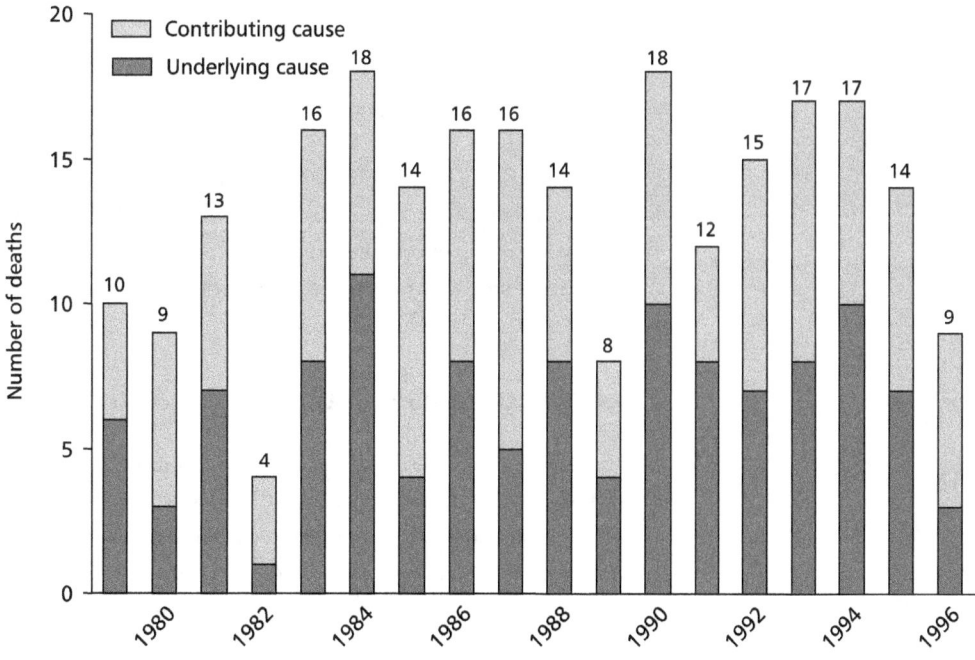

Figure 3-4. Number of deaths recorded with byssinosis as an underlying or contributing cause on the death certificate—U.S. residents aged 15 and older, 1979–1996. (Source: NSSPM [1999].)

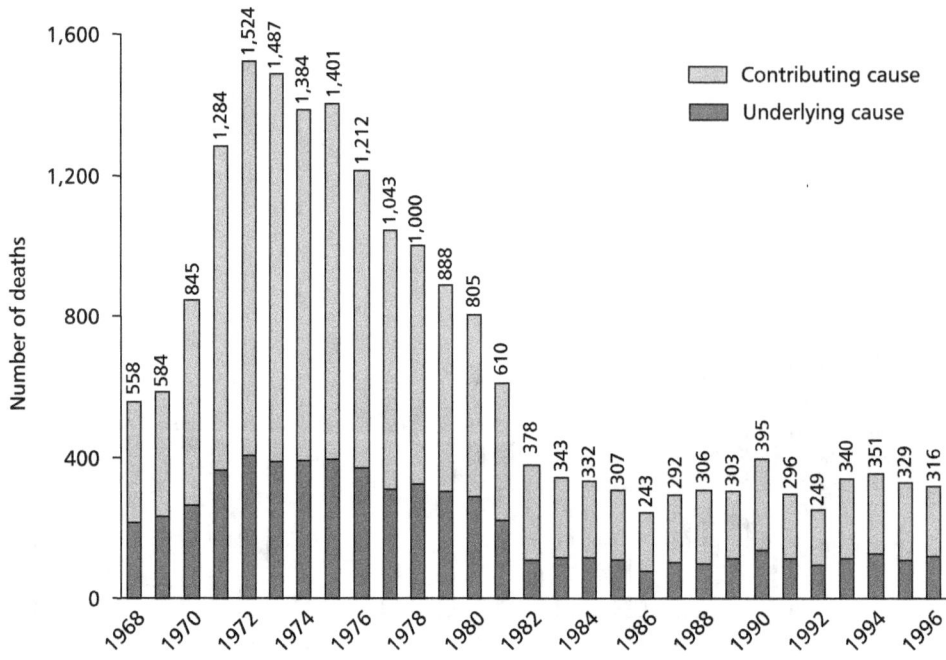

Figure 3–5. Number of deaths recorded with unspecified and other pneumoconiosis as an underlying or contributing cause on the death certificate—U.S. residents aged 15 and older, 1968–1996. (Source: NSSPM [1999].)

Pneumoconiosis Deaths by State

Asbestosis mortality is highest in northeastern, southern, and west coast States (Figure 3–6), and CWP mortality is highest in Appalachian mining areas (Figure 3–7). Silicosis mortality appears less concentrated by geographic region than asbestosis or CWP mortality (Figure 3–8). Byssinosis deaths are concentrated in textile-producing States (Figure 3–9). The pattern of mortality for unspecified and other pneumoconiosis most resembles that of CWP (Figure 3–10).

Pneumoconiosis Deaths by Sex and Race

The distribution of different types of pneumoconiosis deaths varies by sex (Figure 3–11) and race (Figure 3–12). Women accounted for 28% of byssinosis deaths and less than 5% of deaths with all other types of pneumoconiosis. Blacks accounted for 15% of silicosis deaths, 13% of byssinosis deaths, and less than 7% of deaths with all other types of pneumoconiosis.

FATAL ILLNESS

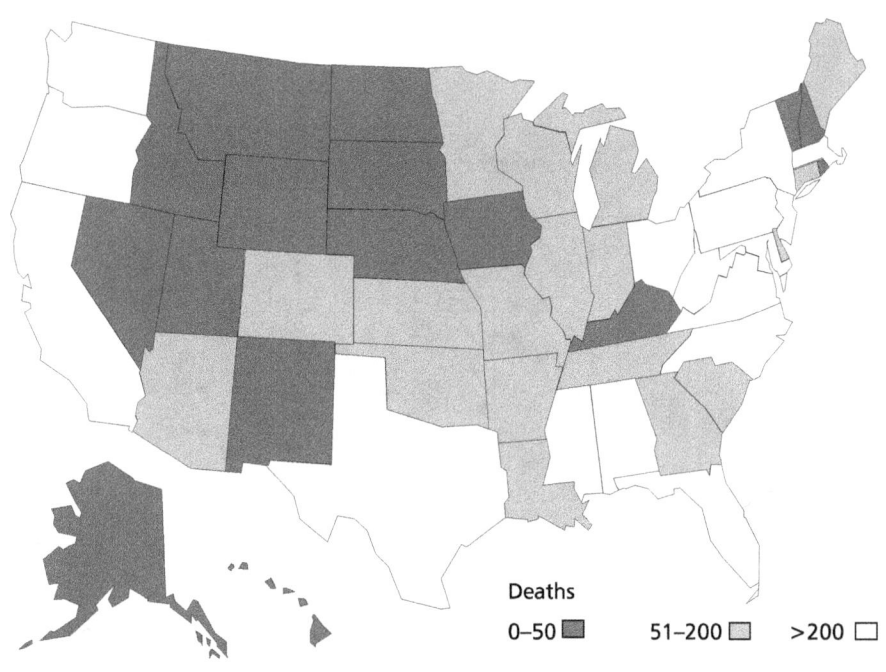

Figure 3–6. Number of asbestosis deaths by State—U.S. residents aged 15 and older, 1987–1996. (Source: NSSPM [1999].)

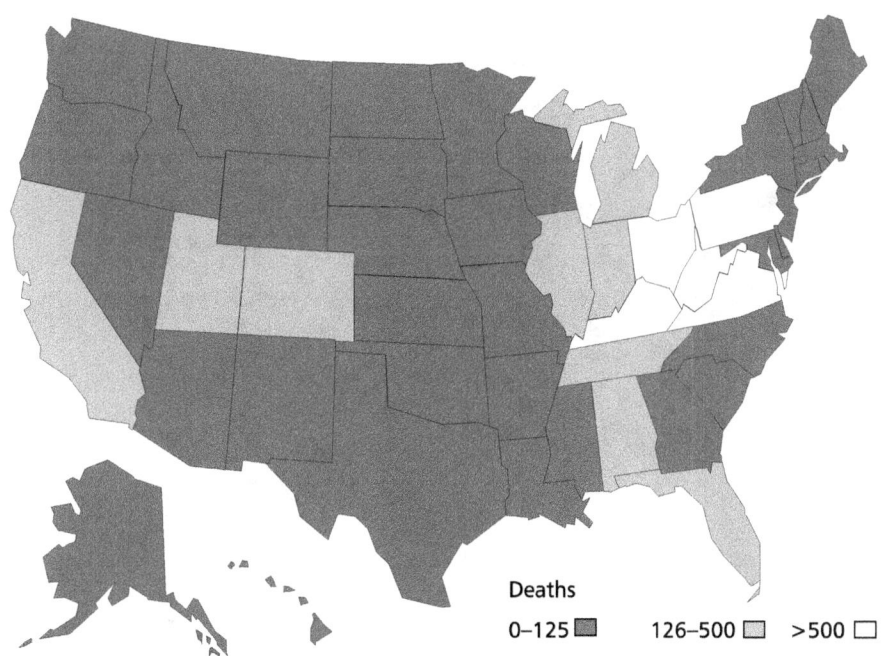

Figure 3–7. Number of CWP deaths by State—U.S. residents aged 15 and older, 1987–1996. (Source: NSSPM [1999].)

Fatal Illness

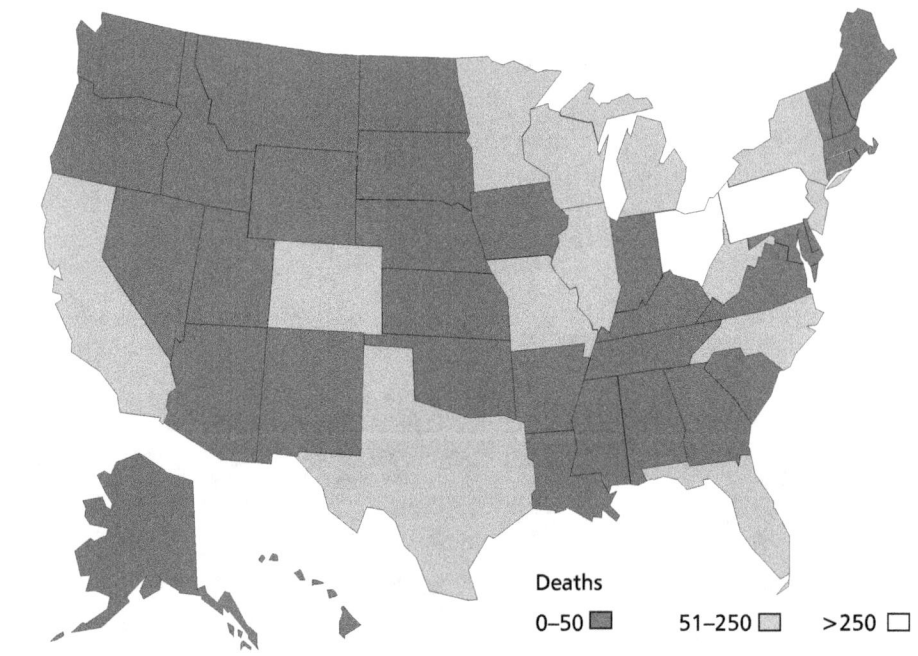

Figure 3–8. Number of silicosis deaths by State—U.S. residents aged 15 and older, 1987–1996. (Source: NSSPM [1999].)

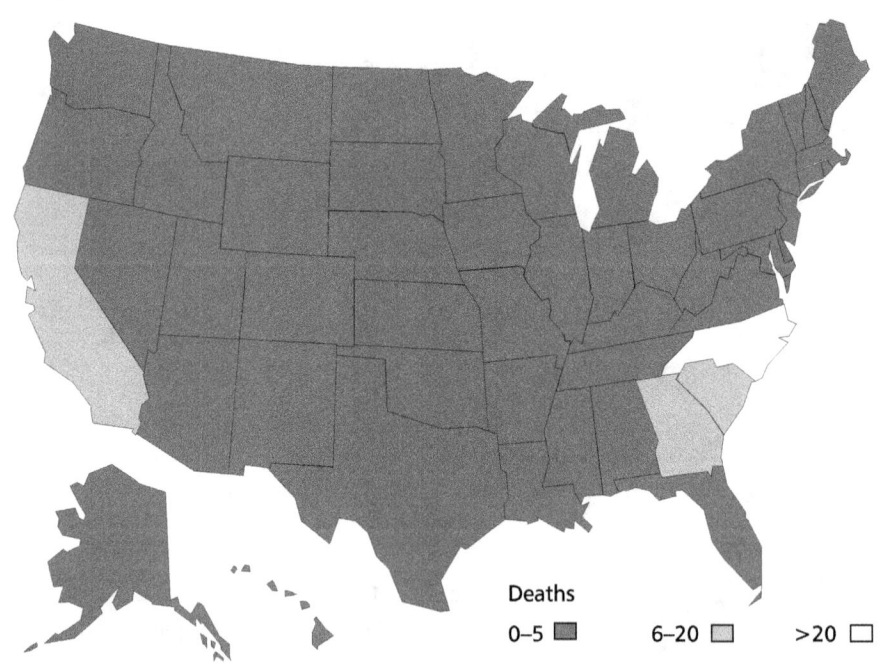

Figure 3–9. Number of byssinosis deaths by State—U.S. residents aged 15 and older, 1987–1996. (Source: NSSPM [1999].)

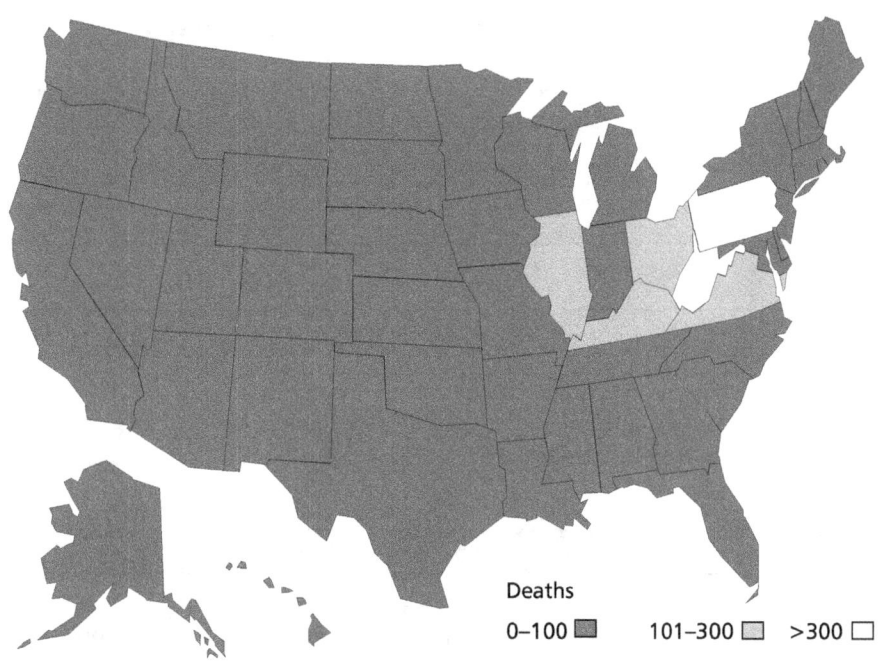

Figure 3–10. Number of unspecified and other pneumoconiosis deaths by State—U.S. residents aged 15 and older, 1987–1996. (Source: NSSPM [1999].)

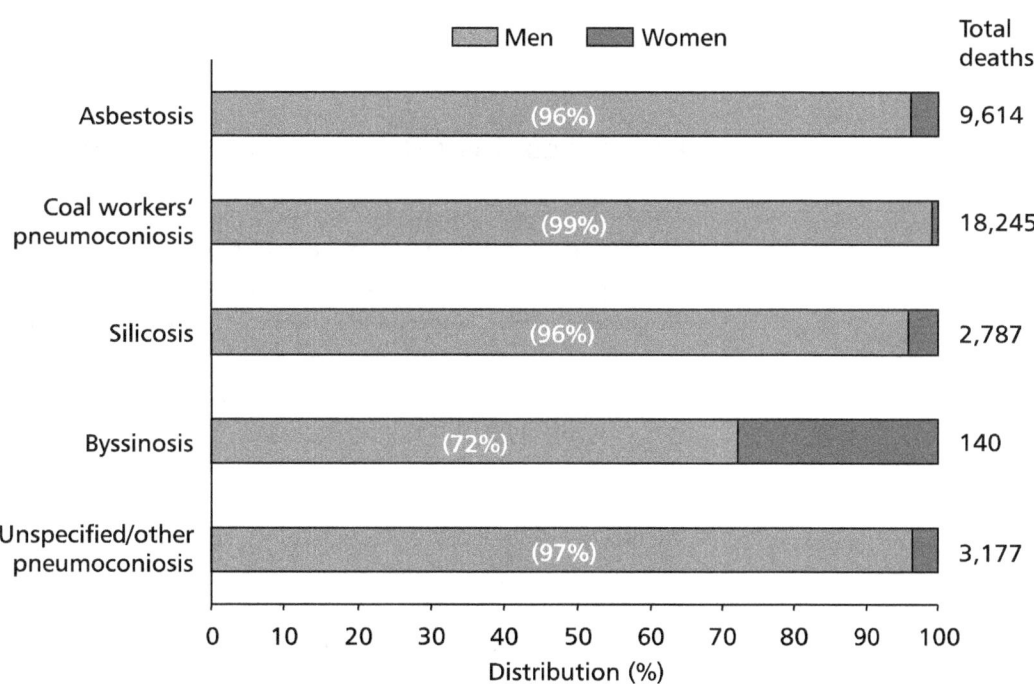

Figure 3–11. Distribution of types of pneumoconiosis deaths by sex—U.S. residents aged 15 and older, 1987–1996. (Source: NSSPM [1999].)

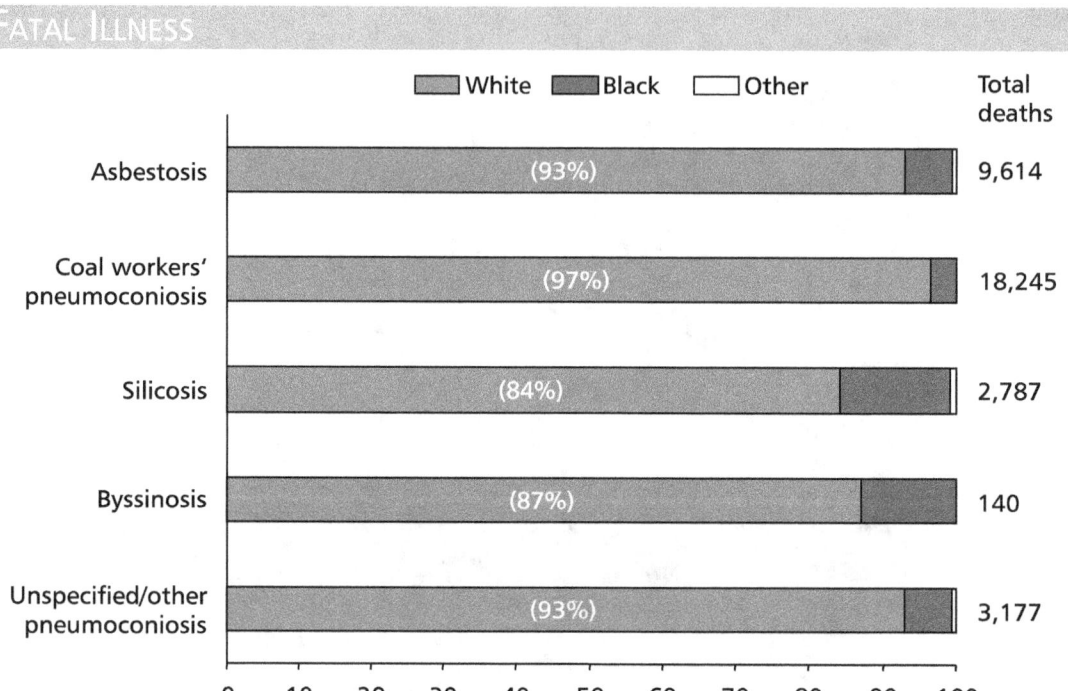

Figure 3–12. Distribution of types of pneumoconiosis deaths by race—U.S. residents aged 15 and older, 1987–1996. (Source: NSSPM [1999].)

Pneumoconiosis Deaths by Occupation

Proportionate mortality ratios (PMRs) associating pneumoconiosis deaths with various occupations are presented in Figures 3–13 through 3–17. A PMR above 1.0 indicates that more deaths occurred with the condition than expected in an occupation or industry. PMRs with lower 95% confidence limits that exceed 1.0 are statistically significant. PMRs calculated from a large subset of national data indicate that mining machine operators have extremely high relative mortality from CWP and from unspecified and other pneumoconioses (Figures 3–13 and 3–14). Insulation workers and related occupations had the highest PMRs for asbestosis (Figure 3–15). Workers in metal and plastic processing, hand molding and shaping, and crushing and grinding in mining occupations had the highest PMRs for silicosis mortality (Figure 3–16). Textile machine operators and repairers had significant PMRs for byssinosis (Figure 3–17).

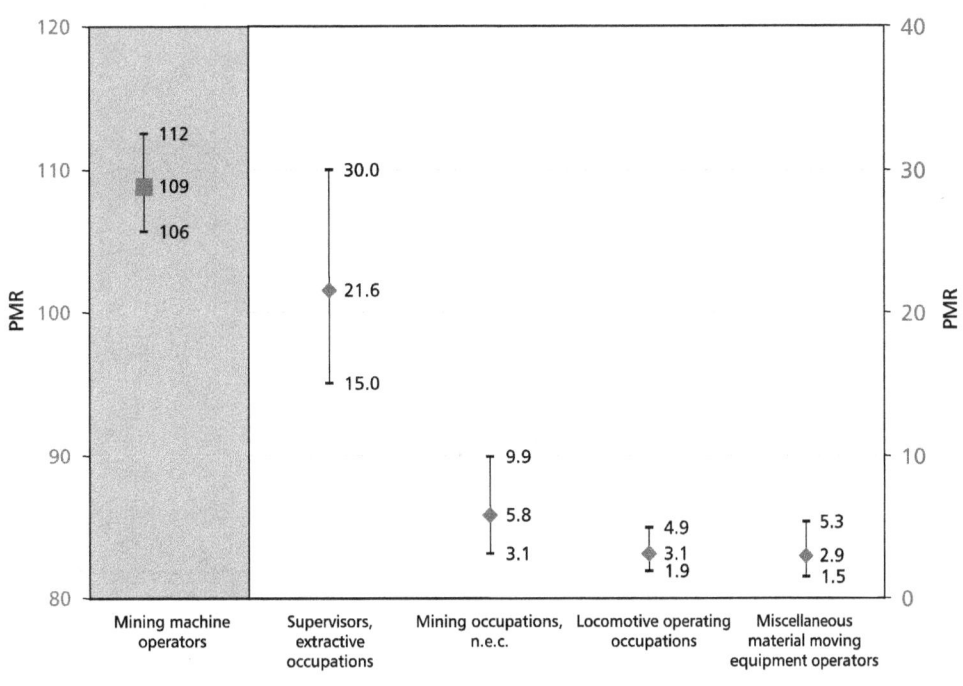

Figure 3–13. PMRs (and 95% CIs) for CWP by occupation—U.S. residents aged 15 and older, 1987–1996. (Source: NSSPM [1999].)

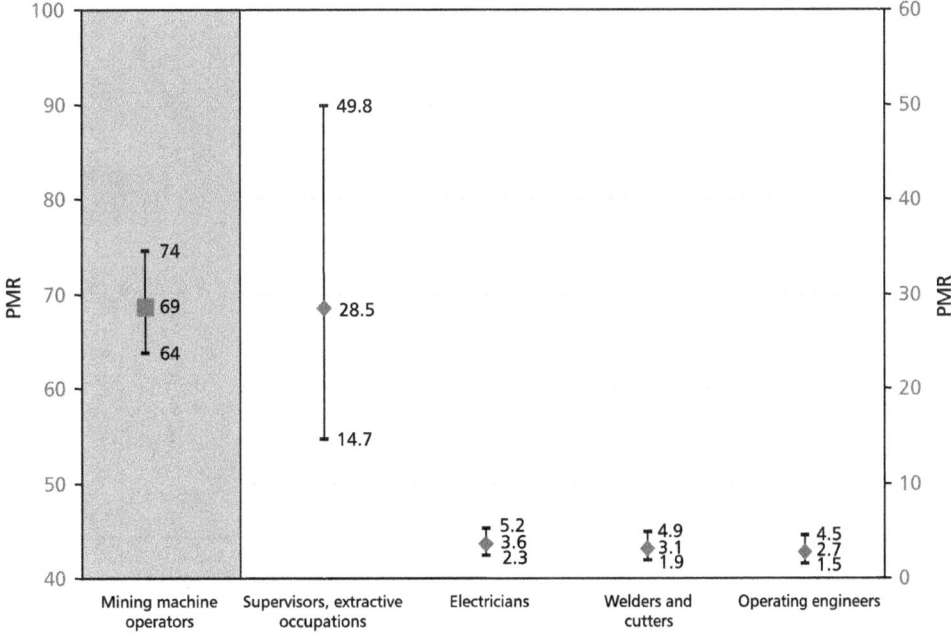

Figure 3–14. PMRs (and 95% CIs) for unspecified and other pneumoconioses by occupation—U.S. residents aged 15 and older, 1987–1996. (Source: NSSPM [1999].)

Fatal Illness

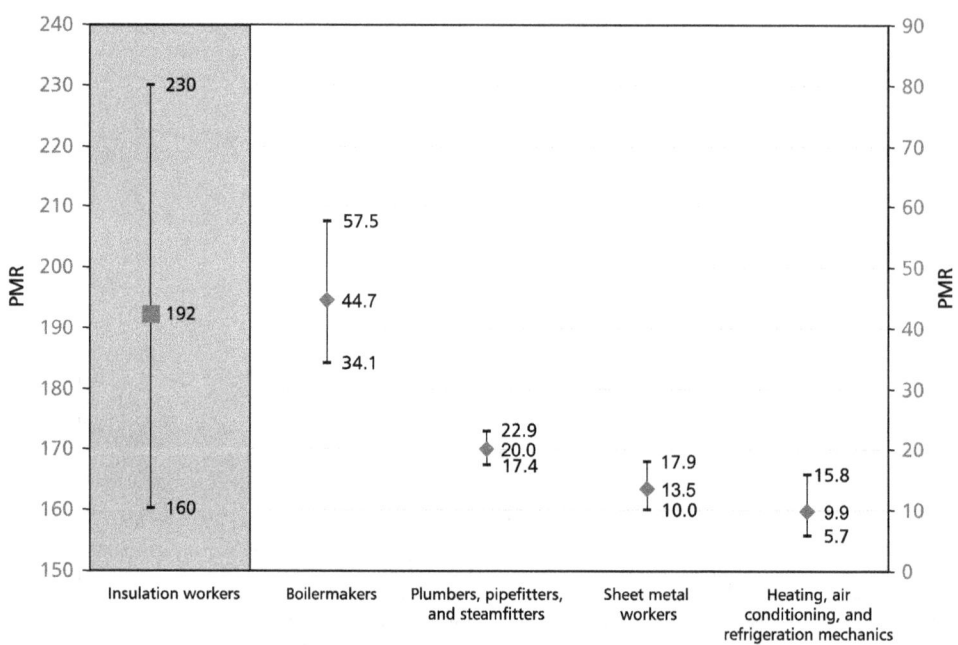

Figure 3–15. PMRs (and 95% CIs) for asbestosis by occupation—U.S. residents aged 15 and older, 1987–1996. (Source: NSSPM [1999].)

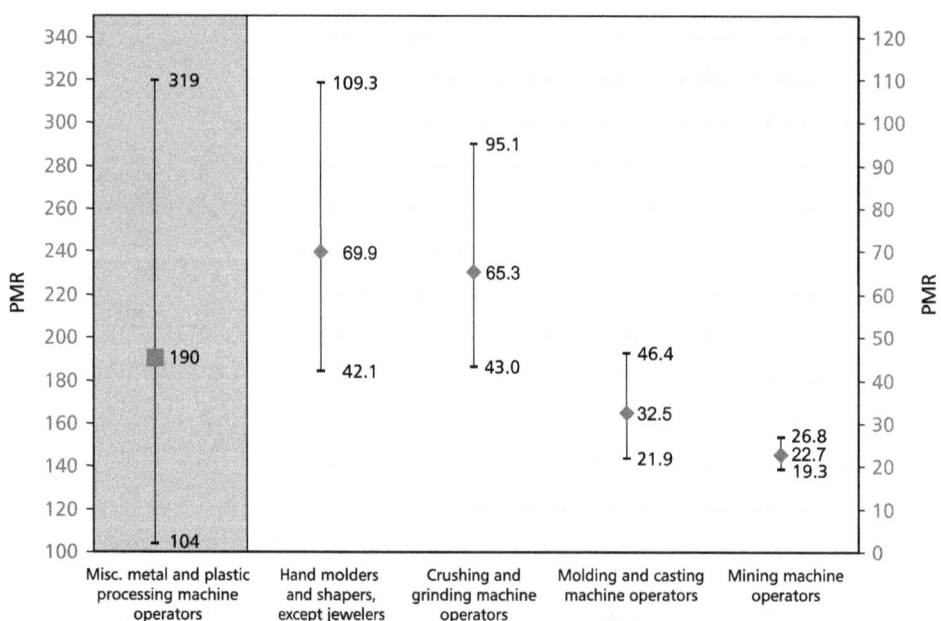

Figure 3–16. PMRs (and 95% CIs) for silicosis by occupation—U.S. residents aged 15 and older, 1987–1996. (Source: NSSPM [1999].)

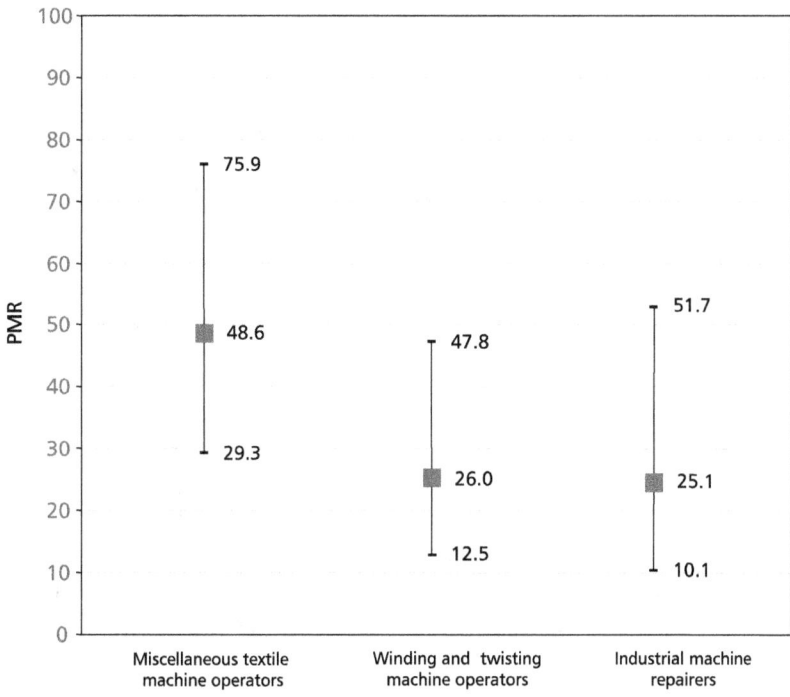

Figure 3–17. PMRs (and 95% CIs) for byssinosis in textile machine operators and repairers—U.S. residents aged 15 and older, 1987–1996. (Source: NSSPM [1999].)

Malignant Pleural Neoplasm

Mortality due to malignant pleural neoplasm (cancer of the lung lining) can serve as a surrogate for mortality due to malignant mesothelioma (often a cancer of the lung lining) because no unique cause-of-death code is currently available for mesothelioma. Asbestos exposure is by far the leading cause of malignant mesothelioma. The number of deaths associated with malignant pleural neoplasm increased during 1968–1996 (Figure 3–18). A geographic distribution of cases is presented in Figure 3–19. From 1987 to 1996, men accounted for 76% of the deaths from malignant pleural neoplasm (Figure 3–20), and white U.S. residents accounted for 94% of these deaths (Figure 3–21). Occupations with the highest PMRs for malignant pleural neoplasm (Figure 3–22) are similar to those with high PMRs for asbestosis (Figure 3–15).

FATAL ILLNESS

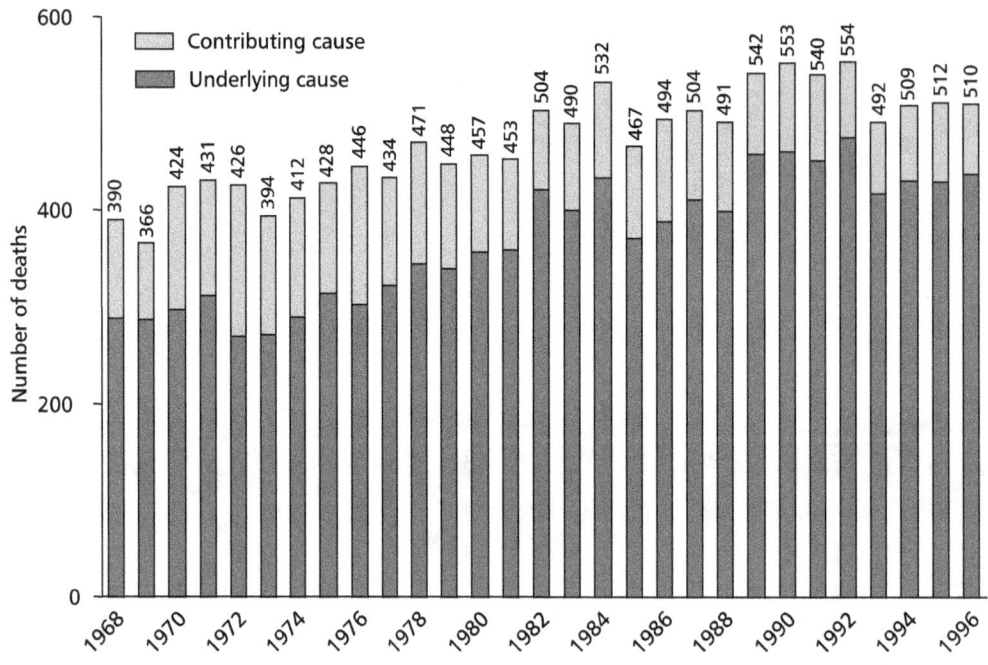

Figure 3–18. Number of deaths with malignant pleural neoplasm recorded as an underlying or contributing cause on the death certificate—U.S. residents aged 15 and older, 1968–1996. (Source: NCHS [1999].)

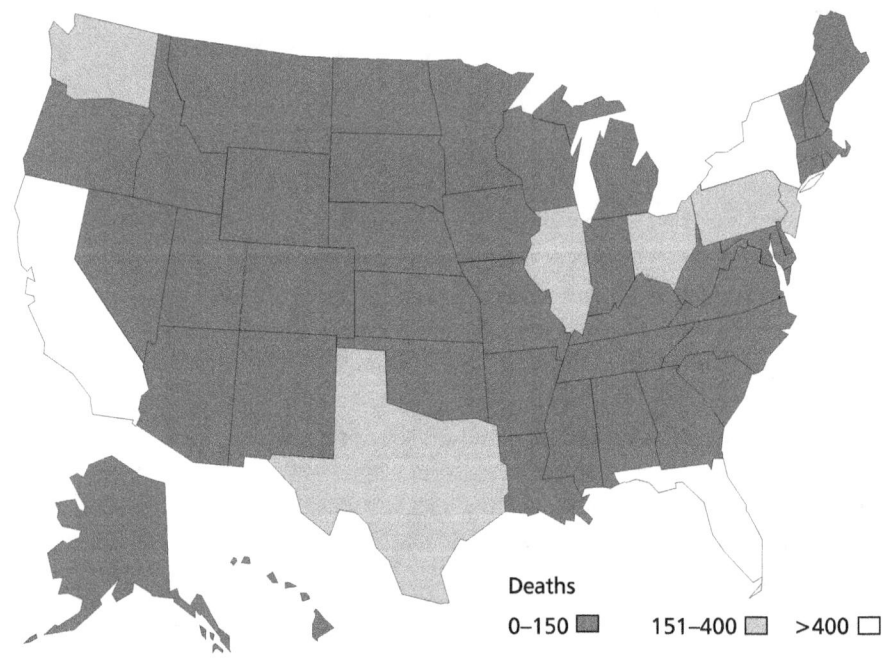

Figure 3–19. Number of deaths due to malignant pleural neoplasm by State—U.S. residents aged 15 and older, 1987–1996. (Source: NCHS [1999].)

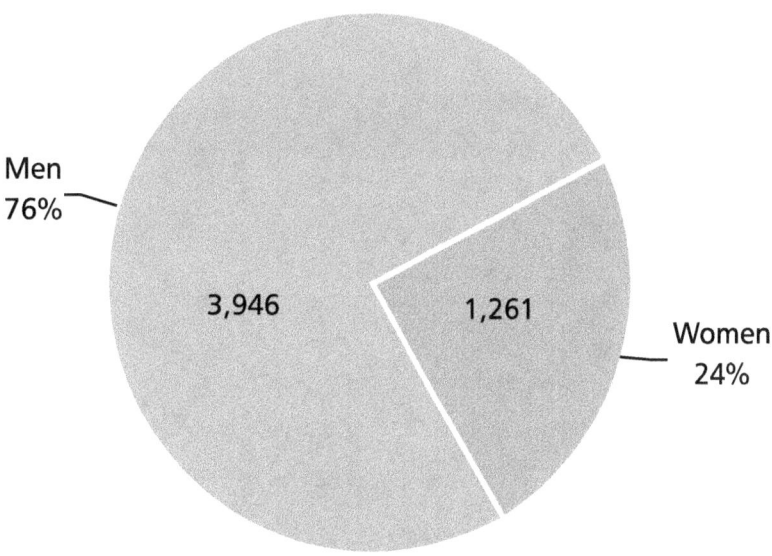

Figure 3–20. Distribution and number of deaths due to malignant pleural neoplasm by sex—U.S. residents aged 15 and older, 1987–1996. (Source: NCHS [1999].)

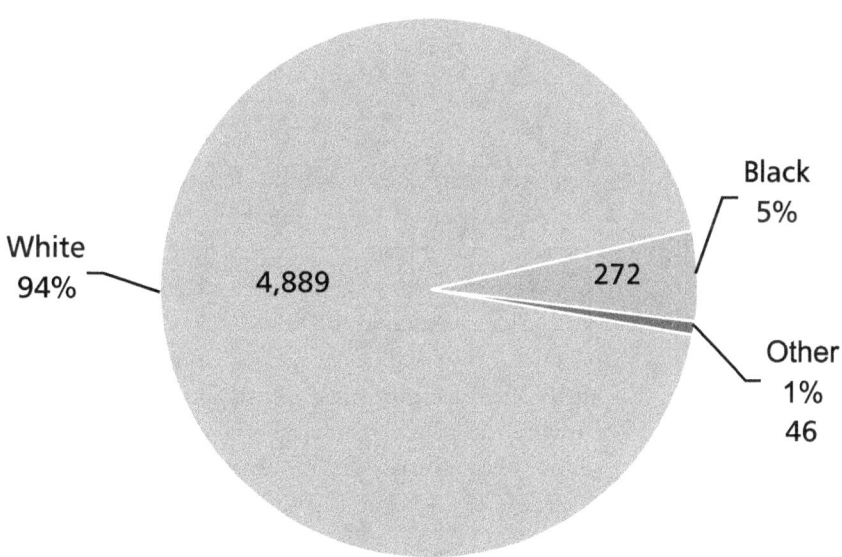

Figure 3–21. Distribution and number of deaths due to malignant pleural neoplasm by race—U.S. residents aged 15 and older, 1987–1996. (Source: NCHS [1999].)

Fatal Illness

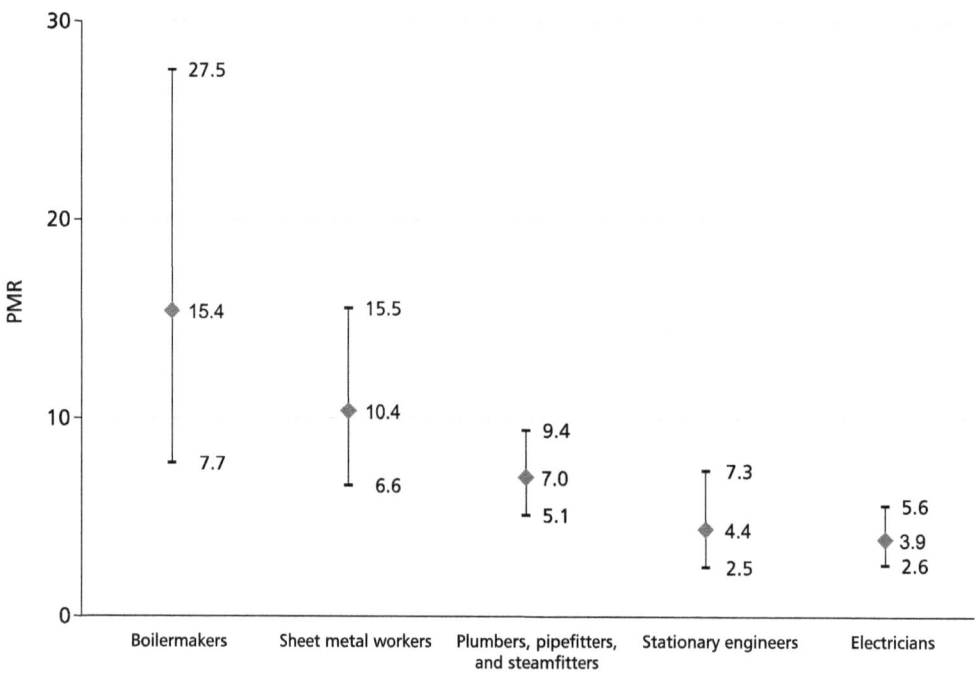

Figure 3–22. PMRs (and 95% CIs) for malignant pleural neoplasm by usual occupation—U.S. residents aged 15 and older, 1987–1996. (Source: NCHS [1999].)

Hypersensitivity Pneumonitis

Hypersensitivity pneumonitis is a lung disease that is often related to occupation. Examples of this disease are farmers' lung, mushroom workers' lung, and bird fanciers' disease. The annual number of deaths with hypersensitivity pneumonitis as an underlying or contributing cause has generally increased since 1979 (Figure 3–23). A geographic distribution of cases is presented in Figure 3–24. Nearly 30% of decedents during 1987–1996 were women (Figure 3–25), and 95% were white U.S. residents (Figure 3–26). The only occupation with a significantly high PMR for this disease was nonhorticultural farmer, with a value of 11.6 (95% confidence interval [CI] = 8.5–15.6).

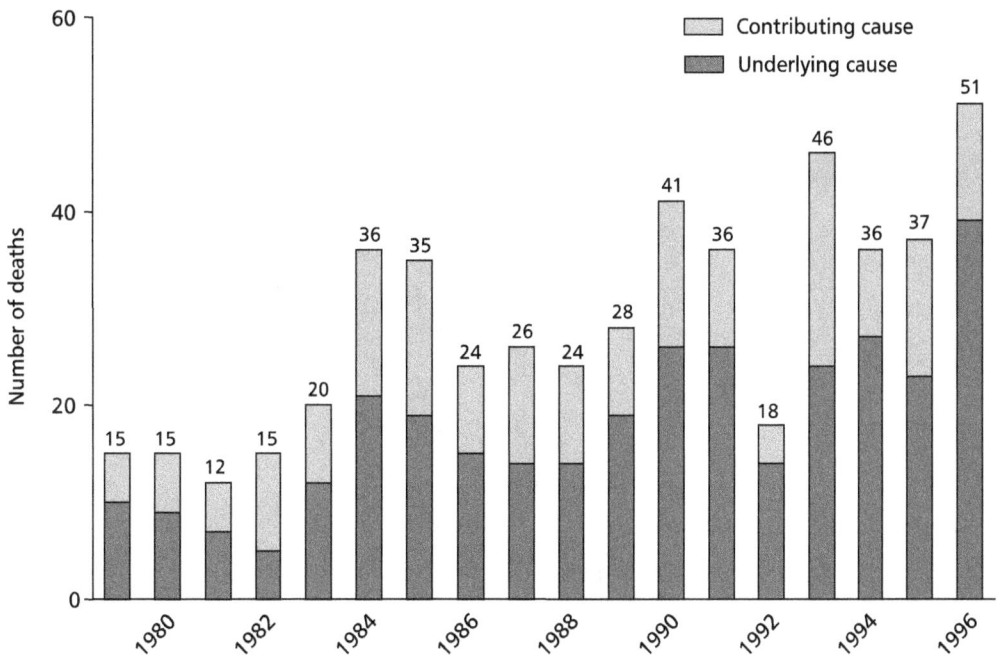

Figure 3–23. Number of deaths recorded with hypersensitivity pneumonitis as an underlying or contributing cause on the death certificate—U.S. residents aged 15 and older, 1979–1996. (Source: NCHS [1999].)

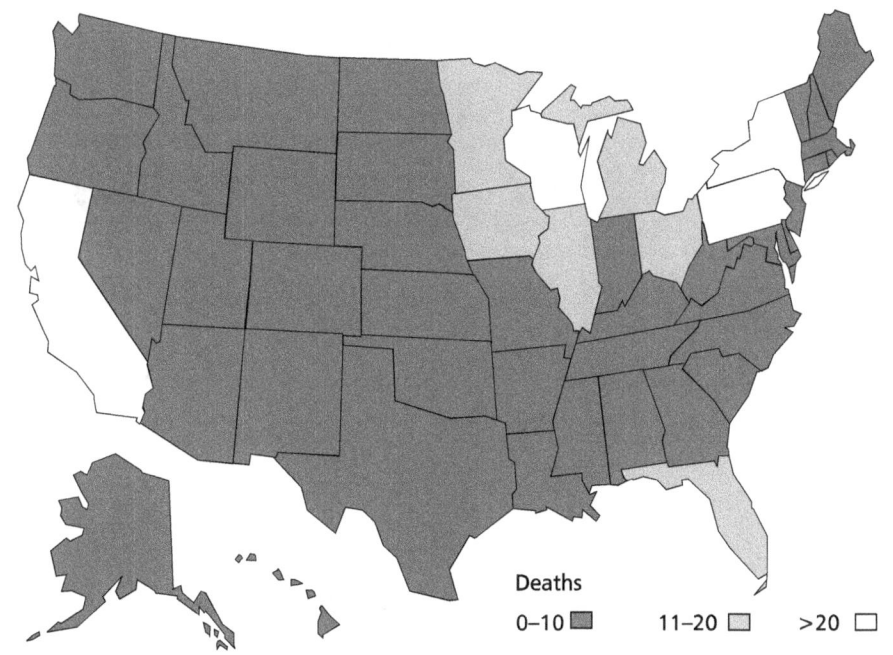

Figure 3–24. Number of hypersensitivity pneumonitis deaths by State—U.S. residents aged 15 and older, 1987–1996. (Source: NCHS [1999].)

FATAL ILLNESS

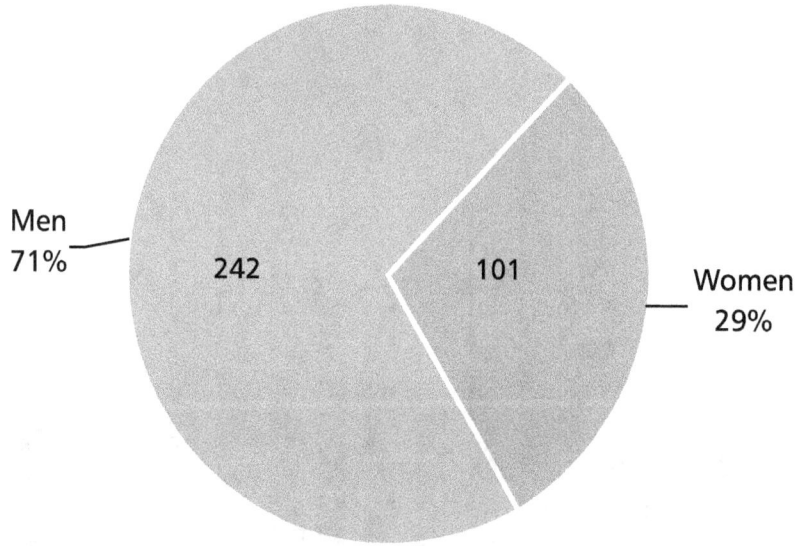

Figure 3–25. Distribution and number of hypersensitivity pneumonitis deaths by sex—U.S. residents aged 15 and older, 1987–1996. (Source: NCHS [1999].)

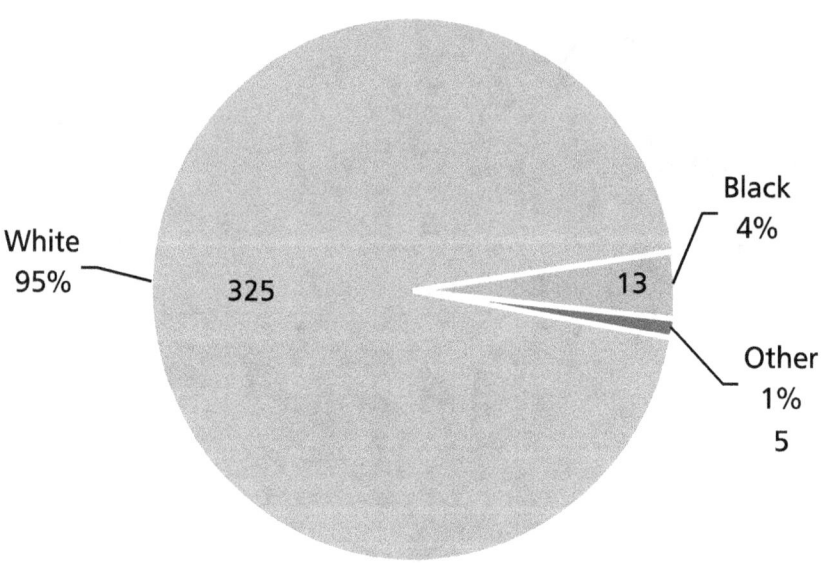

Figure 3–26. Distribution and number of hypersensitivity pneumonitis deaths by race—U.S. residents aged 15 and older, 1987–1996. (Source: NCHS [1999].)

PMRs for Selected Occupations and Causes of Death

PMRs for selected occupations and fatal illnesses are presented in Table 3–1 using data from the National Occupational Mortality Surveillance System (NOMS). Deaths related to diseases other than the pneumoconioses and mesothelioma are difficult to link to the workplace. Therefore, any apparent excess of mortality for an occupation must be verified by more definitive studies. Statistically significant elevations of PMR do not necessarily indicate a causal relationship between an occupation and a fatal illness. Some PMRs may be elevated by chance, which is likely when a very large number of PMRs are tested for statistical significance. Other elevated PMRs may be associated more strongly with confounding factors such as cigarette smoking, alcohol use, or socioeconomic factors such as availability of health care and proper diet. The information in Table 3–1 may be used to identify possible associations between fatal illnesses and occupations that warrant further study. The findings may also be used to describe, to evaluate the results of other studies, or to target occupational groups for health promotion activities. For additional discussion about the use of PMRs from NOMS, see *Mortality by Occupation, Industry, and Cause of Death: 24 Reporting States (1984–1988)* [NIOSH 1997]. Additional information about lung diseases can be obtained from the *Work-Related Lung Disease Surveillance Report* (WoRLD) [NIOSH 1999].

Fatal Illness

Table 3-1. PMRs for white and black men and women aged 18 to 90 for selected occupations and

					Cause of death		
			Malignant neoplasms				
Occupation and census occupation codes‡	Number of deaths	All cancers (140–208)	Liver (155)	Larynx (161)	Lung (162)	Mesothelioma (158.8, 158.9, 163)	Skin (173)
Executive and administrative occupations (003–019):							
White men	269,754	1.10§	1.15§	0.88§	0.99	1.23**	0.95
Black men	9,011	1.06§	1.30	0.50§	0.93**	2.80	0.70
White women	87,200	1.19§	1.17§	1.11	1.29§	1.28	1.12
Black women	4,765	1.30§	1.40	0.52	1.36§	—	0.76
Legislative and public administrative officials (003–005):							
White men	14,933	1.09§	0.98	0.65**	0.95**	0.73	0.57
Black men	763	1.26§	2.45**	0.00**	0.97	—	1.87
White women	5,759	1.21§	1.34	0.37	1.34§	1.08	0.76
Black women	415	1.49§	1.46	—	1.60**	—	—
Professional specialty occupations (043–199):							
White men	234,721	1.06§	1.11§	0.75§	0.75§	0.96	1.14**
Black men	16,050	0.97**	1.10	0.54§	0.74§	0.81	0.73
White women	229,259	1.15§	1.02	0.81**	0.91§	1.38§	0.90
Black women	23,542	1.22§	1.16	0.74	1.05	1.21	0.48
Engineers (044–062):							
White men	67,509	1.12§	1.13**	0.83**	0.90§	1.67§	0.98
Black men	1,246	1.06	0.52	0.54	1.14	—	1.21
White women	973	1.21§	0.56	2.07	1.33**	6.01	1.53
Black women	75	1.48	—	—	2.20	—	—
Natural scientists (069–083):							
White men	11,300	1.07§	1.03	0.64**	0.74§	0.61	1.08
Black men	354	1.19	3.00	0.63	0.65	16.00	—
White women	1,062	1.23§	1.06	2.01	1.14	2.88	—
Black women	67	1.59**	10.02**	—	0.70	—	—
Health diagnosing occupations (084–089):							
White men	21,310	1.06§	1.21	0.66§	0.65§	0.22**	1.60§
Black men	556	0.99	1.25	0.44	0.56§	—	1.65
White women	1,454	1.21§	1.25	0.85	0.89	2.39	—
Black women	75	1.41	—	—	1.32	—	—
Physicians (084):							
White men	13,034	1.05§	1.25	0.74	0.64§	0.18**	1.63**
Black men	347	0.89	1.34	0.73	0.56**	—	2.70
White women	1,017	1.18§	1.46	1.15	0.86	3.31	—
Black women	56	1.42	—	—	1.76	—	—
Health assessment and treating occupations (095–106):							
White men	9,055	0.98	1.04	1.08	0.77§	0.85	0.89
Black men	896	0.89	1.05	0.28	0.55§	—	1.71
White women	65,181	1.07§	0.97	0.84	1.02	1.15	0.90
Black women	8,105	1.11§	0.80	1.05	1.14**	2.54	0.22
Nurses (095, 207):							
White men	2,117	0.83§	1.03	1.46	0.70§	1.29	1.32
Black men	484	0.90	0.49	—	0.64**	—	2.91
White women	75,167	1.03§	0.94	0.80	1.02	0.96	0.93
Black women	10,934	1.09§	0.90	1.11	1.13§	2.25	0.19

See footnotes at end of table.

FATAL ILLNESS

causes of death in 28 States,* 1984–1995

and ICD-9† codes

Bladder (188)	Kidney (189.0–189.2)	Non-Hodgkins lymphoma (200, 202.0–202.2, 202.8, 202.9)	Hodgkins disease (201)	Leukemia (204–208)	Parkinson's disease (332)	IHD (410–414)	COPD (490–496)	Pneumo-conioses (500–505)	Chronic liver disease (571)	Renal failure (584–586)	Suicide (E590–E959)
1.10§	1.21§	1.27§	1.16**	1.22§	1.19§	1.01§	0.80§	0.33§	0.93§	0.93§	0.94§
1.02	1.53§	1.50§	1.04	1.29**	0.81	1.10§	0.87**	0.63	0.73§	0.90	1.45§
1.05	1.13§	1.21§	0.98	1.19§	0.95	0.89§	1.06§	0.22	0.90§	0.90§	1.00
1.18	1.79§	1.12	1.44	1.30	2.01	0.91**	1.14	—	0.63§	0.78	1.25
1.08	1.28§	1.31§	1.27	1.29§	1.15	1.02	0.79§	0.26§	0.99	0.87	0.91
1.18	2.30	1.74	—	1.12	1.28	0.99	0.75	—	0.89	0.52	0.84
1.03	1.25	1.22	0.44	1.23	0.80	0.85§	1.08	—	0.75	0.92	0.95
0.83	0.66	0.98	—	1.60	4.15	0.67§	2.26§	—	0.40	0.31	1.44
1.03	1.24§	1.43§	1.39§	1.33§	1.68§	0.97§	0.68§	0.18§	0.88§	0.92§	1.03**
1.39§	1.03	1.49§	1.42	1.41§	1.93§	1.07§	0.75§	0.49	0.80§	1.06	1.06
0.95	1.01	1.20§	1.35§	1.23§	1.46§	0.89§	0.90§	0.26§	0.88§	0.81§	1.27§
0.93	1.20	1.32§	1.52	1.26§	1.62**	0.92§	0.97	1.14	0.62§	0.82§	1.21
1.11**	1.30§	1.34§	1.42§	1.37§	1.56§	0.97§	0.75§	0.36§	0.90§	0.88§	1.03
1.73	0.55	1.07	—	1.39	—	1.13	0.75	—	1.10	1.19	1.32
1.46	1.38	1.06	1.70	0.47	0.91	0.81**	1.05	—	1.15	0.67	1.61§
—	14.47§	5.13	—	4.07	—	0.42	0.73	—	0.65	—	—
1.29**	1.23	1.62§	1.55	1.50§	1.81§	0.93§	0.74§	0.28§	1.07	0.88	1.30§
2.42	1.94	1.29	5.13	1.46	—	1.10	0.80	3.67	1.05	0.42	1.49
2.14	0.38	1.02	2.11	1.51	0.34	0.67§	1.01	—	0.92	0.33**	1.88§
—	—	2.92	—	—	—	0.59	0.83	—	0.74	—	—
0.94	1.35§	1.61§	1.43	1.43§	1.69§	0.99	0.54§	0.00§	0.75§	0.83**	1.73§
1.40	0.86	1.77	—	1.83	2.56	1.23**	0.54	—	0.54	0.83	2.72§
0.66	1.34	0.80	1.20	1.31	1.83	0.82§	1.08	—	0.88	0.77	2.33§
—	—	2.69	11.43	—	—	1.08	0.77	—	0.62	—	—
0.90	1.36§	1.60§	1.44	1.36§	1.72§	0.99	0.52§	0.00§	0.68§	0.94	1.82§
1.52	0.69	2.12	—	0.98	2.05	1.17	0.48	—	0.21	0.98	2.88**
0.62	1.44	0.84	1.81	1.25	2.32	0.89	1.11	—	0.78	0.55	2.83§
—	—	—	—	—	—	1.36	1.01	—	0.81	—	—
0.89	1.23	1.26**	0.97	1.18	1.39**	0.98	0.77§	0.00§	0.60§	1.14	1.13**
1.24	1.12	0.48	4.26	2.24**	2.69	0.99	1.00	—	0.92	1.22	0.97
0.98	1.04	1.15§	1.11	1.14§	1.17**	0.92§	1.12§	0.00	0.91**	0.89§	1.37§
0.69	1.29	1.17	1.51	1.04	0.80	0.97	1.10	3.45	0.71§	0.90	1.38
0.61	1.50	1.29	0.72	0.85	1.68	0.98	0.94	—	0.77	0.77	0.98
3.70**	1.05	0.42	4.61	1.94	3.43	0.90	0.87	—	0.92	1.33	0.45
0.95	1.06	1.13§	1.21	1.11§	1.15**	0.94§	1.13§	0.00**	0.94	0.93**	1.34§
0.70	1.43**	1.06	1.15	1.09	0.83	0.98	1.18**	3.04	0.78**	0.92	1.54**

(Continued)

FATAL ILLNESS

Table 3–1 (Continued). PMRs for white and black men and women aged 18 to 90 for selected

					Cause of death	
				Malignant neoplasms		
Occupation and census occupation codes‡	Number of deaths	All cancers (140–208)	Liver (155)	Larynx (161)	Lung (162)	Mesothelioma (158.8, 158.9, 163)	Skin (173)
Teachers (113–159):							
White men	42,189	1.03§	1.02	0.48§	0.62§	0.83	0.82
Black men	3,795	0.98	0.87	0.60	0.65§	—	1.02
White women	119,036	1.18§	1.02	0.67§	0.76§	1.37	0.87
Black women	10,934	1.28§	1.13	0.55	0.89**	0.68	0.36
Social workers (174):							
White men	3,237	0.96	1.03	0.79	0.83§	—	0.83
Black men	955	1.01	0.73	0.26	1.03	—	—
White women	6,236	1.14§	1.02	1.64	1.05	1.56	1.68
Black women	1,628	1.30§	2.31**	0.91	1.52§	—	1.07
Clergy (176):							
White men	19,879	0.96§	1.03	0.36§	0.47§	0.96	1.11
Black men	4,378	0.91§	1.11	0.42**	0.59§	—	0.24
White women	1,299	1.17§	1.92	—	0.44§	—	—
Black women	665	1.18**	1.40	—	0.41§	—	—
Lawyers and judges (178, 179):							
White men	15,224	1.09§	1.25	1.07	0.77§	0.32	1.57**
Black men	292	1.17	1.51	—	0.93	—	—
White women	1,127	1.22§	0.79	1.10	1.14	9.10§	—
Black women	53	1.47	—	—	1.85	—	—
Technicians and related support occupations (203–235):							
White men	42,132	1.04§	1.02	0.92	0.92§	1.40	1.08
Black men	3,130	1.04	0.38**	0.33**	1.03	—	0.95
White women	32,519	1.03§	1.01	0.45§	1.07§	0.59	0.92
Black women	5,401	1.18§	1.33	0.72	1.20§	—	0.32
Engineering and related technologists and technicians (213–218):							
White men	20,528	1.05§	1.15	0.98	0.92§	1.85**	0.81
Black men	1,078	1.03	0.43	0.23	0.95	—	0.67
White women	2,851	1.12§	1.51	—	1.26§	0.99	1.00
Black women	383	1.25**	0.84	—	1.23	—	—
Sales occupations (243–285):							
White men	297,253	1.03§	1.06**	0.91§	0.95§	0.69§	0.93
Black men	8,552	1.00	1.08	0.72	0.89§	3.26	0.50
White women	150,884	1.11§	1.01	0.89	1.16§	1.17	0.76**
Black women	6,158	1.21§	1.25	1.07	1.15**	2.21	1.00
Administrative support occupations, including clerical (303–389):							
White men	134,837	1.01**	1.01	0.85§	0.96§	0.86	1.06
Black men	15,091	1.07§	1.25	0.84	0.96	1.25	1.47
White women	289,646	1.13§	0.93**	0.91	1.10§	0.96	0.99
Black women	14,714	1.30§	1.30	0.48**	1.54§	1.72	0.97

See footnotes at end of table.

FATAL ILLNESS

occupations and causes of death in 28 States,* 1984–1995

and ICD-9† codes

Bladder (188)	Kidney (189.0–189.2)	Non-Hodgkins lymphoma (200, 202.0–202.2, 202.8, 202.9)	Hodgkins disease (201)	Leukemia (204–208)	Parkinson's disease (332)	IHD (410–414)	COPD (490–496)	Pneumo-conioses (500–505)	Chronic liver disease (571)	Renal failure (584–586)	Suicide (E590–E959)
0.89	1.21§	1.61§	1.60§	1.33§	1.99§	0.99	0.56§	0.09§	0.79§	0.90	0.95
1.69**	1.34	1.79§	0.46	1.14	2.92§	0.99	0.69§	—	0.86	0.98	1.10
0.89**	0.99	1.28§	1.51§	1.29§	1.61§	0.88§	0.73§	0.36	0.82§	0.81§	0.94
0.83	1.20	1.30	1.36	1.44§	2.31§	0.89§	0.87	—	0.54§	0.74§	1.00
0.56	1.04	0.98	1.04	1.20	2.20§	0.93	0.86	0.60	1.05	1.34	1.08
1.94	0.79	2.55§	3.45	1.61	6.03**	1.09	0.84	—	0.66	1.28	0.73
1.21	0.82	1.19	1.04	1.15	2.06§	0.84§	1.23§	—	0.83	0.83	1.58§
1.09	0.69	1.64	0.94	1.41	—	0.88	0.96	—	0.63**	0.92	0.74
0.83**	1.35§	1.64§	1.80**	1.44§	1.76§	1.12§	0.44§	0.16§	0.43§	1.02	0.52§
1.02	0.94	1.36	1.37	1.23	1.39	1.19§	0.59§	0.85	0.42§	1.06	0.87
2.05	1.19	0.93	3.29	1.27	1.52	0.90	0.42§	—	0.87	0.95	1.06
2.53	1.43	2.46	—	0.82	—	1.16	0.59	—	0.62	0.73	1.07
1.00	1.13	1.34§	1.06	1.42§	1.88§	0.94§	0.63§	0.00§	1.04	0.86	1.35§
—	3.22	0.76	5.21	1.80	—	0.95	0.65	—	0.57	1.82	1.57
1.77	0.96	1.10	1.32	1.27	1.85	0.81§	0.89	—	0.68	1.12	1.78§
8.30	—	—	—	—	—	0.78	3.17	—	—	—	2.89
1.07	1.15**	1.22§	1.38§	1.10	1.32§	0.99	0.89§	0.30§	0.90§	0.98	1.08§
1.22	1.29	1.66**	1.39	1.35	0.49	1.02	1.08	0.69	0.84	0.74	1.56§
0.85	1.06	1.14**	1.27	1.16§	1.03	0.95§	1.11§	0.58	0.94	0.98	1.28§
1.23	1.43	1.48**	1.28	1.52§	0.53	0.99	1.17	—	0.83	0.85	1.70§
1.15	1.10	1.19§	1.56**	1.11	1.33**	1.00	0.84§	0.26§	0.92	1.02	1.11§
1.56	1.15	1.93	1.03	1.37	—	1.02	0.91	—	1.02	0.52	1.63**
0.55	0.98	1.18	0.00	1.36	1.73	0.92	1.13	6.20	0.80	0.92	1.28
5.24§	0.72	3.09**	—	2.49	—	0.97	1.14	—	0.89	0.92	2.92**
1.07§	1.16§	1.14§	1.07	1.08§	1.16§	1.03§	0.84§	0.18§	0.99	0.99	1.04§
1.10	0.84	1.43**	1.04	1.03	1.39	1.04	0.92	0.18	0.60§	1.18	1.25§
1.15§	1.13§	1.09§	1.03	1.08§	1.06	0.96§	1.01	0.51	0.91§	0.88§	1.06**
1.10	1.08	1.21	0.30	1.16	1.35	0.95	1.14	—	0.67§	0.83	1.34**
1.02	0.97	1.13§	1.06	1.03	1.26§	1.02§	0.90§	0.25§	0.97	0.95	0.90§
1.07	1.13	1.67§	0.74	1.13	1.52	1.01	0.93	0.29**	0.93	1.00	0.94
1.08§	1.04	1.07§	0.99	1.12§	1.19§	0.90§	1.06§	0.58	0.94§	0.86§	1.00
1.20	1.02	1.57§	1.34	1.28§	2.04**	0.85§	1.16**	—	0.65§	0.76§	1.21

(Continued)

Fatal Illness

Table 3-1 (Continued). PMRs for white and black men and women aged 18 to 90 for selected

		Cause of death					
		Malignant neoplasms					
Occupation and census occupation codes‡	Number of deaths	All cancers (140–208)	Liver (155)	Larynx (161)	Lung (162)	Mesothelioma (158.8, 158.9, 163)	Skin (173)
Secretaries, stenographers, and typists (313–315):							
White men	2,363	0.98	1.85**	0.50	0.80§	2.10	2.51**
Black men	177	1.13	—	—	0.61	—	4.71
White women	112,996	1.16§	0.85§	1.08	1.09§	0.78	0.99
Black women	3,540	1.28§	1.17	0.46	1.40§	1.72	1.54
Mail and message distributing occupations (354–357):							
White men	30,933	1.04§	1.05	0.88	0.98	0.98	0.95
Black men	5,045	1.10§	0.86	1.06	0.98	—	0.83
White women	5,141	1.08§	1.36	0.66	1.05	1.88	1.46
Black women	863	1.25§	0.38	—	1.27	—	2.01
Material recording, scheduling, and distributing clerks, n.e.c. (359–374):							
White men	40,807	1.01	0.98	0.88	1.00	0.76	0.92
Black men	4,799	1.06**	1.40	0.91	0.94	2.63	2.01**
White women	9,183	1.10§	1.02	0.86	1.22§	0.32	0.63
Black women	997	1.31§	0.97	0.75	1.42**	6.50	—
General office clerks (379):							
White men	19,559	0.95§	1.08	0.92	0.89§	0.71	1.05
Black men	1,842	1.03	1.82	0.71	0.96	3.48	0.93
White women	44,791	1.09§	0.96	0.94	1.05§	1.20	0.77
Black women	3,108	1.22§	1.05	0.80	1.54§	4.11	1.20
Service occupations (403–469):							
White men	213,546	0.96§	1.11§	1.11§	1.01**	0.76**	1.09
Black men	62,865	1.02§	1.10	0.98	0.99	0.98	1.08
White women	237,600	0.97§	0.98	1.21§	1.11§	0.74**	0.89
Black women	110,346	1.00	1.00	1.11	1.00	0.92	0.92
Protective service occupations (413–427):							
White men	65,823	0.98§	1.11	0.87	0.99	0.70	0.96
Black men	6,859	1.02	1.10	0.83	0.96	0.96	0.89
White women	3,928	1.03	1.10	0.73	1.23§	—	0.73
Black women	1,049	1.21§	2.23	2.10	1.02	—	—
Food preparation and service occupations (433–444):							
White men	31,189	0.90§	1.29§	1.83§	0.98	0.59	1.50§
Black men	13,495	0.98	1.33**	1.11	0.90§	1.09	1.03
White women	83,294	0.98§	1.04	1.39§	1.22§	0.50§	0.93
Black women	16,394	0.98	0.88	1.50	0.94	1.27	1.17
Cleaning and building service occupations, excluding private household (448–455):							
White men	90,769	0.98§	1.07	1.14**	1.06§	0.88	1.00
Black men	33,255	1.04§	1.00	1.01	1.04**	1.20	1.07
White women	24,958	0.96§	0.73**	1.42	1.10§	0.77	0.71
Black women	15,075	1.04§	0.94	1.09	1.12§	1.41	1.14

See footnotes at end of table.

Fatal Illness

occupations and causes of death in 28 States,* 1984–1995

and ICD–9† codes

Bladder (188)	Kidney (189.0–189.2)	Non-Hodgkins lymphoma (200, 202.0–202.2, 202.8, 202.9)	Hodgkins disease (201)	Leukemia (204–208)	Parkinson's disease (332)	IHD (410–414)	COPD (490–496)	Pneumo-conioses (500–505)	Chronic liver disease (571)	Renal failure (584–586)	Suicide (E590–E959)
0.90	1.09	1.55**	0.50	1.17	1.08	0.99	0.83	0.18**	0.74	0.77	0.72
2.77	3.27	1.16	5.19	1.09	—	0.71	0.64	—	1.06	1.60	0.70
1.09	1.00	1.07**	1.00	1.15§	1.28§	0.87§	1.05§	0.17**	1.02	0.79§	1.04
1.72	0.64	1.85§	1.06	1.41	2.74	0.79§	1.13	—	0.73**	0.68**	0.95
1.11	0.94	1.21§	1.23	1.07	1.35§	1.01	0.88§	0.21§	1.04	0.93	1.07
0.92	1.23	1.52**	1.13	1.27	1.24	0.98	0.89	0.00**	1.00	0.93	1.28
1.23	1.09	0.98	1.04	1.21	1.26	0.98	0.91	3.62	0.89	1.05	1.08
1.85	0.66	1.63	—	1.45	—	0.88	1.59**	—	0.71	0.57	1.49
1.03	0.91	1.04	0.97	0.97	1.14	1.04§	0.92§	0.42§	0.96	1.04	0.88§
1.35	1.17	1.75§	0.57	1.08	2.06	1.01	0.91	0.32	1.00	1.01	1.06
1.13	1.07	1.21**	0.65	0.96	1.08	0.95**	1.05	2.02	0.78**	0.92	0.98
0.76	0.28	1.43	2.50	1.27	7.39§	0.79§	1.69§	—	0.59	0.75	1.17
0.99	1.03	1.13	1.17	0.90	1.25**	1.02	0.92**	0.23§	1.07	0.84**	0.95
1.04	1.10	1.36	0.72	1.23	2.07	0.96	1.00	—	0.99	0.99	0.58**
1.08	1.06	0.99	1.09	1.10	1.17**	0.95§	1.01	1.65	0.89**	0.89**	0.94
0.98	1.38	1.18	1.49	0.71	1.10	0.98	1.02	—	0.64§	0.83	1.40
0.94**	0.92§	0.89§	1.02	0.90§	0.79§	1.01	1.03§	0.45§	1.10§	1.04	0.99
0.99	1.01	1.01	1.04	1.01	1.08	0.97§	0.97	0.32§	1.04	1.02	1.04
1.02	0.98	0.92§	0.94	0.90§	0.80§	1.03§	1.07§	0.87	1.01	1.00	0.95**
1.07	0.97	0.99	0.96	0.96	0.80**	1.00	1.00	0.94	1.01	0.98	0.91
0.89**	1.04	0.98	1.40§	0.94	0.74§	1.06§	0.92§	0.31§	0.99	1.03	1.17§
0.94	1.44**	1.19	1.35	1.31	1.61	1.12§	0.76§	0.24	0.82**	1.10	1.77§
0.90	0.95	1.10	0.60	0.80	0.90	1.02	0.98	4.71	0.93	0.86	1.46§
1.58	2.10	1.14	2.06	1.47	—	0.97	0.99	—	0.63	0.62	3.16§
0.93	0.67§	0.72§	0.63§	0.66§	0.63§	0.88§	1.22§	0.29§	1.60§	1.13	0.84§
1.28	1.04	0.87	0.99	0.96	1.26	0.89§	1.03	0.12§	1.22§	1.21§	0.89
0.95	1.03	0.86§	0.80	0.86§	0.72§	1.02§	1.14§	1.11	1.09**	1.05	0.90§
1.00	0.99	0.96	0.66	1.10	1.19	1.00	1.01	3.36	1.06	0.99	0.81
0.97	0.92**	0.85§	0.99	0.95	0.80§	1.01**	1.07§	0.63§	1.02	0.99	1.04**
0.99	0.88	1.08	1.01	1.01	0.98	0.98	0.98	0.40§	1.02	0.94	1.00
1.17	0.95	0.91	1.03	0.84**	0.62§	1.11§	0.92§	2.22	0.91	0.89	0.94
1.23	0.87	1.08	1.35	0.88	1.02	1.01	0.99	3.72	0.97	0.85**	0.72

(Continued)

Fatal Illness

Table 3-1 (Continued). PMRs for white and black men and women aged 18 to 90 for selected

						Cause of death	
					Malignant neoplasms		
Occupation and census occupation codes‡	Number of deaths	All cancers (140–208)	Liver (155)	Larynx (161)	Lung (162)	Mesothelioma (158.8, 158.9, 163)	Skin (173)
Personal service occupations (456–469):							
White men	19,790	0.94§	1.02	0.70**	0.96	0.79	1.57§
Black men	4,852	1.02	1.10	0.97	0.91	—	1.20
White women	33,756	1.08§	0.97	1.34	1.19§	1.16	0.80
Black women	7,784	1.14§	1.29	1.32	1.20§	1.78	0.50
Farming and other agricultural occupations (473–489):							
White men	269,546	0.89§	0.86§	0.75§	0.81§	0.53§	1.22§
Black men	38,103	0.88§	0.68§	0.92	0.86§	0.45	0.77
White women	9,124	0.99	0.84	0.78	0.89**	1.25	1.22
Black women	7,751	0.79§	0.76	0.99	0.53§	2.00	2.33**
Forestry, fishing, and hunting occupations (494–499):							
White men	16,031	0.94§	1.01	1.28	1.09§	0.15**	0.87
Black men	4,564	0.93§	0.61	0.82	0.96	—	1.00
White women	156	0.97	—	—	1.26	—	—
Black women	45	0.49	—	—	1.13	—	—
Precision production, craft, and repair occupations (503–699):							
White men	766,839	1.03§	0.94§	1.10§	1.12§	1.54§	0.95
Black men	61,703	1.06§	1.09	1.08	1.10§	1.45	1.03
White women	43,881	1.07§	1.05	0.85	1.14§	1.17	0.79
Black women	4,989	1.16§	1.39	0.70	1.34§	1.39	0.38
Mechanics and repairers (503–549):							
White men	170,950	1.03§	0.90§	1.09**	1.11§	1.08	0.94
Black men	14,087	1.05§	0.83	1.14	1.07**	2.53	0.77
White women	3,185	1.09§	0.91	—	1.19§	2.03	0.47
Black women	333	1.25**	1.90	—	1.53	21.88	5.04
Vehicle and mobile equipment mechanics and repairers (505–517):							
White men	80,025	1.01	0.97	1.20§	1.13§	0.79	0.97
Black men	8,912	1.03	0.85	1.28	1.05	2.69	0.88
White women	390	1.04	1.46	—	1.48**	8.17	—
Black women	69	1.07	—	—	0.97	—	—
Millwrights (544):							
White men	10,058	1.09§	0.79	1.22	1.17§	1.09	0.79
Black men	232	1.01	0.89	2.66	0.88	—	—
White women	56	1.05	—	—	1.24	57.94**	—
Black women	6	0.61	—	—	—	—	—
Construction trades (553–599):							
White men	281,961	1.04§	0.97	1.22§	1.18§	2.08§	1.04
Black men	30,348	1.05§	1.09	1.13	1.09§	0.60	1.24
White women	1,667	1.03	2.12**	1.40	1.30§	3.98	—
Black women	192	1.18	1.78	—	1.17	—	—

See footnotes at end of table.

76

FATAL ILLNESS

occupations and causes of death in 28 States,* 1984–1995

and ICD–9† codes

Bladder (188)	Kidney (189.0–189.2)	Non-Hodgkins lymphoma (200, 202.0–202.2, 202.8, 202.9)	Hodgkins disease (201)	Leukemia (204–208)	Parkinson's disease (332)	IHD (410–414)	COPD (490–496)	Pneumo-conioses (500–505)	Chronic liver disease (571)	Renal failure (584–586)	Suicide (E590–E959)
0.97	0.92	1.03	1.00	0.99	1.12	0.96§	0.96	0.21§	0.98	1.16**	0.79§
0.90	1.32	0.99	0.68	0.99	1.04	0.93	1.01	0.00**	1.06	1.11	0.71**
1.26§	1.10	1.15§	1.07	1.06	0.85	0.94§	1.11§	0.56	1.05	0.88**	0.95
1.25	1.22	1.20	1.69	1.24	1.04	0.96	1.04	—	1.09	1.03	0.57**
0.81§	0.96	1.01	0.89	1.05**	0.93§	1.03§	1.01	0.12§	0.73§	1.01	1.11§
0.88	0.71§	0.63§	0.62	0.82§	0.75**	1.04§	1.02	0.11§	0.90**	0.97	0.69§
0.82	1.19	1.08	1.03	1.14	1.07	1.00	0.96	—	1.02	0.96	1.43§
0.47§	1.26	0.77	0.36	0.67**	0.97	0.99	0.76§	3.08	1.14	1.07	0.78
0.84	0.78**	0.76§	0.64	0.83**	0.61§	0.92§	1.31§	0.21§	1.00	0.83**	1.01
1.05	0.74	0.81	—	0.56**	0.91	0.96	0.85	0.00	0.70§	0.94	0.66**
—	1.24	1.14	—	0.68	—	0.85	1.12	—	2.25	0.71	1.95
—	7.13	—	—	—	—	1.38	2.58	—	1.75	—	7.41
1.04§	0.98	0.93§	0.89§	0.95§	0.83§	0.98§	1.09§	2.87§	1.06§	0.97§	1.04§
0.95	1.14**	1.02	1.26	0.97	0.80	0.99	1.01	3.33§	1.03	0.91§	1.12§
1.07	0.94	1.06	1.08	1.03	0.85	1.01	1.02	5.12§	0.88**	0.96	1.05
0.96	1.20	1.31	0.37	0.62**	1.77	0.97	1.14	—	0.83	0.61§	1.39
1.06	0.97	0.92§	0.90	0.98	0.81§	0.99	1.09§	0.50§	1.01	0.97	1.05§
0.83	1.19	1.22	1.24	1.07	0.71	1.02	0.97	1.50	0.95	0.98	1.32§
1.15	0.77	1.03	1.24	0.88	1.31	0.98	1.04	—	1.08	0.59**	1.10
2.36	0.85	1.81	—	0.47	—	0.89	0.98	—	0.79	0.00**	1.83
1.07	0.93	0.84§	0.87	0.92**	0.74§	0.98§	1.19§	0.45§	1.06**	0.97	1.06§
0.92	1.21	0.99	1.41	0.82	0.81	1.01	0.98	1.11	0.97	0.95	1.16
—	—	0.93	1.86	0.77	0.93	1.05	0.68	—	1.72	0.91	1.14
—	—	—	—	2.35	—	1.12	—	—	1.97	—	—
1.12	1.28**	0.94	1.26	0.96	0.78	1.02	0.93	0.71	1.09	1.15	0.96
—	0.93	2.03	—	1.43	—	1.00	0.41	5.04	0.89	0.91	3.17
—	—	—	—	1.86	5.40	1.31	0.85	—	3.44	—	1.40
—	—	—	—	—	—	4.13**	—	—	—	—	—
1.03	0.92§	0.91§	0.77§	0.90§	0.81§	0.94§	1.18§	0.78§	1.19§	0.90§	1.07§
0.98	1.02	0.91	0.83	0.83**	0.69	0.95§	0.97	0.78	1.06	0.87§	0.97
1.39	1.27	0.90	0.41	0.66	1.08	0.97	1.19	11.91	0.98	0.70	1.84§
—	—	1.06	—	—	—	1.11	0.60	—	1.35	0.37	0.84

(Continued)

Fatal Illness

Table 3–1 (Continued). PMRs for white and black men and women aged 18 to 90 for selected

						Cause of death	
			Malignant neoplasms				
Occupation and census occupation codes‡	Number of deaths	All cancers (140–208)	Liver (155)	Larynx (161)	Lung (162)	Mesothelioma (158.8, 158.9, 163)	Skin (173)
Carpenters (554, 567, 569):							
White men	87,576	1.02§	0.88**	1.14**	1.18§	1.33	1.22**
Black men	6,687	1.02	0.95	1.03	1.07	—	0.88
White women	227	0.93	1.32	—	1.60	—	—
Black women	27	1.54	—	—	3.33	—	—
Electricians and power transmission installers (555, 575–577):							
White men	46,173	1.05§	0.92	1.09	1.10§	2.12§	0.77
Black men	1,983	1.04	1.04	0.86	0.98	3.26	0.78
White women	285	0.89	0.99	—	1.21	—	—
Black women	36	1.26	—	—	1.65	—	—
Painters, paperhangers, and plasterers (556, 579–584):							
White men	35,274	1.02	1.16	1.53§	1.27§	2.15§	1.21
Black men	4,770	1.11§	1.39	1.36	1.15§	1.30	0.83
White women	570	1.17**	3.25**	4.41	1.24	6.37	—
Black women	55	1.11	—	—	0.95	—	—
Plumbers, pipefitters, and steamfitters (557, 585, 587):							
White men	32,975	1.07§	1.00	1.30§	1.18§	4.31§	0.96
Black men	2,481	0.98	1.59	0.75	0.98	—	0.99
White women	108	0.86	—	—	1.21	—	—
Black women	16	1.23	—	—	—	—	—
Extractive occupations (613–617):							
White men	58,959	0.90§	0.92	1.04	1.11§	0.44§	0.86
Black men	2,289	0.94	0.83	1.04	1.14	—	1.72
White women	166	0.72	—	—	1.40	—	—
Black women	14	0.85	—	—	1.68	—	—
Precision production occupations (633–699):							
White men	254,969	1.05§	0.92**	0.97	1.07§	1.52§	0.87**
Black men	14,979	1.10§	1.38§	0.92	1.14§	2.31	0.77
White women	38,863	1.08§	1.02	0.91	1.12§	0.97	0.86
Black women	4,450	1.15§	1.34	0.80	1.33§	—	—
Precision metal working occupations (634–655):							
White men	104,650	1.04§	0.93	1.10	1.09§	1.67§	0.82**
Black men	4,645	1.10§	1.45	0.88	1.21§	4.98**	1.00
White women	3,896	1.08§	1.08	1.35	1.34§	1.57	0.75
Black women	448	1.25**	1.43	—	1.55**	—	—
Tool and die makers (634, 635):							
White men	15,716	1.08§	0.82	0.96	1.09§	1.28	0.51**
Black men	188	1.17	1.12	—	1.36	—	—
White women	90	1.14	3.15	12.37	1.10	—	—
Black women	9	0.44	—	—	2.21	—	—
Machinists (637, 639):							
White men	68,053	1.02**	0.96	1.07	1.07§	1.07	0.82
Black men	3,741	1.10§	1.44	1.02	1.24§	6.13§	1.24
White women	1,750	1.00	0.47	1.72	1.24**	3.37	—
Black women	339	1.27**	0.94	—	1.69**	—	—

See footnotes at end of table.

FATAL ILLNESS

occupations and causes of death in 28 States,* 1984–1995

and ICD–9† codes

Bladder (188)	Kidney (189.0–189.2)	Non-Hodgkins lymphoma (200, 202.0–202.2, 202.8, 202.9)	Hodgkins disease (201)	Leukemia (204–208)	Parkinson's disease (332)	IHD (410–414)	COPD (490–496)	Pneumo-conioses (500–505)	Chronic liver disease (571)	Renal failure (584–586)	Suicide (E590–E959)
0.97	0.84§	0.94	0.62§	0.86§	0.74§	0.94§	1.25§	0.46§	1.18§	0.89§	1.15§
0.77	1.11	0.97	0.25	0.72	0.72	1.00	0.92	0.39	0.91	0.92	1.07
—	0.97	1.23	—	0.82	5.47**	1.05	0.96	—	0.70	0.55	1.61
—	—	7.59	—	—	—	1.20	—	—	—	—	—
1.05	1.03	1.06	0.78	0.97	0.93	0.96§	1.07§	1.06	1.07	0.99	1.04
1.00	1.34	1.12	1.39	1.60	0.52	1.03	1.07	0.79	0.90	0.70	1.31
2.15	—	0.64	—	0.73	2.21	1.14	1.43	—	1.94	0.39	2.51**
—	—	—	—	—	—	1.00	—	—	—	1.97	—
1.02	0.74§	0.73§	0.90	0.71§	0.75§	0.91§	1.35§	0.51§	1.35§	0.92	1.06**
1.03	1.16	0.76	0.64	0.91	0.39	0.89§	0.96	0.00	1.20	0.95	0.90
1.72	1.16	1.20	—	1.10	—	0.93	1.13	—	0.97	0.20	1.55
—	—	—	—	—	—	1.31	1.11	—	1.85	—	2.91
1.11	0.99	0.89	0.91	0.98	0.78§	0.96§	1.10§	1.16	1.19§	0.79§	1.04
0.88	1.31	1.04	0.74	0.61	1.26	1.01	0.91	1.01	1.29	0.82	0.75
3.07	—	—	—	0.94	—	1.03	1.22	—	0.82	1.07	2.05
—	—	—	—	—	—	1.03	3.49	—	—	—	—
0.78§	0.80§	0.76§	1.01	0.79§	0.46§	0.94§	1.33§	25.85§	0.86§	0.97	0.92**
0.90	0.88	1.31	1.54	1.12	0.62	1.08	1.78§	39.07§	1.10	1.03	1.83**
1.95	—	—	—	1.28	2.06	0.94	1.24	744.21§	1.10	1.35	0.87
—	—	—	—	—	—	0.99	4.20	—	—	—	16.88
1.09§	1.09§	1.01	1.03	1.02	0.96	1.03§	0.95§	0.56§	0.98	1.04	1.00
1.02	1.36§	1.02	2.24§	1.12	1.15	1.03	0.95	1.07	1.04	0.88	1.23§
1.04	0.95	1.07	1.12	1.06	0.81**	1.01	1.01	2.40	0.86§	0.99	0.95
0.91	1.29	1.29	0.44	0.66	1.91	0.97	1.17	—	0.80	0.66§	1.32
1.05	1.01	0.97	1.08	0.97	0.96	1.02§	0.97**	0.59§	1.04	1.05	1.10§
1.07	1.53**	1.12	2.23	0.89	0.74	1.03	1.11	1.18	0.98	0.88	1.48§
0.98	0.81	1.11	1.45	0.69	0.79	1.09**	1.04	9.14**	0.85	1.06	1.07
—	1.87	2.27	—	1.10	—	0.80	1.57	—	1.10	0.74	1.93
1.13	1.17	1.13	0.89	0.96	0.98	1.04§	0.84§	0.13§	0.97	1.02	1.16**
3.54	2.40	1.14	—	0.95	—	0.69	0.64	—	1.50	0.43	1.48
—	—	1.03	—	2.41	6.99	0.98	0.25	—	—	1.20	—
—	—	—	—	—	—	0.59	—	—	9.42	—	—
1.03	0.95	0.92**	1.19	0.96	0.96	1.03§	0.99	0.50§	1.01	1.05	1.11§
0.85	1.58**	1.09	1.86	0.87	0.69	1.06	1.19	0.73	0.98	0.75	1.63§
1.02	0.77	0.81	—	0.54	0.56	1.09	1.28**	10.02	0.82	1.21	1.40
—	2.47	1.80	—	1.46	—	0.88	1.73	—	0.66	0.39	2.70

(Continued)

FATAL ILLNESS

Table 3–1 (Continued). PMRs for white and black men and women aged 18 to 90 for selected

				Cause of death			
				Malignant neoplasms			
Occupation and census occupation codes‡	Number of deaths	All cancers (140–208)	Liver (155)	Larynx (161)	Lung (162)	Mesothelioma (158.8, 158.9, 163)	Skin (173)
Sheet metal workers (653, 654):							
White men	10,948	1.10§	1.05	1.38	1.16§	4.65§	1.18
Black men	369	1.10	2.46	0.65	1.06	—	—
White women	290	1.16	2.78	3.24	2.00§	—	—
Black women	28	1.89**	12.34	—	0.84	—	—
Precision textile, apparel, and furnishings machine workers (666–674):							
White men	10,760	0.99	1.00	0.95	0.99	0.68	1.25
Black men	1,402	0.99	1.42	1.31	0.97	—	—
White women	14,444	1.05§	1.11	0.65	0.91**	1.15	0.67
Black women	2,001	1.16§	1.11	0.56	1.31**	—	—
Precision workers, assorted materials (675–684):							
White men	8,631	1.01	0.85	0.72	0.97	0.80	0.42**
Black men	508	1.19	1.84	1.46	1.42**	—	—
White women	7,008	1.07§	0.99	1.28	1.15§	0.85	1.03
Black women	469	1.14	—	1.57	1.54**	—	—
Precision food production occupations (686–688):							
White men	22,121	1.00	0.89	1.11	1.01	0.75	1.04
Black men	2,976	1.01	1.59	1.16	0.98	—	0.82
White women	3,943	1.01	0.46	0.64	1.06	1.79	—
Black women	603	0.96	1.53	—	1.05	—	—
Plant and system operators (694–699):							
White men	22,238	1.08§	0.84	0.79	1.05**	2.55§	0.72
Black men	1,328	1.27§	1.94**	1.18	1.20**	4.32	1.79
White women	360	1.07	1.54	2.89	1.06	—	4.10
Black women	103	1.27	6.08	—	0.65	—	—
Operators, fabricators, and laborers (703–889):							
White men	765,521	0.97§	0.96§	1.17§	1.07§	0.83§	0.93**
Black men	178,087	1.01§	0.98	1.08**	1.05§	0.83	1.06
White women	247,323	0.98§	1.01	1.11	1.03§	0.90	0.94
Black women	34,052	1.09§	1.01	0.81	1.18§	1.21	0.93
Metalworking and plastic machine operators (703–725):							
White men	30,311	1.03§	1.06	1.06	1.10§	0.66	0.69**
Black men	3,398	1.07**	1.33	0.85	1.28§	1.57	0.95
White women	3,549	1.05	0.87	1.16	1.37§	1.73	1.22
Black women	300	1.41§	1.04	—	2.13§	—	—
Woodworking machine operators (726–733):							
White men	7,445	0.92§	1.03	0.91	1.06	0.32	0.40
Black men	1,656	0.92	0.81	0.84	0.98	—	0.50
White women	375	1.07	1.52	2.76	1.07	8.19	—
Black women	82	1.02	—	—	1.05	—	—
Printing machine operators (734–737):							
White men	19,559	1.04§	0.87	1.15	1.04	0.83	1.08
Black men	926	1.23§	0.76	0.53	1.22	—	1.68
White women	2,762	1.13§	0.64	0.41	1.31§	3.58	1.08
Black women	144	1.47**	—	—	1.38	—	—

See footnotes at end of table.

FATAL ILLNESS

occupations and causes of death in 28 States,* 1984–1995

and ICD-9† codes

Bladder (188)	Kidney (189.0–189.2)	Non-Hodgkins lymphoma (200, 202.0–202.2, 202.8, 202.9)	Hodgkins disease (201)	Leukemia (204–208)	Parkinson's disease (332)	IHD (410–414)	COPD (490–496)	Pneumo-conioses (500–505)	Chronic liver disease (571)	Renal failure (584–586)	Suicide (E590–E959)
1.14	1.08	1.11	0.57	1.09	0.82	0.94§	1.05	0.95	1.19**	0.99	1.02
2.62	1.95	1.21	—	0.47	2.72	0.98	0.42	4.17	0.90	1.95	1.01
1.06	1.91	1.77	—	—	—	1.03	0.90	—	1.39	0.80	0.89
—	—	14.71**	—	—	—	0.26	—	—	1.67	2.57	—
1.11	0.61§	1.04	1.31	1.04	0.88	1.03	0.91**	0.26§	1.00	1.19	1.01
0.97	1.16	0.34	1.35	0.97	—	0.95	1.03	—	1.48**	0.82	0.75
1.12	1.05	0.99	1.05	1.07	0.84	1.00	0.99	2.76	0.90	1.03	0.58§
1.12	1.66	0.96	—	0.64	2.41	0.95	1.46**	—	0.50**	0.78	1.27
1.25	1.06	1.17	1.37	1.06	1.27	0.98	0.96	1.31	0.97	1.17	1.15**
—	1.45	2.22	—	0.69	—	0.81	1.18	6.06	1.03	0.64	1.34
1.22	0.85	1.29**	0.78	0.79	0.70	1.01	1.09	2.50	0.87	1.09	1.20
0.80	—	1.27	—	0.34	—	1.03	0.45	—	1.41	0.57	0.42
1.18**	0.94	0.87	0.79	0.89	0.88	1.01	1.04	0.20§	1.03	1.12	0.97
1.03	1.26	0.31§	3.02	0.80	2.21	0.96	0.80	0.42	1.26	1.20	1.08
0.74	0.72	1.04	0.65	1.25	0.92	1.05	0.89	—	0.75	0.84	1.16
0.52	0.48	2.86**	—	0.28	4.36	1.04	0.54	—	1.18	0.59	2.04
1.08	1.09	1.20§	0.94	1.16**	0.95	1.01	0.93§	0.54§	1.03	0.99	1.00
0.62	1.37	1.92	4.23	1.65	1.19	1.05	0.82	—	1.06	0.77	1.14
0.86	0.54	1.24	—	1.15	—	1.02	0.82	—	0.93	0.64	0.67
—	2.80	—	17.55	—	—	0.88	0.51	—	—	0.58	—
0.99	0.89§	0.83§	0.86§	0.86§	0.81§	1.01§	1.14§	0.81§	1.01	1.03§	0.99
1.01	1.03	0.94	0.92	0.99	0.96	0.99	1.03§	1.01	1.03	1.00	0.96**
1.06	1.10§	0.99	0.86**	0.96**	0.75§	1.05§	0.96§	3.95§	0.91§	1.05**	0.82§
0.75§	0.94	0.97	0.90	1.18**	1.03	0.98	0.96	1.71	0.87§	0.94	1.01
1.07	0.95	0.95	0.91	0.93	0.86	1.03§	1.09§	1.13	1.01	0.92	1.00
0.86	0.87	1.04	1.18	1.24	2.04	1.11**	1.32§	6.46§	1.00	1.08	1.09
1.40	0.94	1.25	1.03	0.77	0.62	1.07**	1.30§	—	0.77	1.01	0.93
1.15	0.91	2.07	5.34	0.55	—	0.95	1.06	—	1.17	0.64	2.55
0.75	0.80	0.92	0.30	0.85	0.85	1.02	1.16§	0.29§	0.69§	0.98	1.04
0.79	1.12	0.58	—	1.02	—	1.05	0.89	2.14	0.71	0.64	1.22
0.83	1.58	0.48	—	1.10	1.93	0.98	1.17	—	0.70	0.92	0.64
—	—	—	—	2.05	—	1.09	1.32	—	3.07	0.73	—
1.10	0.96	1.09	0.81	0.92	1.31§	0.99	0.98	0.11§	1.16**	0.90	0.99
1.83	1.07	0.96	—	1.34	1.14	1.00	0.80	1.73	1.15	1.07	1.21
1.22	1.20	1.06	1.07	0.86	0.46	1.04	0.96	—	1.00	0.80	0.95
—	—	1.38	—	2.21	—	0.84	0.39	—	0.35	1.00	—

(Continued)

Fatal Illness

Table 3–1 (Continued). PMRs for white and black men and women aged 18 to 90 for selected

					Cause of death	
				Malignant neoplasms		
Occupation and census occupation codes‡	Number of deaths	All cancers (140–208)	Liver (155)	Larynx (161)	Lung (162)	Mesothelioma (158.8, 158.9, 163)	Skin (173)
Textile, apparel, and furnishing machine operators (738–749):							
White men	38,837	0.89§	0.94	1.00	0.97	0.79	1.25
Black men	6,922	1.02	0.99	1.17	0.98	—	0.85
White women	105,775	0.95§	1.00	0.79	0.89§	0.52§	1.07
Black women	13,468	1.05§	1.14	1.08	1.10**	1.02	1.12
Machine operators, assorted materials (753–779):							
White men	111,490	1.00	0.97	1.14**	1.05§	0.77	0.97
Black men	19,429	1.07§	0.97	0.97	1.14§	1.49	0.60**
White women	33,842	1.00	1.03	1.48**	1.08§	1.14	1.18
Black women	5,005	1.12§	0.83	0.64	1.06	—	—
Fabricators, assemblers, and hand working occupations (783–795):							
White men	61,029	1.01	1.03	1.10	1.13§	1.22	0.85
Black men	6,793	1.06§	1.12	0.89	1.09**	0.91	0.68
White women	22,617	1.03§	1.07	1.26	1.21§	1.64	0.39**
Black women	2,164	1.23§	0.59	1.33	1.37§	—	0.78
Assemblers (785):							
White men	24,366	0.99	0.85	0.92	1.02	0.74	1.01
Black men	2,962	1.03	0.76	0.96	1.10	2.14	0.79
White women	19,472	1.03**	1.10	1.13	1.19§	1.42	0.45**
Black women	1,805	1.22§	0.54	1.57	1.30§	—	0.93
Production inspectors, testers, samplers, and weighers (796–799):							
White men	20,942	1.01	0.95	1.20	1.01	1.68	0.69
Black men	1,476	1.19§	2.54§	0.93	1.15	—	—
White women	19,876	1.03§	0.95	1.27	1.08§	0.96	0.89
Black women	1,599	1.27§	1.18	0.46	1.11	4.13	—
Transportation and material moving occupations (803–859):							
White men	237,073	1.00	0.95	1.15§	1.14§	0.77**	0.93
Black men	39,706	1.02	0.85**	0.90	1.07§	0.58	0.89
White women	5,235	1.03	1.16	0.52	1.14§	1.05	0.54
Black women	981	1.23§	1.68	—	1.41**	—	—
Motor vehicle operators (803–814):							
White men	153,852	0.99§	0.93	1.18§	1.14§	0.70§	0.92
Black men	29,635	1.01	0.83**	0.99	1.06§	0.39	0.89
White women	4,066	1.04**	1.20	0.64	1.10	0.65	0.68
Black women	821	1.23§	1.60	—	1.47§	—	—
Rail transportation occupations (823–826):							
White men	23,349	1.03**	1.06	0.94	1.11§	1.08	0.89
Black men	841	1.05	0.82	0.31	1.03	—	1.13
White women	115	0.93	—	—	0.90	—	—
Black women	15	1.34	—	—	—	—	—

See footnotes at end of table.

FATAL ILLNESS

occupations and causes of death in 28 States,* 1984–1995

and ICD-9† codes

Bladder (188)	Kidney (189.0–189.2)	Non-Hodgkins lymphoma (200, 202.0–202.2, 202.8, 202.9)	Hodgkins disease (201)	Leukemia (204–208)	Parkinson's disease (332)	IHD (410–414)	COPD (490–496)	Pneumo-conioses (500–505)	Chronic liver disease (571)	Renal failure (584–586)	Suicide (E590–E959)
0.85§	0.83§	0.73§	0.78	0.79§	0.77§	1.04§	1.14§	0.39§	0.95	0.90	0.92**
1.45**	1.14	0.87	0.90	1.26	0.61	1.02	1.04	0.39	0.90	1.01	1.30§
1.00	1.15§	0.93**	0.95	0.97	0.71§	1.07§	0.86§	4.97§	1.00	1.04	0.68§
0.87	0.72	0.69**	0.26**	1.16	0.66	0.98	0.96	—	0.90	0.87**	0.85
1.06	1.00	0.93**	0.99	0.99	0.87§	1.02§	1.05§	0.76§	0.99	1.03	1.01
0.97	0.85	1.09	0.90	1.36§	0.80	1.09§	1.01	1.65**	0.91	0.94	1.11
1.26§	1.01	0.98	0.75	0.92	0.66§	1.05§	1.02	5.37§	0.94	1.02	0.85**
0.68	1.29	1.36	1.56	1.23	2.00	0.95	0.89	5.84	0.74**	1.05	1.23
0.93	0.91	0.83§	1.07	0.90**	0.86**	1.00	1.13§	0.88	0.95	0.99	1.09§
0.95	1.39	1.01	0.43	1.18	1.18	1.03	0.99	1.34	0.88	1.03	1.17
1.13	1.20**	1.07	1.24	1.01	0.77**	1.06§	1.10§	2.35	0.83§	1.16**	0.85**
1.02	1.03	1.42	1.26	1.19	1.59	1.01	1.09	—	0.57§	0.80	0.78
0.96	0.98	0.83**	0.99	1.03	0.93	1.02**	1.00	0.58§	0.91	1.14**	1.00
0.55	1.29	1.13	0.96	1.18	0.98	0.98	0.97	0.51	0.97	1.03	1.25
1.16	1.22**	1.05	1.37	1.05	0.75**	1.06§	1.08**	0.91	0.79§	1.14	0.87
1.25	1.08	1.59	1.48	1.24	1.98	1.08	0.99	—	0.57**	0.67	0.90
1.05	0.95	1.00	0.96	1.03	1.05	1.07§	0.93**	0.50§	0.87**	0.92	0.96
1.18	0.63	2.19**	1.16	1.39	—	0.97	0.75	0.90	0.59**	1.13	1.04
1.08	1.04	1.10	0.68	1.07	0.91	1.03**	0.95	4.45**	0.90	0.98	0.88
0.21	1.58	1.43	4.87§	1.02	—	0.97	0.86	—	0.65	0.44§	1.10
1.03	0.92§	0.83§	0.92	0.88§	0.74§	1.01§	1.18§	0.69§	0.93§	1.05**	0.95§
1.02	1.29§	0.99	1.57**	0.94	1.06	1.05§	0.94**	0.69	0.87§	1.02	1.05
0.78	1.33	1.03	0.67	0.79	0.97	1.05	0.94	3.75	0.85	0.92	1.02
1.82	0.85	1.60	2.33	1.10	—	0.92	0.96	—	0.42§	1.00	0.82
1.02	0.92§	0.79§	0.94	0.85§	0.75§	1.01§	1.22§	0.44§	0.90§	1.08§	0.93§
1.08	1.19**	0.93	1.75§	0.94	1.09	1.04§	0.94	0.42§	0.86§	1.03	1.04
0.78	1.49**	1.12	0.64	0.76	1.03	1.03	0.94	—	0.84	0.87	0.99
2.21	0.67	1.89	2.75	1.31	—	0.88	1.01	—	0.38§	1.21	0.78
1.07	0.97	0.93	1.12	0.95	0.78**	1.02	1.06**	1.11	1.14**	0.99	1.05
1.47	1.44	1.16	2.72	1.40	3.07	1.14	0.93	—	0.88	1.24	1.38
—	—	—	10.84	3.05	2.38	1.13	1.51	—	—	—	2.28
—	—	—	—	—	—	1.74	—	—	—	—	—

(Continued)

Fatal Illness

Table 3–1 (Continued). PMRs for white and black men and women aged 18 to 90 for selected

			Cause of death				
			Malignant neoplasms				
Occupation and census occupation codes‡	Number of deaths	All cancers (140–208)	Liver (155)	Larynx (161)	Lung (162)	Mesothelioma (158.8, 158.9, 163)	Skin (173)
Water transportation occupations (828–834):							
White men	6,141	1.01	1.14	1.50	1.17§	1.45	1.54
Black men	511	0.96	1.64	0.42	0.98	—	—
White women	42	0.39**	—	—	0.72	—	—
Black women	9	1.18	—	—	3.78	—	—
Material moving equipment operators (843–859):							
White men	53,731	1.02**	0.94	1.10	1.14§	0.77	0.92
Black men	8,719	1.05§	0.86	0.70	1.12§	1.33	0.94
White women	1,012	1.00	1.13	—	1.36§	3.17	—
Black women	136	1.23	2.42	—	1.05	—	—
Handlers, equipment cleaners, helpers, and laborers (863–889):							
White men	238,835	0.92§	0.93**	1.27§	1.01	0.79**	0.94
Black men	97,781	0.99	1.00	1.22§	1.02**	0.86	1.29§
White women	53,292	0.97§	1.01	1.37**	1.07§	0.79	0.81
Black women	10,309	1.02	0.94	0.62	1.27§	2.00	1.51
Construction laborers (869):							
White men	56,765	0.91§	0.99	1.43§	1.08§	0.83	1.06
Black men	27,394	1.01	0.88	1.50§	1.06§	1.12	1.31
White women	570	0.90	0.52	—	1.07	—	2.91
Black women	222	0.97	1.61	—	1.90**	—	—
Laborers, excluding construction (889):							
White men	146,536	0.91§	0.92**	1.19§	0.99	0.79	0.94
Black men	54,459	0.98**	1.02	1.07	1.03	0.98	1.22
White women	39,082	0.96§	1.05	1.42**	1.06§	0.42**	0.73
Black women	8,511	1.01	0.94	0.65	1.23§	1.61	1.83
Military (905):							
White men	52,541	1.05§	1.10	1.11	1.09§	0.81	1.22
Black men	6,004	1.05**	1.05	0.73	1.01	1.08	1.63
White women	1,188	1.00	0.51	2.09	1.26	—	1.35
Black women	194	1.08	—	—	0.24	—	—

Source: NOMS [1999].

*Alaska, Colorado, Georgia, Hawaii, Idaho, Indiana, Kansas, Kentucky, Maine, Missouri, Nebraska, Nevada, New Hampshire, New Jersey, New Mexico, New York except New York City, North Carolina, Ohio, Oklahoma, Pennsylvania, Rhode Island, South Carolina, Tennessee, Utah, Vermont, Washington, West Virginia, and Wisconsin.

†Abbreviations: COPD = chronic obstructive pulmonary disease; ICD–9 = International Classification of Diseases, 9th Revision [WHO 1977]; IHD = ischemic heart disease.

‡Bureau of the Census [1992].

§$P<0.01$.

**$P<0.05$.

occupations and causes of death in 28 States,* 1984–1995

and ICD-9† codes

Bladder (188)	Kidney (189.0–189.2)	Non-Hodgkins lymphoma (200, 202.0–202.2, 202.8, 202.9)	Hodgkins disease (201)	Leukemia (204–208)	Parkinson's disease (332)	IHD (410–414)	COPD (490–496)	Pneumo-conioses (500–505)	Chronic liver disease (571)	Renal failure (584–586)	Suicide (E590–E959)
0.95	0.75	0.78	0.39	0.91	0.88	0.95**	1.07	0.21§	1.24**	0.96	1.02
0.99	0.87	1.41	—	1.27	2.81	0.87	1.33	—	0.91	1.23	1.67
—	—	—	—	—	8.92	0.63	2.23	—	1.90	—	2.90
—	—	—	—	—	—	—	6.93	—	—	—	15.02
1.07	0.92	0.89**	0.85	0.93	0.66§	1.02§	1.14§	1.20**	0.95	1.04	0.98
0.76	1.66§	1.16	0.97	0.89	0.54	1.11§	0.91	1.80	0.90	0.95	1.04
0.90	0.79	0.82	—	0.73	0.33	1.13	0.85	18.30	0.98	1.22	1.07
—	2.10	—	—	—	—	1.10	0.42	—	0.71	—	—
0.93§	0.80§	0.76§	0.73§	0.76§	0.77§	0.98§	1.18§	1.09**	1.15§	1.09§	1.00
0.99	0.93	0.86§	0.73**	0.88§	0.95	0.94§	1.07§	0.79	1.15§	1.01	0.86§
1.02	1.04	1.02	0.70**	0.94	0.83**	1.04§	1.05**	2.07	0.80§	1.07	0.87§
0.57**	0.94	0.87	0.31	1.20	1.24	1.00	0.97	2.73	1.06	1.10	1.14
0.88**	0.79§	0.68§	0.63§	0.67§	0.66§	0.93§	1.27§	0.91	1.41§	1.02	1.02
0.99	0.91	0.69§	0.53**	0.93	0.80	0.89§	1.15§	0.53**	1.03	0.98	0.76§
2.75	0.78	0.96	—	0.60	—	0.92	1.39	—	0.54	0.24	0.90
—	2.65	0.92	—	1.99	—	0.88	0.54	—	1.41	1.01	—
0.92**	0.77§	0.74§	0.76§	0.79§	0.80§	1.00	1.16§	1.26§	1.05**	1.10§	0.98
1.02	0.92	0.93	0.72	0.82§	0.90	0.96§	1.07§	1.00	1.19§	1.07	0.89§
1.00	1.10	1.01	0.78	0.96	0.84	1.05§	1.06**	1.88	0.81§	1.08	0.86§
0.64	0.90	0.91	0.20	1.16	1.47	1.01	0.94	3.24	1.07	1.13	1.25
0.95	1.03	0.87§	1.01	1.03	0.97	0.93§	1.13§	0.13§	1.31§	0.99	1.07§
0.97	1.10	1.15	1.17	1.61§	1.24	1.15§	1.03	—	1.18	1.02	1.59§
0.58	0.56	0.71	1.71	0.40**	1.56	0.73§	1.69§	—	1.35	0.71	1.90§
—	3.41	3.92**	—	0.61	21.53	1.04	1.74	—	0.25	1.62	1.27

4 Nonfatal Injury

4 Nonfatal Injury

Three surveillance systems provide information about the characteristics of nonfatal occupational injuries: the Survey of Occupational Injuries and Illnesses (SOII), the National Electronic Injury Surveillance System (NEISS), and the National Hospital Ambulatory Medical Care Survey (NHAMCS). SOII is based on employer-generated workplace incident logs, and NEISS and NHAMCS are based on visits to emergency departments in hospitals. NEISS and NHAMCS both collect data on occupational injuries, but they use different methods.

Nonfatal occupational injuries constitute more than 90% of the events recorded by SOII. In 1997, more than 5.7 million nonfatal occupational injuries were estimated to have occurred in the United States, resulting in a rate of 6.6 cases per 100 full-time, private-sector workers. Among industry divisions, incidence rates for the total number of nonfatal injuries ranged from a low of 2.0 cases per 100 full-time workers in finance, insurance, and real estate to a high of 9.3 cases per 100 full-time workers in construction (Figure 4–1). Rates for four of the eight industry divisions are above the average for all industries.

Injuries treated in emergency departments* are usually more urgent or severe than those treated in physicians' offices or walk-in clinics. NEISS estimates that approximately 3.6 million nonfatal occupational injuries were treated in U.S. hospital emergency departments in 1998. The average rate for all nonfatal occupational injuries treated in emergency departments that year was 2.8 per 100 full-time workers. The rate for men (3.4 per 100 full-time workers) was nearly twice the rate for women (2 per 100 full-time workers) (Figure 4–2). Rates were higher in younger workers (aged 16 to 19), with steady declines in both male and female workers aged 20 and older (Figure 4–2). Hands and fingers were the most commonly injured parts of the body, accounting for 30% of the total (Figure 4–3). Lacerations and punctures (26%), sprains and strains (25%), and contusions, abrasions, and hematomas (19%) were the most frequent types of injuries recorded in NEISS in 1998.

---

*The term *emergency departments* is used in this chapter to refer to hospital emergency rooms (NEISS data) as well as to hospital outpatient departments and hospital emergency departments (NHAMCS data).

NONFATAL INJURY

Figures 4–4 and 4–5 present the average annual rates of emergency department visits related to nonfatal occupational injuries recorded in NHAMCS for 1995–1997. Male workers aged 16–17 had the highest rate (nearly 10 per 100 full-time workers). The rate for black male workers was higher than the average rate for all workers. Overall, the rate for men exceeded the rate for women.

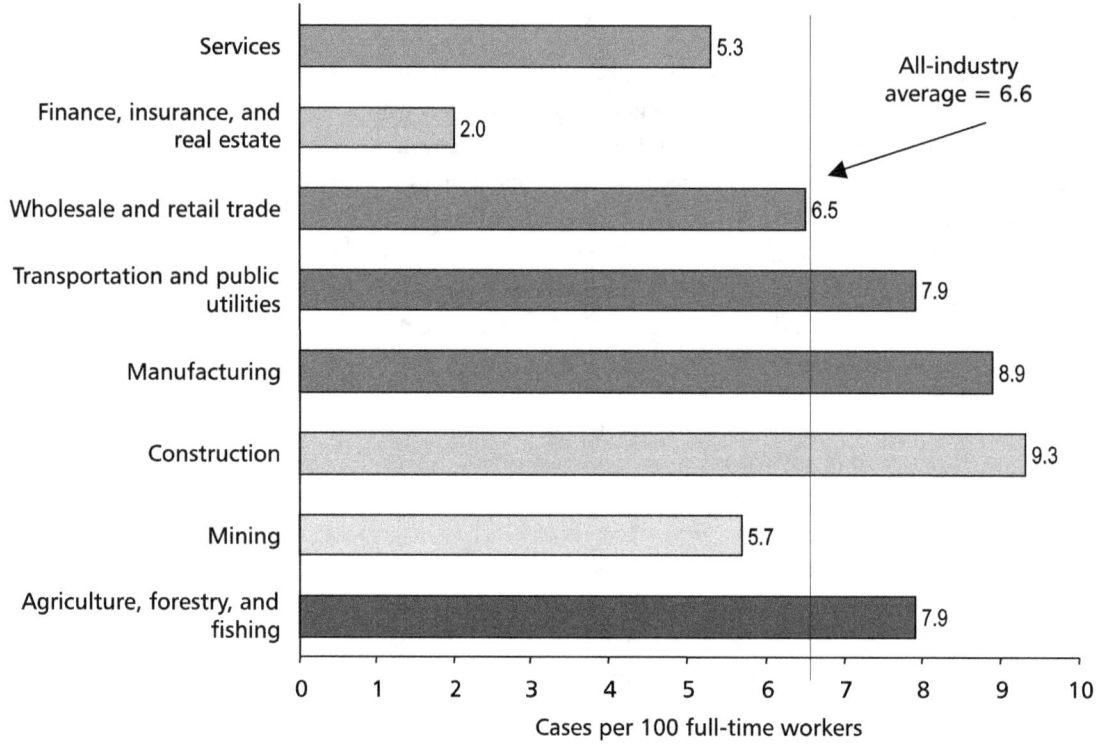

Figure 4–1. Incidence rates for nonfatal occupational injuries in private industry by major industry division, 1997. (Source: SOII [1999].)

NONFATAL INJURY

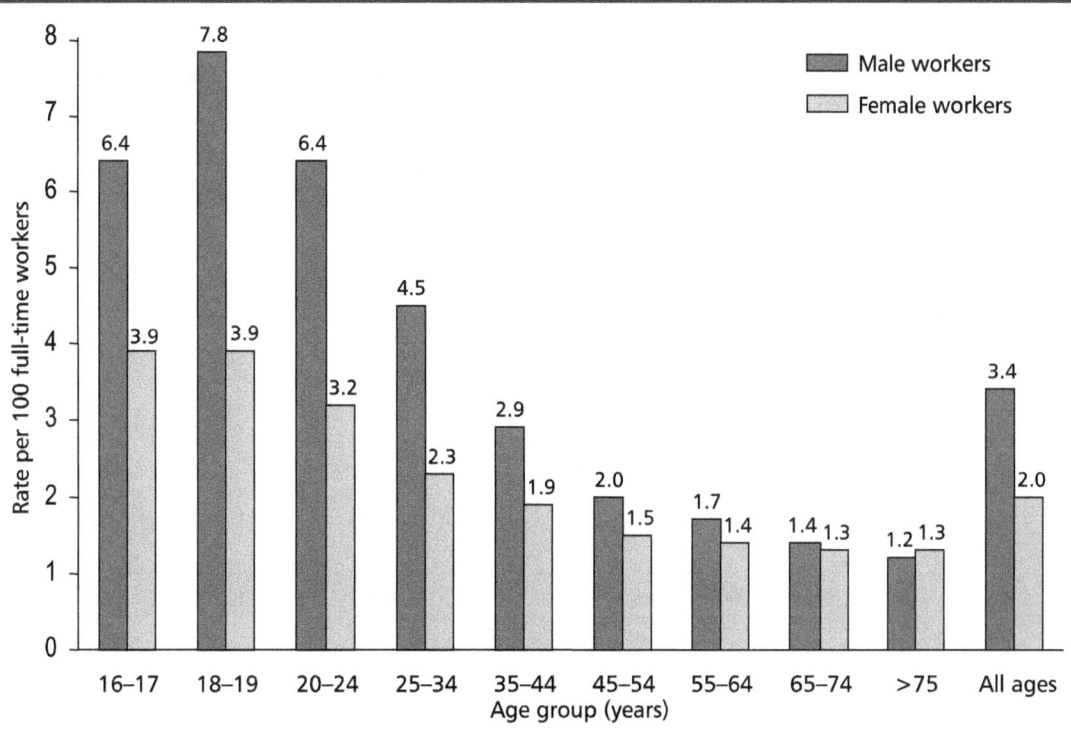

Figure 4–2. Rate of nonfatal occupational injuries treated in emergency departments, by age and sex, 1998. (Source: NEISS [1999].)

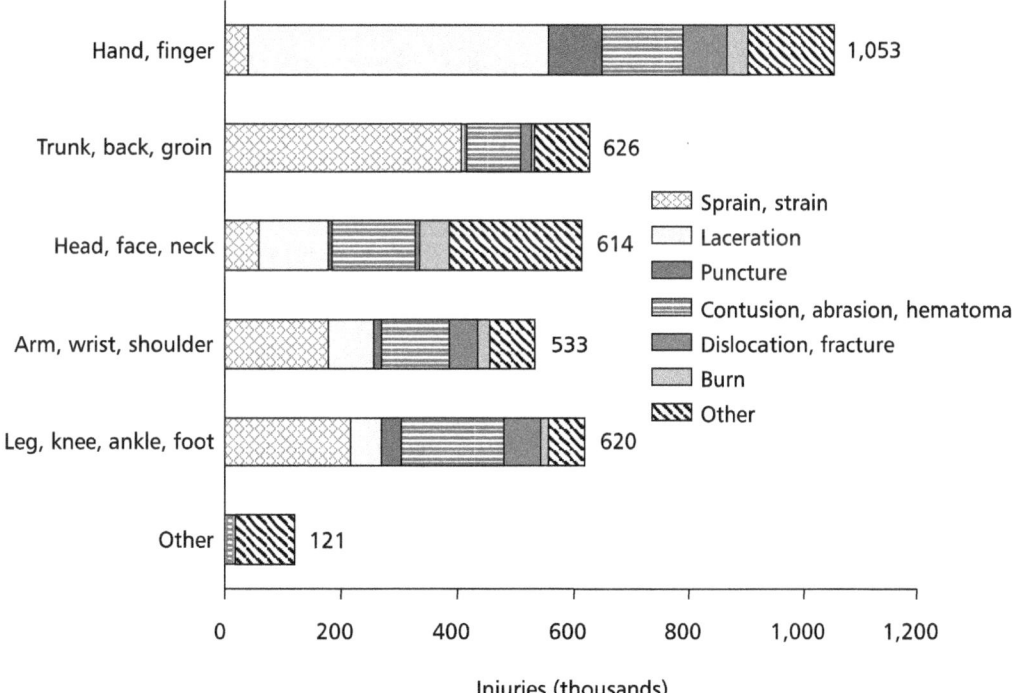

Figure 4–3. Number of nonfatal occupational injuries treated in emergency departments, by anatomic site and type of injury, 1998. (Source: NEISS [1999].)

NONFATAL INJURY

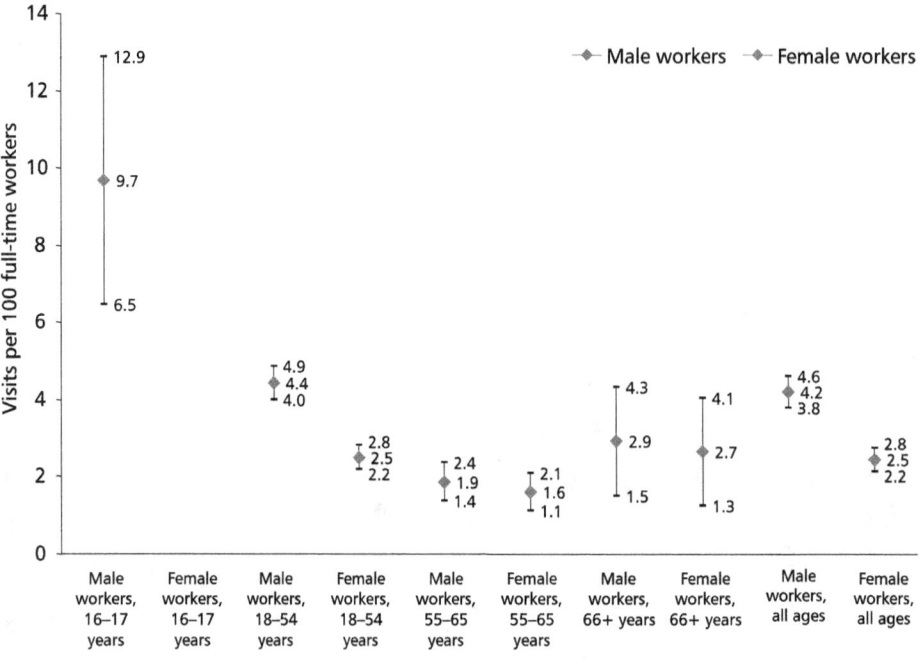

Figure 4–4. Annual rates (and 95% CIs) of emergency department visits related to nonfatal occupational injuries in male and female workers aged 16 and older, by age group—averaged for 1995–1997. (The rate for female workers aged 16–17 does not meet the standards of reliability or precision.) (Source: NHAMCS [1999].)

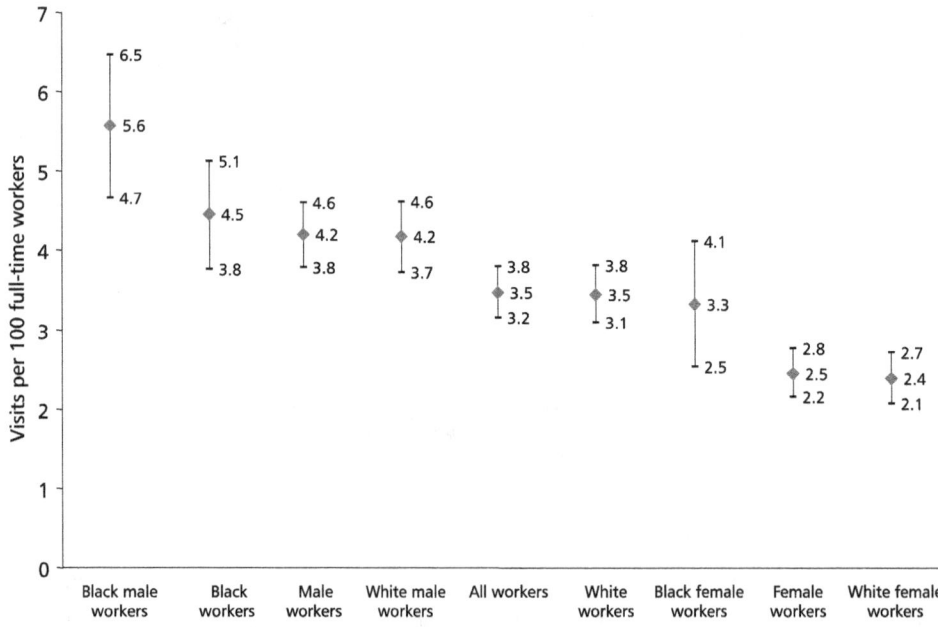

Figure 4–5. Annual rates (and 95% CIs) of emergency department visits related to nonfatal occupational injuries in black and white male and female workers aged 16 and older—averaged for 1995–1997. (Source: NHAMCS [1999].)

Nonfatal Occupational Injuries by Industry and Cases with Lost Workdays

The total number of nonfatal occupational injuries recorded by the Bureau of Labor Statistics (BLS) in SOII has fluctuated between 4.7 and 6.4 million per year over the last two decades. Many of these cases involved lost workdays.† The number of cases with days away from work fluctuated during that period; however, there was a 10-fold increase in cases with restricted work activity only (Figure 4–6). By 1997, 53% of cases involved no time away from work, 31% required at least 1 day away from work, and 16% involved restricted work activity only.

The incidence rate for total nonfatal occupational injuries over the past two decades ranged from a high of 9.2 cases per 100 full-time workers in 1978–1979 to a low of 6.6 cases per 100 full-time workers in 1997 (Figure 4–7). The incidence rate for cases with days away from work declined steadily from 1988 to 1997, and the incidence rate rose 120% for cases involving restricted work activity only.

Incidence rates for lost-workday cases of nonfatal occupational injury are shown for 1992–1997 by industry division in Figure 4–8. For all private industry during this period, the incidence rate declined 14% to 3.1 cases per 100 full-time workers. Finance, insurance, and real estate had the largest relative decline (27%), and construction had the largest absolute decline (1.3 cases per 100 full-time workers). Transportation and public utilities showed the least decline, both relatively (4%) and absolutely (0.2 cases per 100 full-time workers). Injury cases with and without lost workdays in 1997 (including days away from work and days of restricted activity only) are shown by industry division in Figure 4–9. The number of injuries ranged from a low of 46,000 in mining to a high of 1.7 million in manufacturing. The percentage of injury cases involving lost workdays ranged from a low of 38% in finance, insurance, and real estate to a high of 73% in mining.

The increasing incidence rate for cases involving restricted work activity only (Figure 4–7) is presented by industry division in Figure 4–10 for 1992–1997. The percentage of cases with restricted work activity only is shown for each industry division in Figure 4–11 for 1992 and 1997. In both years, manufacturing had the largest percentage of lost-workday cases with restricted activity only (32% and 48%, respectively).

---

†Lost-workday cases include cases with days away from work and cases with restricted work activity only (i.e., cases in which workers report to their jobs for limited duty).

NONFATAL INJURY

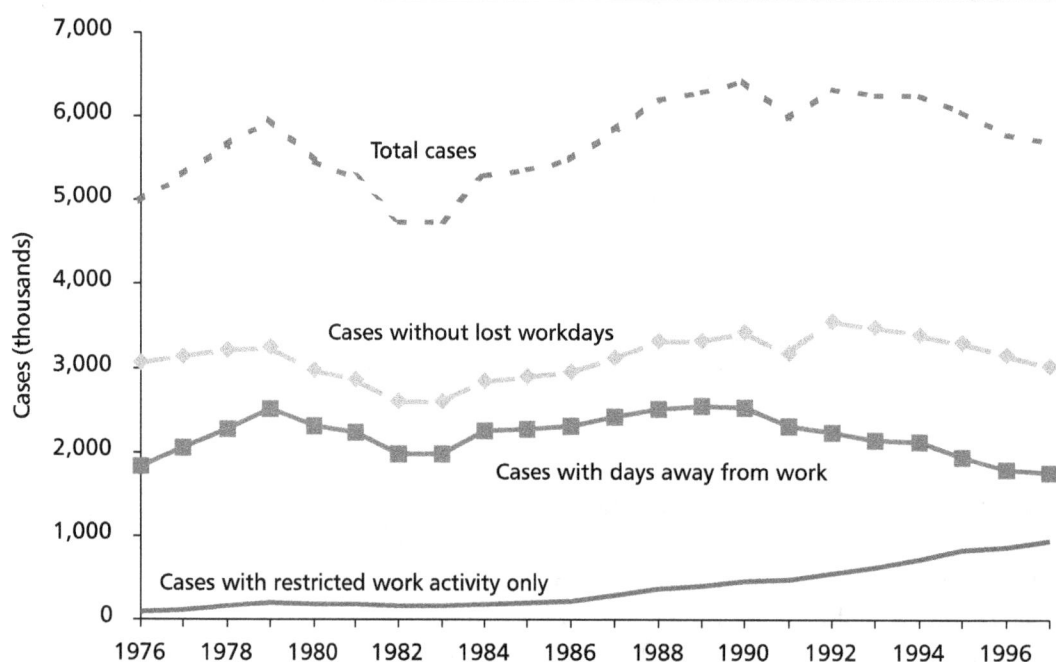

Figure 4–6. Number of nonfatal occupational injury cases in private industry by type of case, 1976–1997. (Source: SOII [1999].)

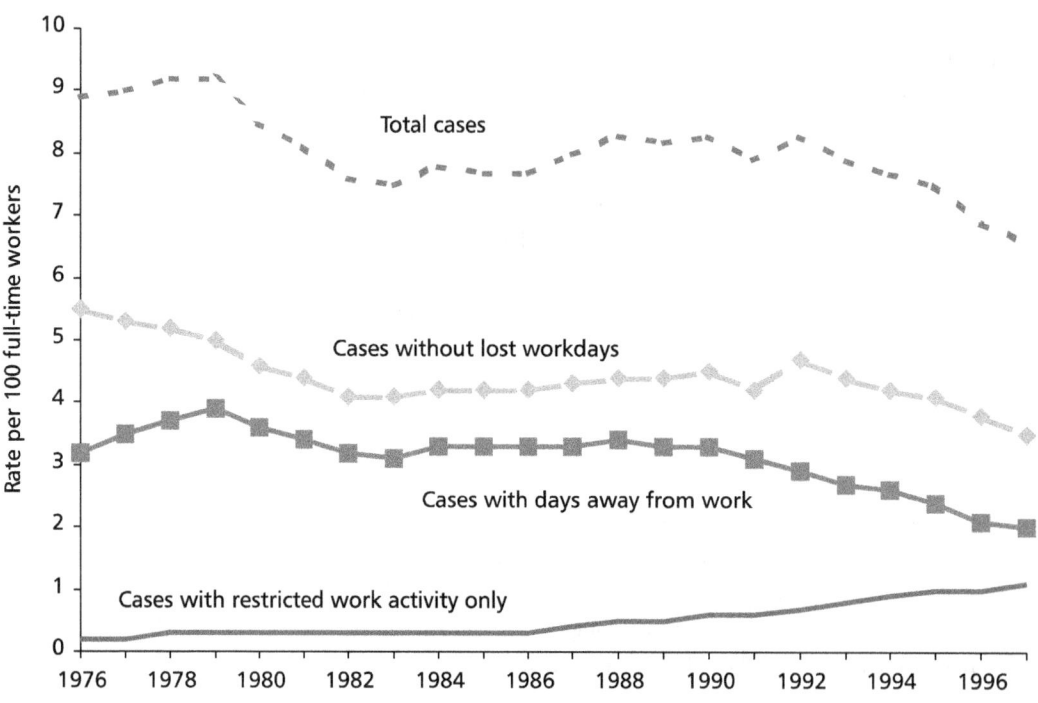

Figure 4–7. Incidence rate of nonfatal occupational injury cases in private industry by type of case, 1976–1997. (Source: SOII [1999].)

NONFATAL INJURY

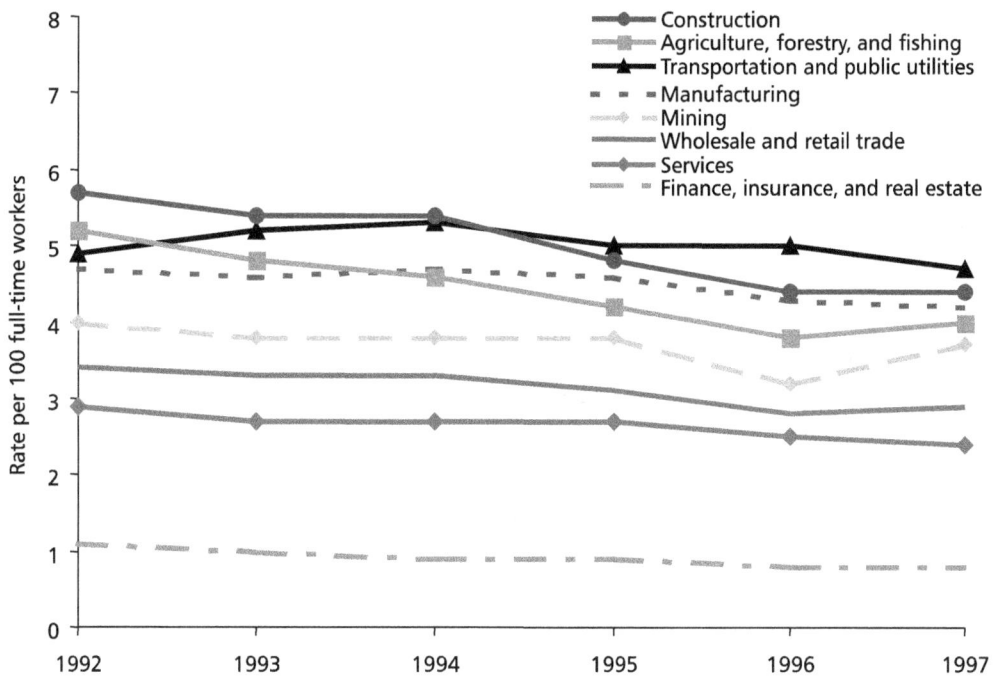

Figure 4–8. Incidence rates for lost-workday cases of nonfatal occupational injury in private industry by industry division, 1992–1997. (Source: SOII [1999].)

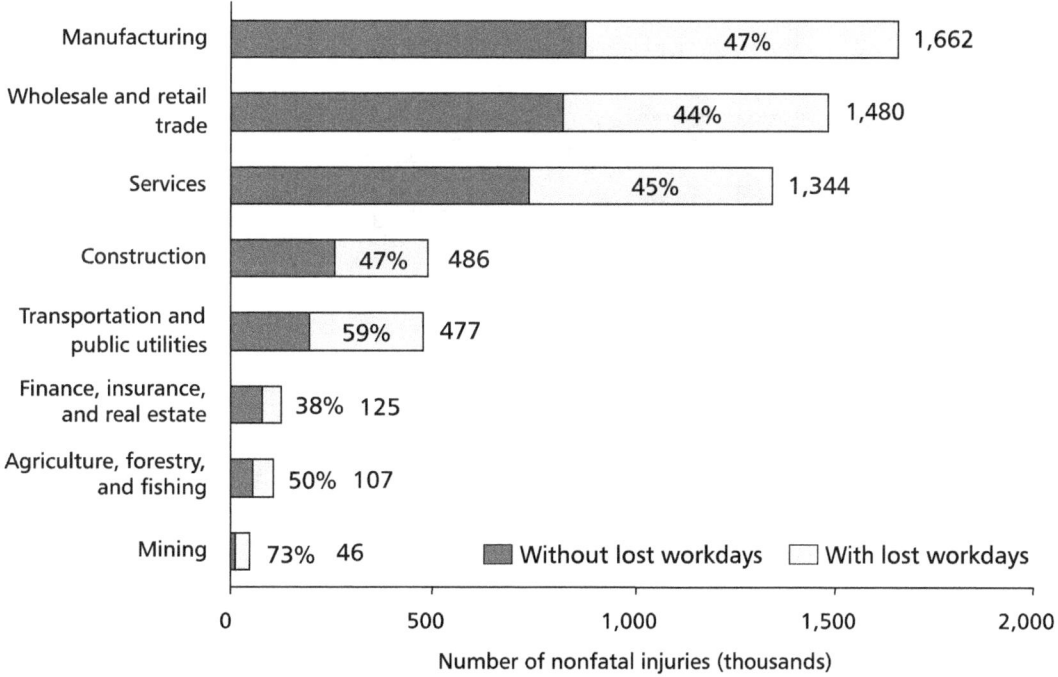

Figure 4–9. Number of nonfatal occupational injury cases in private industry without and with lost workdays by industry division, 1997. Percentage of cases with lost workdays also is shown. (Source: SOII [1999].)

NONFATAL INJURY

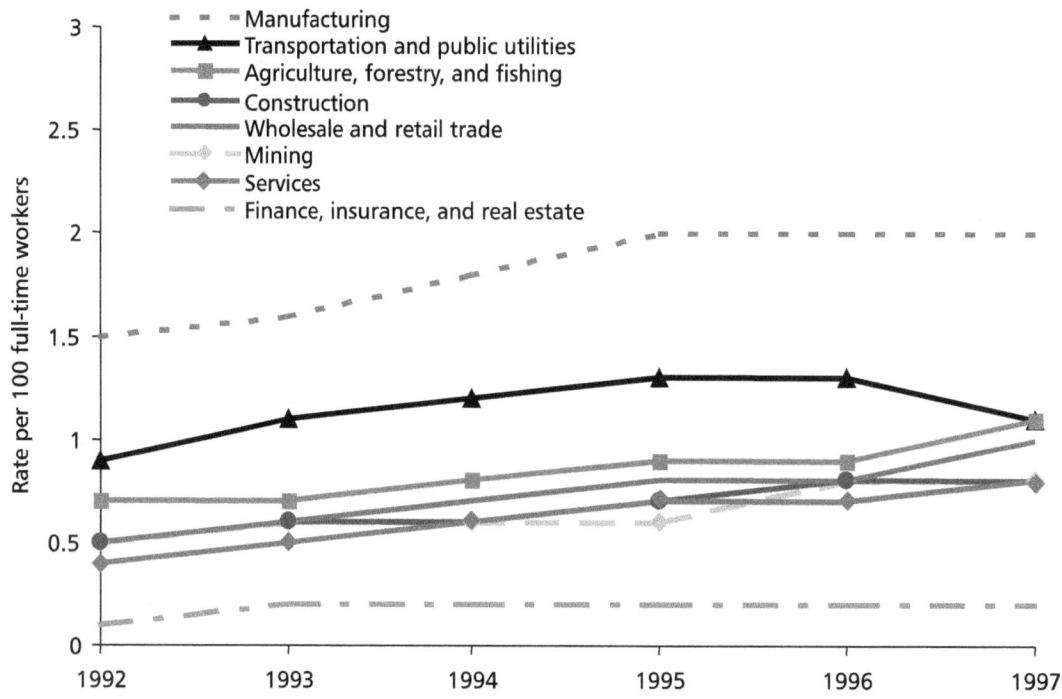

Figure 4–10. Incidence rates in private industry for nonfatal occupational injury cases involving days of restricted work activity only, by industry division, 1992–1997. (Source: SOII [1999].)

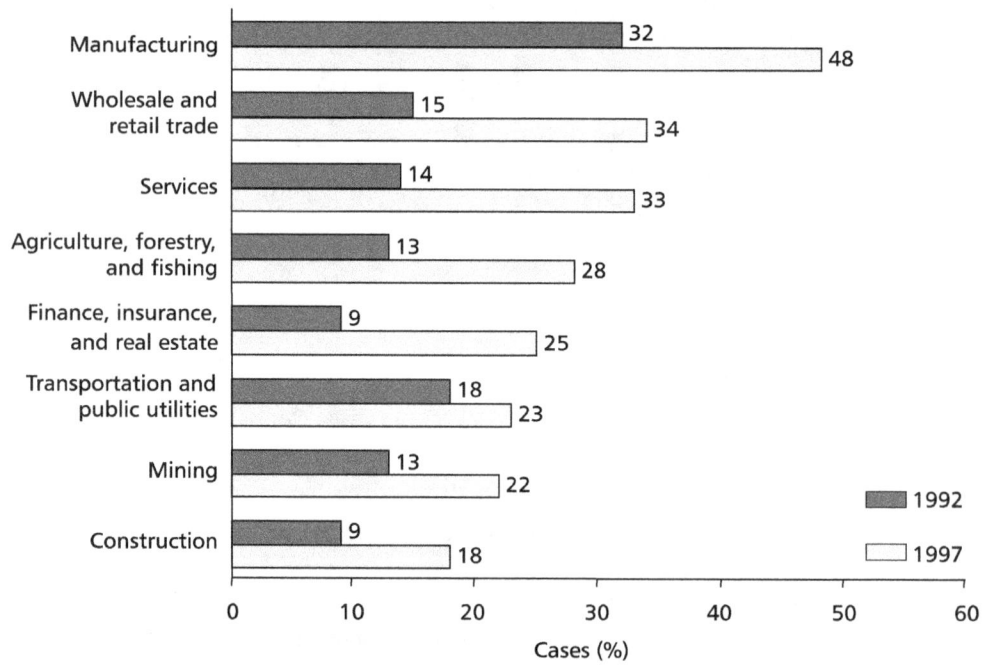

Figure 4–11. Percentage of nonfatal occupational injury cases with lost workdays involving restricted work activity only, by industry division, 1992 and 1997. (Source: SOII [1999].)

Characteristics of Injury Cases with Days away from Work

The total number of nonfatal occupational injury cases involving days away from work for 1992–1997 is shown in Figure 4–12 for seven injury categories. Sprains, strains, and tears accounted for the largest number of events, with approximately 799,000 cases in 1997. Nearly half those cases (about 385,000) involved the back, accounting for more than 80% of all traumatic injuries and disorders to the back. Other categories accounting for many days away from work included bruises and contusions (with nearly 166,000 cases in 1997), cuts and lacerations (with approximately 134,000 cases), and fractures (with approximately 119,000 cases). Presented separately for each of the seven injury categories are charts showing the distributions of cases by (1) major industries, (2) occupational groups, and (3) the sources of the disorder, events or exposures leading to the disorder, or the body parts affected.

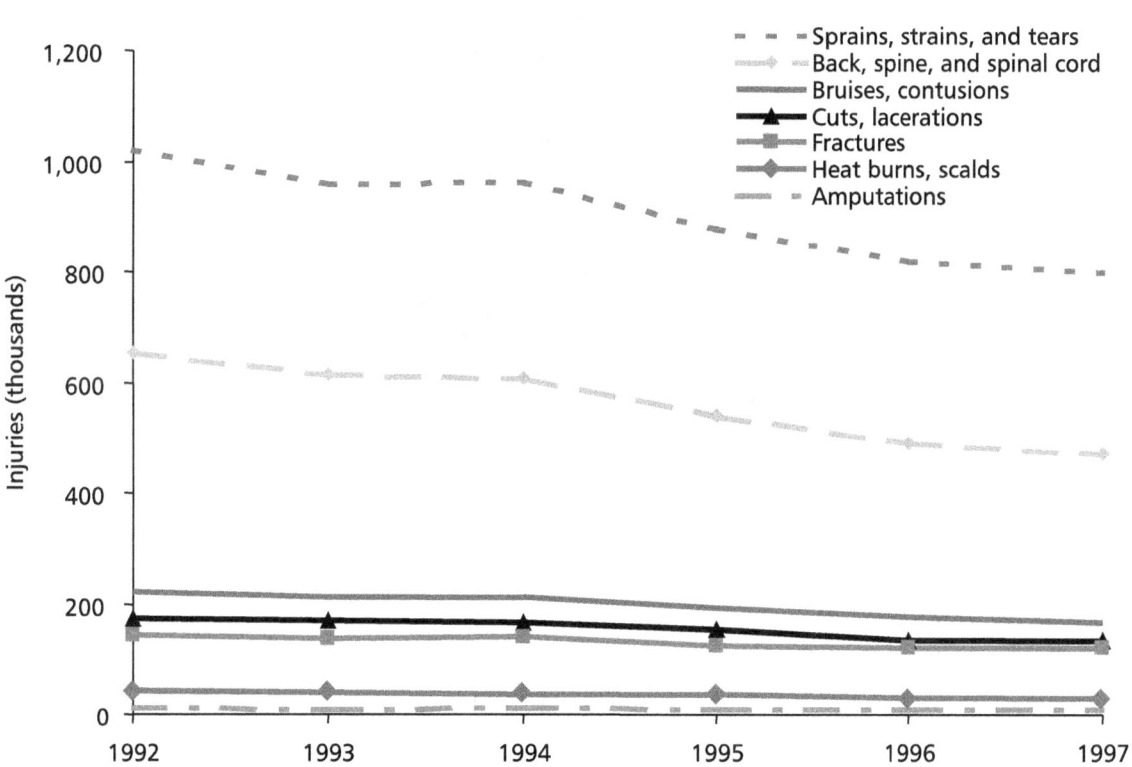

Figure 4–12. Number of nonfatal occupational injury cases with days away from work in private industry by type of injury, 1992–1997. (Source: SOII [1999].)

NONFATAL INJURY

Sprain, Strain, and Tear Cases with Days away from Work, 1997

Nearly half of the approximately 799,000 cases of sprains, strains, and tears involving days away from work in 1997 occurred in services (27%) and manufacturing (21%) (Figure 4–13). Most of these injuries were experienced by operators, fabricators, and laborers (42%) and service personnel (19%) (Figure 4–14). Overexertion was the most common event leading to a sprain, strain, or tear (Figure 4–15). Men accounted for nearly two-thirds of the sprain, strain, and tear cases. Half of the cases required 6 or more days away from work.

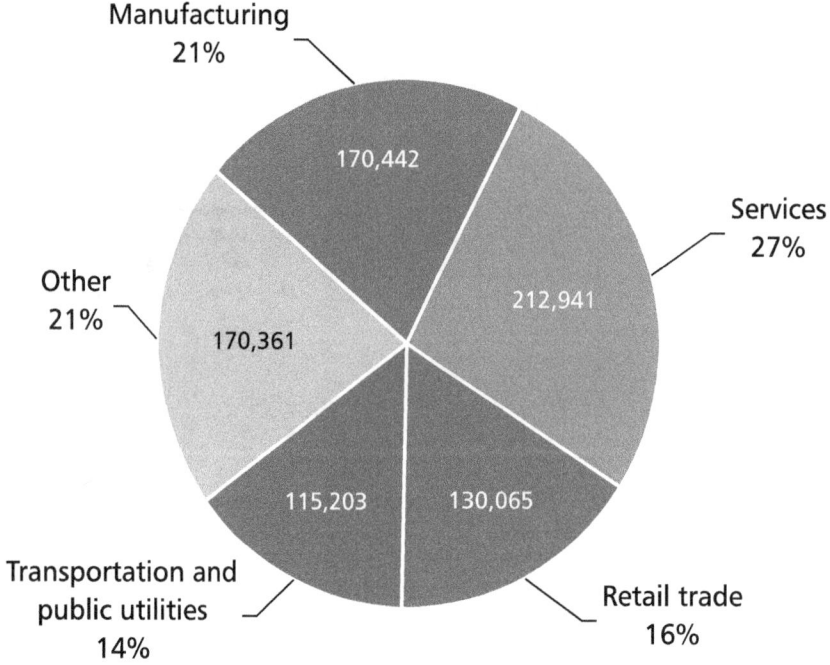

Figure 4–13. Number and distribution of sprain, strain, and tear cases with days away from work in private industry by industry division, 1997. (Source: SOII [1999].)

NONFATAL INJURY

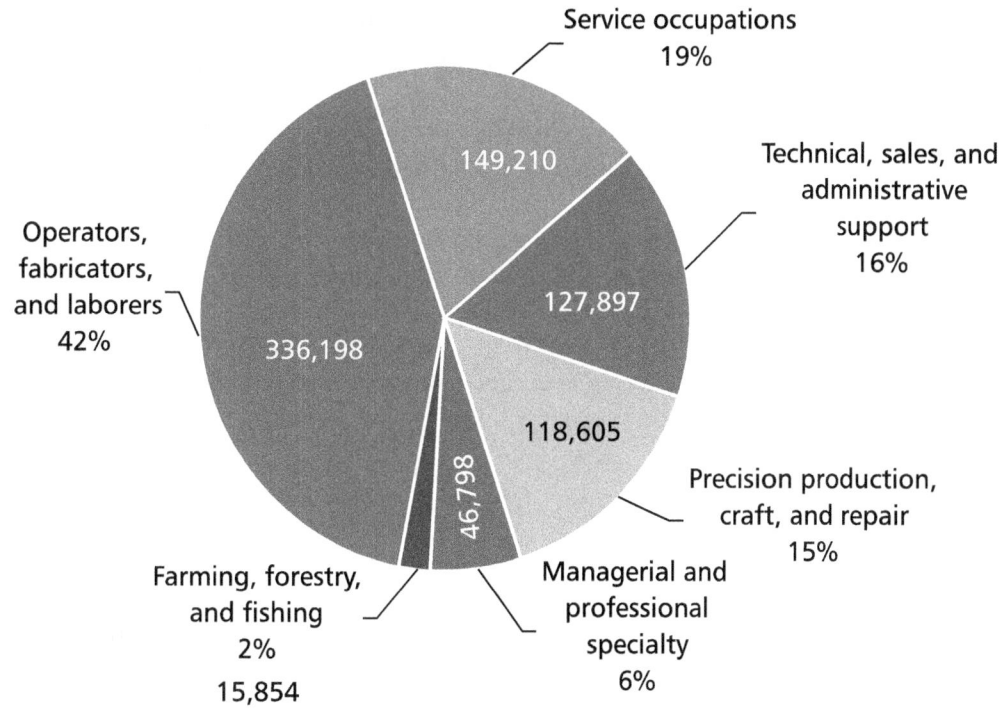

Figure 4–14. Number and distribution of sprain, strain, and tear cases with days away from work in private industry by occupational group, 1997. (Source: SOII [1999].)

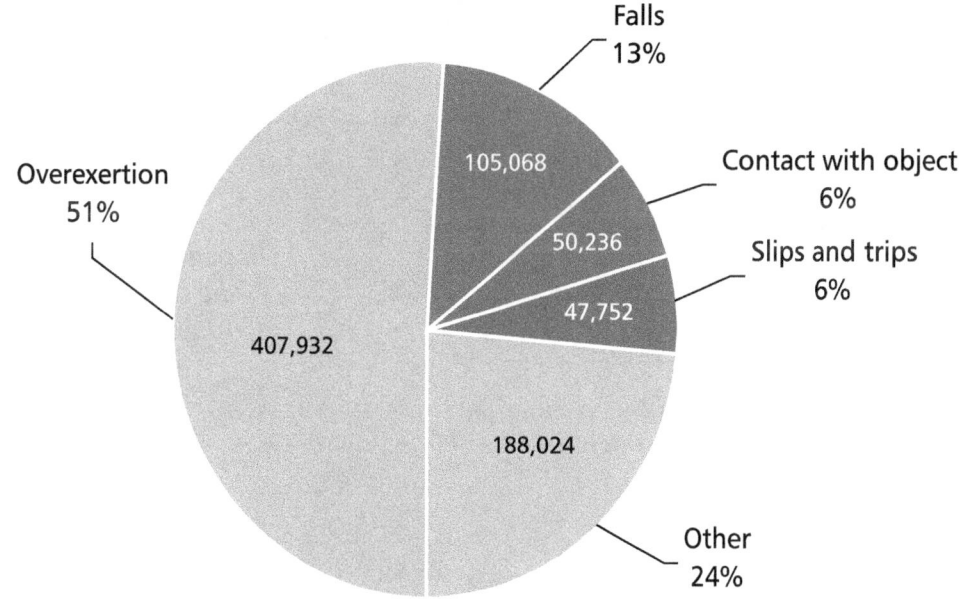

Figure 4–15. Number and distribution of sprain, strain, and tear cases with days away from work in private industry by event or exposure, 1997. (Source: SOII [1999].)

Nonfatal Injury

Back, Spine, or Spinal Cord Cases with Days away from Work, 1997

Nearly two-thirds of the approximately 472,000 back, spine, and spinal cord cases in 1997 occurred in services (28%), manufacturing (21%), and retail trade (16%) (Figure 4–16). Most of the back, spine, and spinal cord disorders were experienced by operators, fabricators, and laborers (41%) and service personnel (19%) (Figure 4–17). The most common sources of cases were containers (26%), worker motion or position (17%), and parts and materials (12%) (Figure 4–18). The event associated with most cases was overexertion, which accounted for 63% of the cases.

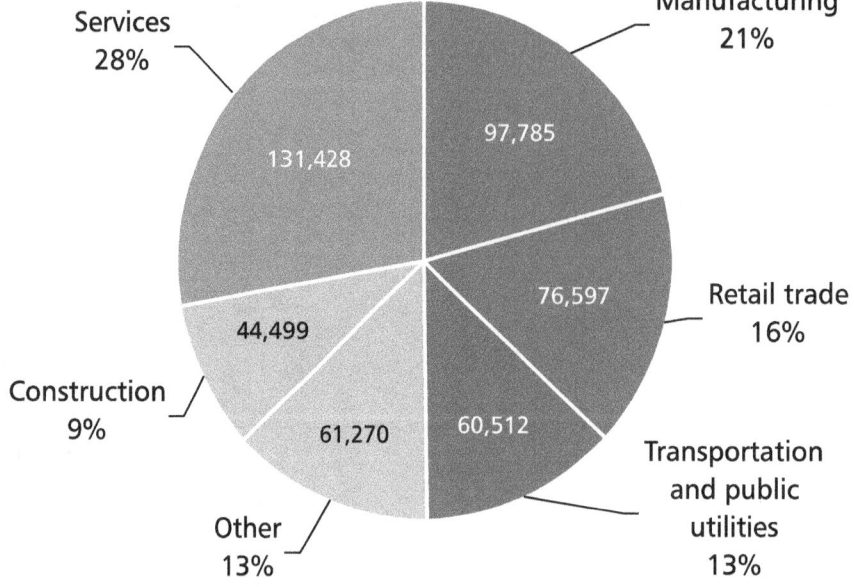

Figure 4–16. Number and distribution of back, spine, and spinal cord cases with days away from work in private industry by industry division, 1997. (Source: SOII [1999].)

NONFATAL INJURY

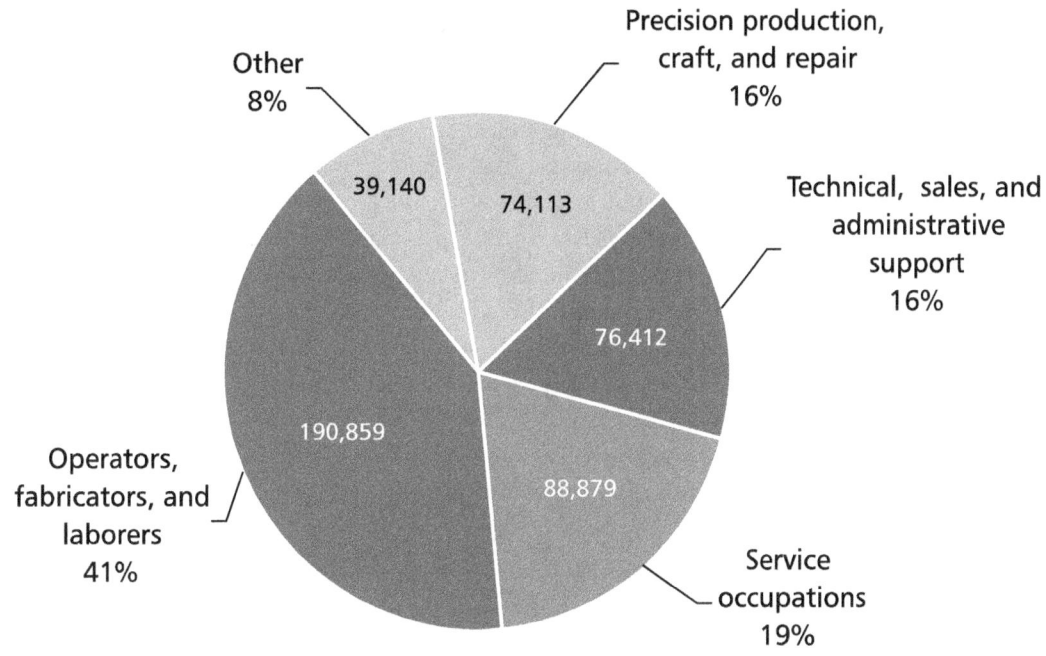

Figure 4–17. Number and distribution of back, spine, and spinal cord cases with days away from work in private industry by occupational group, 1997. (Source: SOII [1999].)

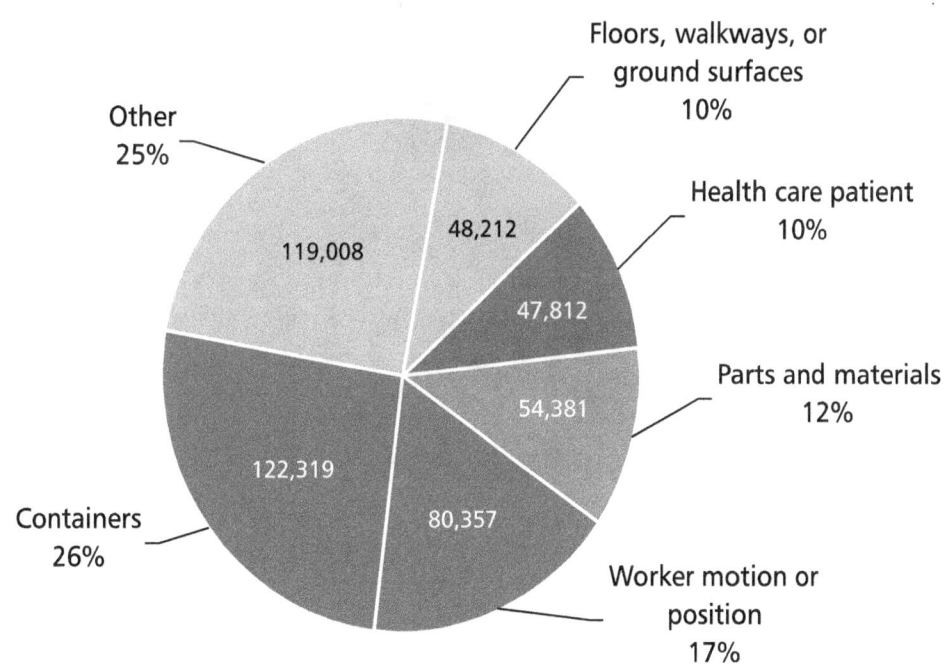

Figure 4–18. Number and distribution of back, spine, and spinal cord cases with days away from work in private industry by source of disorder, 1997. (Source: SOII [1999].)

Nonfatal Injury

Bruise and Contusion Cases with Days away from Work, 1997

Most of the approximately 166,000 bruise and contusion cases with days away from work in 1997 occurred in manufacturing (24%), services (22%), and retail trade (19%) (Figure 4–19). Together, operators, fabricators, and laborers and service personnel experienced more than half of these injuries (Figure 4–20). The most common sources of injury were floors and ground surfaces (26%), vehicles (15%), and parts and materials (13%) (Figure 4–21). Most job-related bruises and contusions resulted from workers being struck by, struck against, or caught in objects, equipment, or materials. In 1997, a median of 3 lost workdays resulted from bruises and contusions. Nearly 9% of these injuries required 31 or more days away from work.

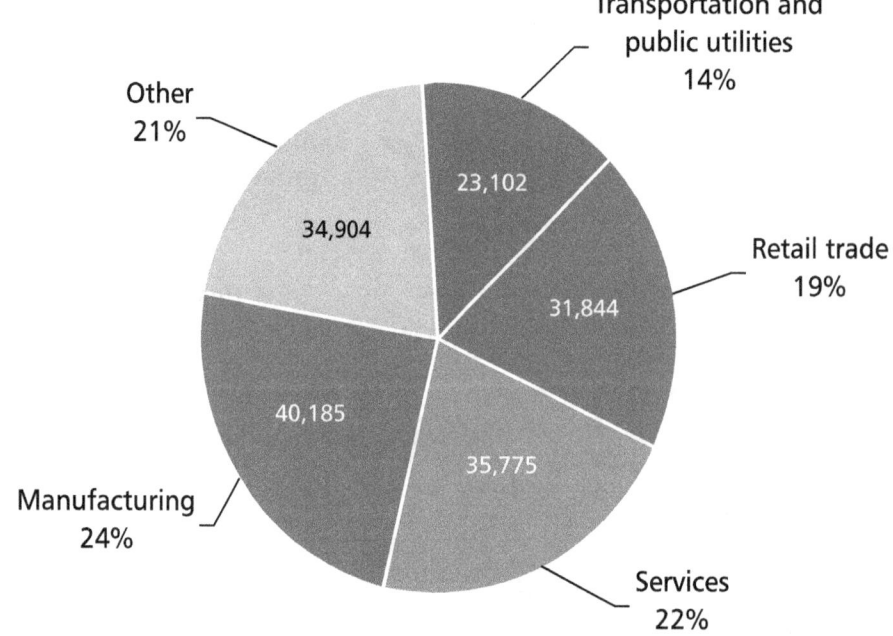

Figure 4–19. Number and distribution of bruise and contusion cases with days away from work in private industry by industry division, 1997. (Source: SOII [1999].)

NONFATAL INJURY

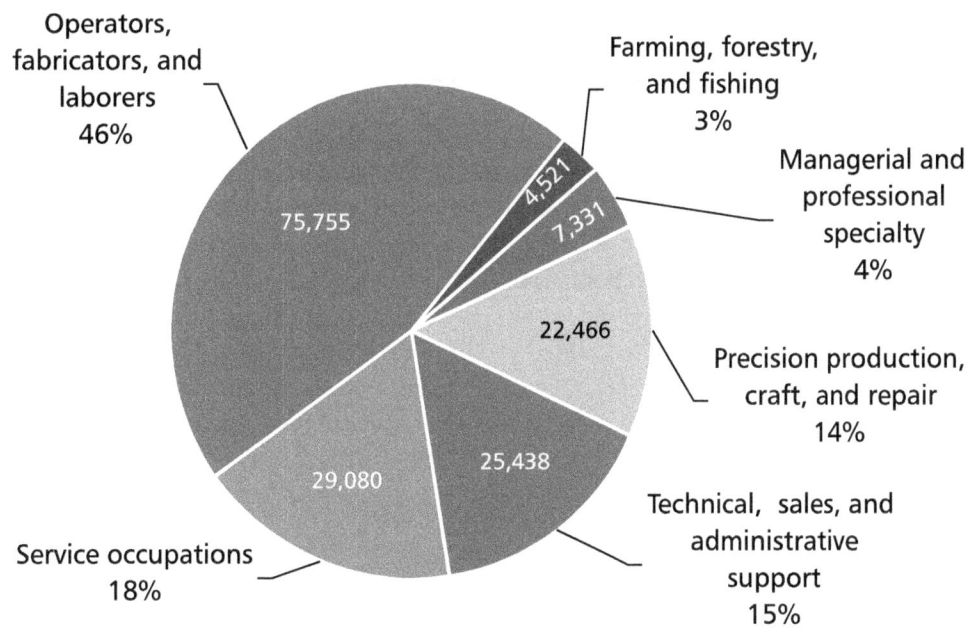

Figure 4–20. Number and distribution of bruise and contusion cases with days away from work in private industry by occupational group, 1997. (Source: SOII [1999].)

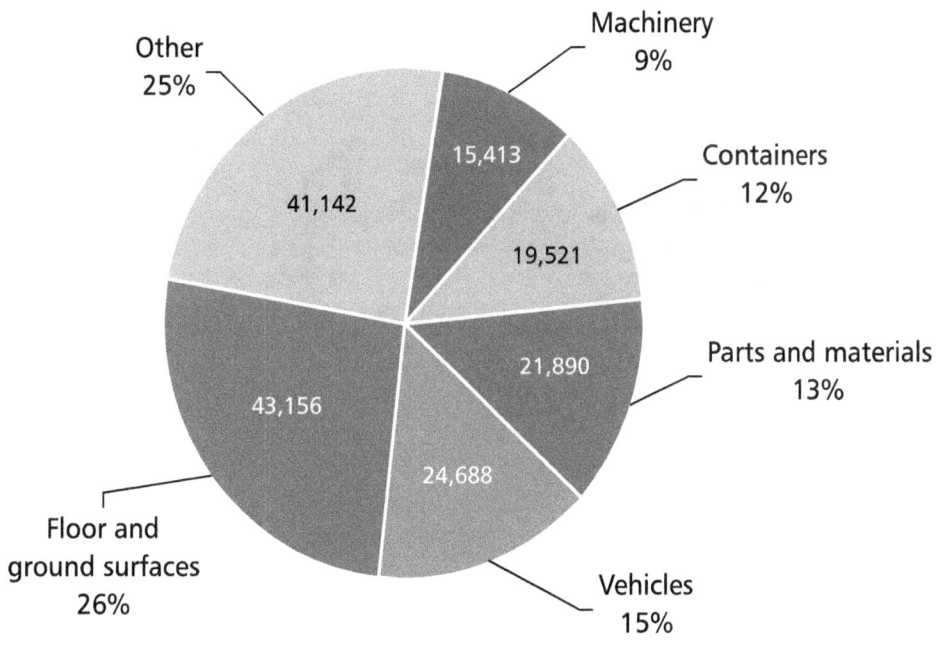

Figure 4–21. Number and distribution of bruise and contusion cases with days away from work in private industry by source of disorder, 1997. (Source: SOII [1999].)

Nonfatal Injury

Cut and Laceration Cases with Days away from Work, 1997

More than half of the approximately 134,000 cut and laceration cases with days away from work in 1997 were in manufacturing (28%) or retail trade (26%) (Figure 4–22). Operators, fabricators, and laborers experienced 42% of cuts and lacerations, and precision production, craft, and repair personnel experienced 24% (Figure 4–23). The most common sources of injury were floors and ground surfaces (25%), machinery (21%), and parts and materials (20%) (Figure 4–24). Finger cuts and lacerations accounted for half of all cuts and lacerations involving days away from work. A median of 3 days away from work resulted from cuts and lacerations.

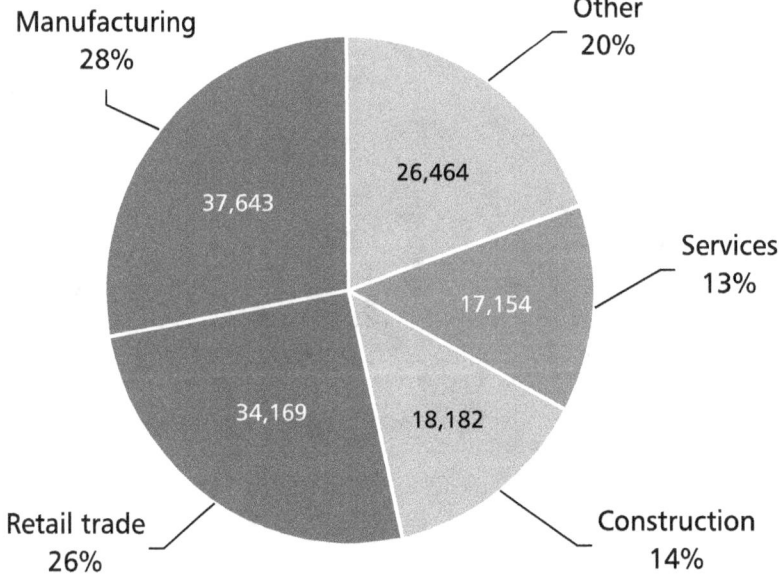

Figure 4–22. Number and distribution of cut and laceration cases with days away from work in private industry by industry division, 1997. (Source: SOII [1999].)

NONFATAL INJURY

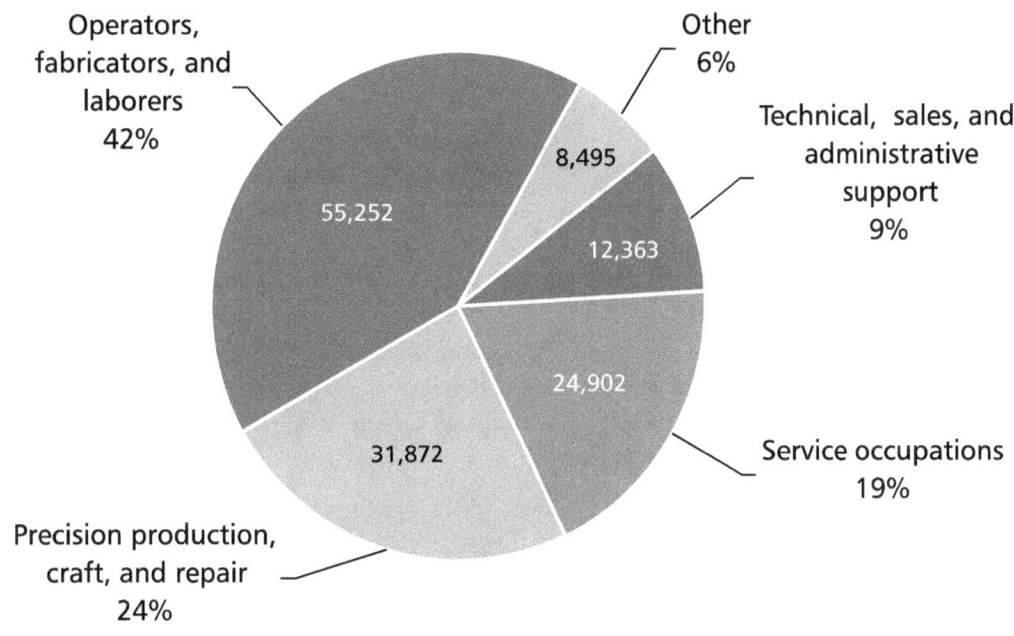

Figure 4–23. Number and distribution of cut and laceration cases with days away from work in private industry by occupational group, 1997. (Source: SOII [1999].)

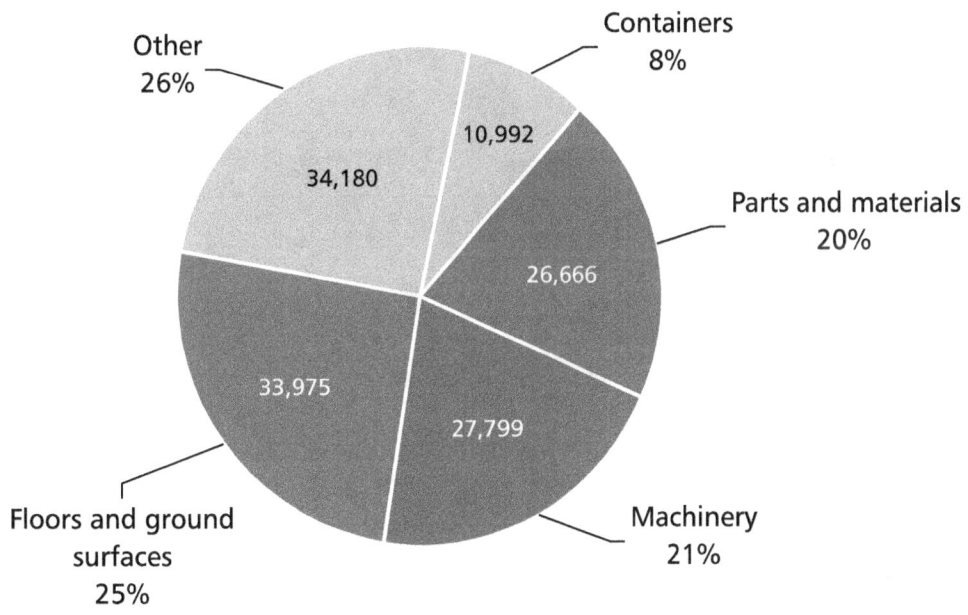

Figure 4–24. Number and distribution of cut and laceration cases with days away from work in private industry by source of disorder, 1997. (Source: SOII [1999].)

NONFATAL INJURY

Fracture Cases with Days away from Work, 1997

Most of the approximately 119,000 fracture cases with days away from work in 1997 occurred in manufacturing (25%), services (18%), and construction (16%) (Figure 4–25). Most of these injuries were experienced by operators, fabricators, and laborers (43%) and precision production, craft, and repair personnel (23%) (Figure 4–26). The most common sources of injury were floor and ground surfaces (43%) and parts and materials (14%) (Figure 4–27). Half of the occupational fractures in 1997 required 21 or more days away from work for recuperation. The categories *struck by object* and *falls on the same level* each accounted for more than 30,000 fractures.

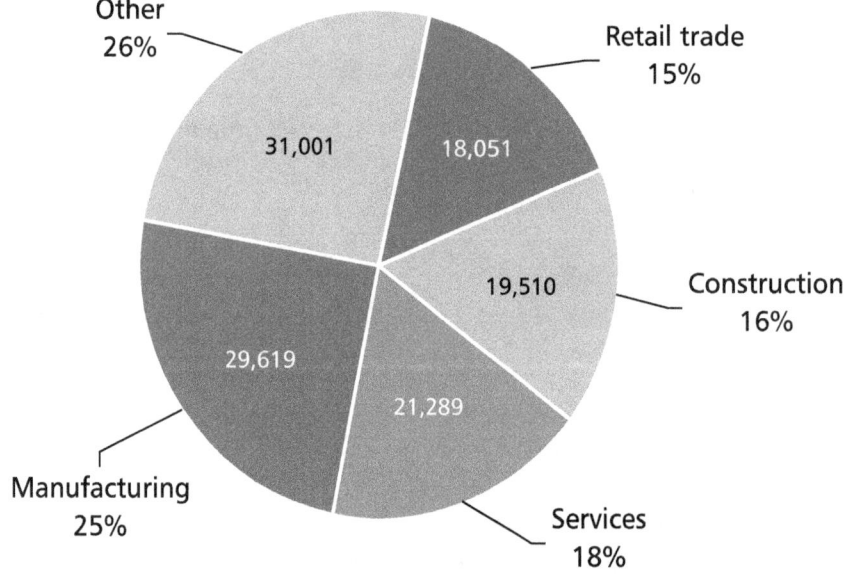

Figure 4–25. Number and distribution of fracture cases with days away from work in private industry by industry division, 1997. (Source: SOII [1999].)

Nonfatal Injury

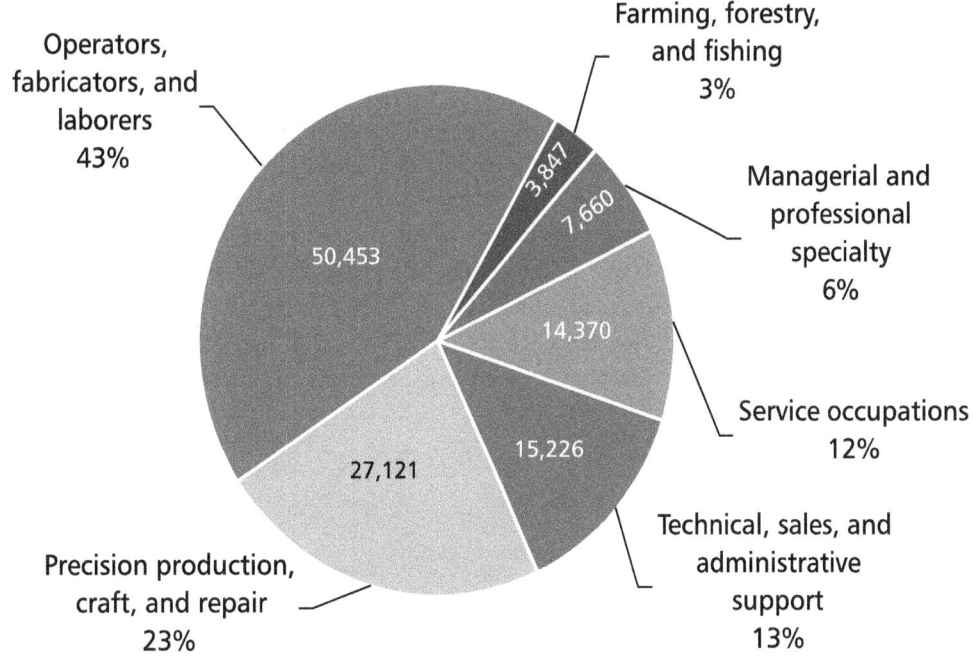

Figure 4–26. Number and distribution of fracture cases with days away from work in private industry by occupational group, 1997. (Source: SOII [1999].)

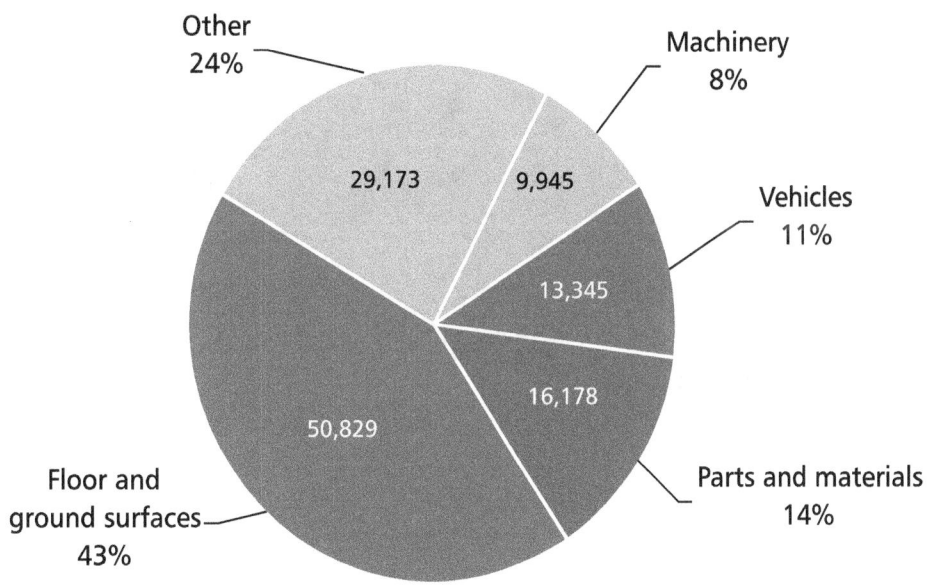

Figure 4–27. Number and distribution of fracture cases with days away from work in private industry by source of disorder, 1997. (Source: SOII [1999].)

NONFATAL INJURY

Heat Burn and Scald Cases with Days away from Work, 1997

More than half of the approximately 30,000 heat burn and scald cases with days away from work in 1997 occurred in retail trade (39%) and manufacturing (26%) (Figure 4–28). Most of these injuries were experienced by service personnel (44%) and operators, fabricators, and laborers (30%) (Figure 4–29). Twenty-four percent of heat burn and scald cases affected the hand (except fingers), 14% affected multiple body parts, and 12% affected the foot or toe (Figure 4–30). A median number of 4 days away from work resulted from heat burns and scalds.

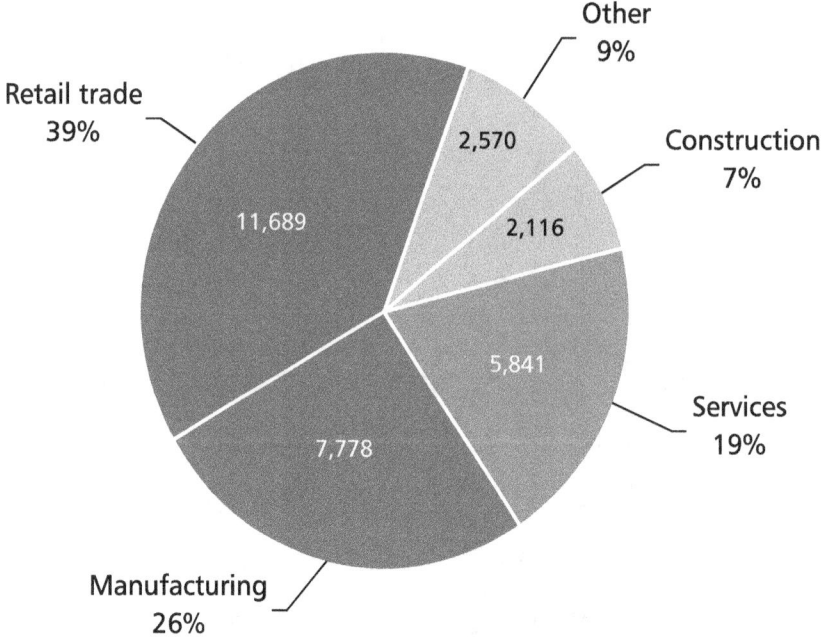

Figure 4–28. Number and distribution of heat burn and scald cases with days away from work in private industry by industry division, 1997. (Source: SOII [1999].)

NONFATAL INJURY

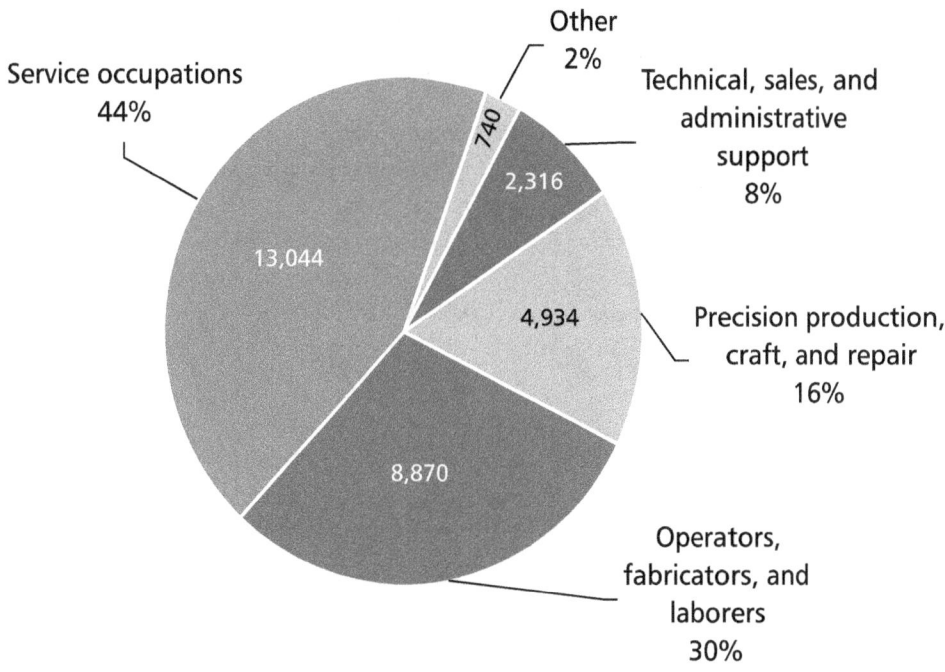

Figure 4–29. Number and distribution of heat burn and scald cases with days away from work in private industry by occupational group, 1997. (Source: SOII [1999].)

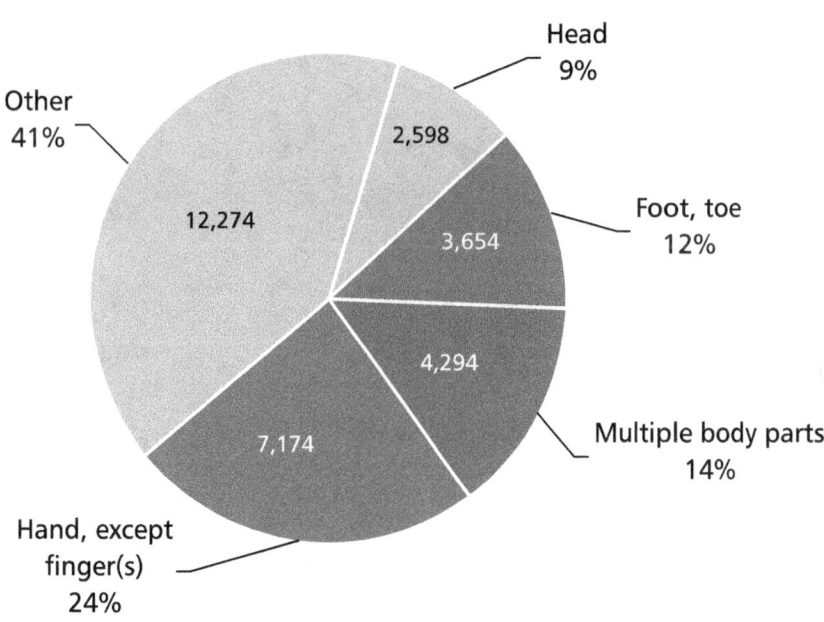

Figure 4–30. Number and distribution of heat burn and scald cases with days away from work in private industry by part of body affected, 1997. (Source: SOII [1999].)

Nonfatal Injury

Amputation Cases with Days away from Work, 1997

More than half of the approximately 10,850 amputation cases with days away from work in 1997 occurred in manufacturing (51%) (Figure 4–31). Operators, fabricators, and laborers experienced 60% of amputations (Figure 4–32). Machinery was the major source of amputation injury (57%) (Figure 4–33). Men accounted for 87% of occupational amputations. Nearly 10,200 amputations (93.8%) were to fingers. A median number of 18 days away from work resulted from amputations.

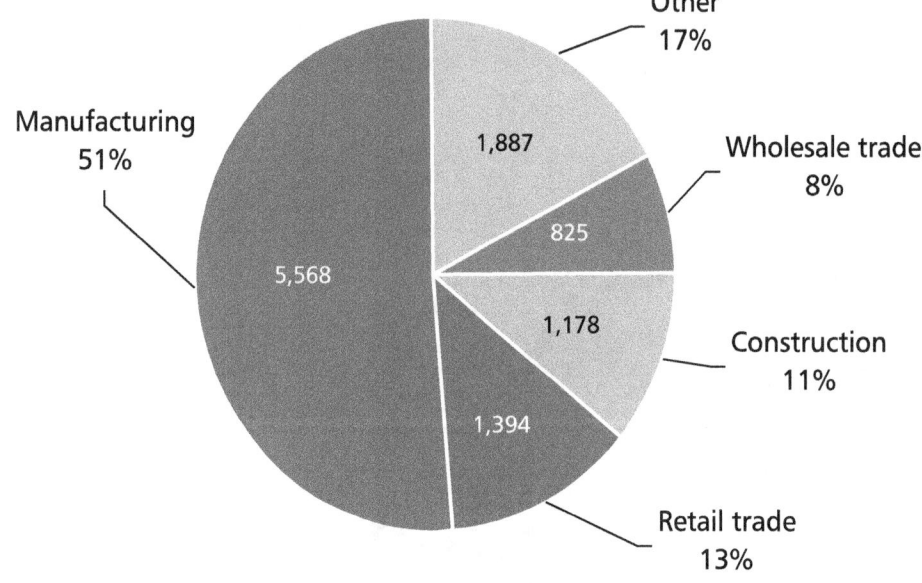

Figure 4–31. Number and distribution of amputation cases with days away from work in private industry by industry division, 1997. (Source: SOII [1999].)

NONFATAL INJURY

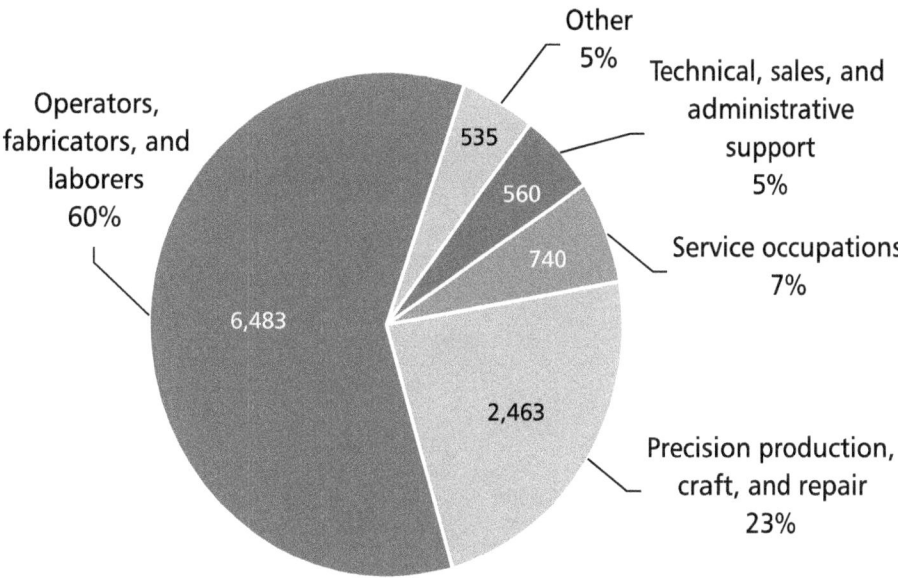

Figure 4–32. Number and distribution of amputation cases with days away from work in private industry by occupational group, 1997. (Source: SOII [1999].)

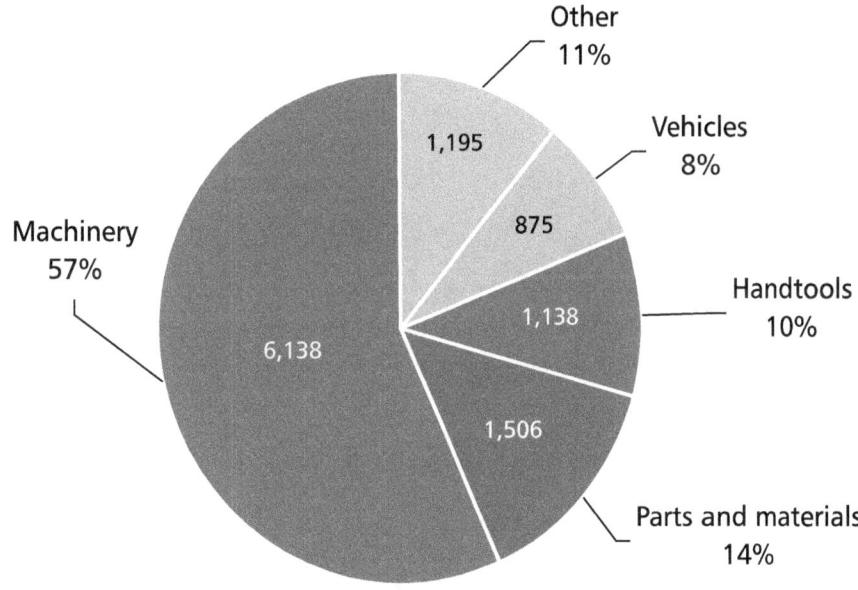

Figure 4–33. Number and distribution of amputation cases with days away from work in private industry by source of disorder, 1997. (Source: SOII [1999].)

5 Nonfatal Illness

5 Nonfatal Illness

Illnesses are often more difficult to link with work than injuries. Illnesses related to occupational exposures (e.g., tuberculosis [TB], cancers, central nervous system disorders, and asthma) appear no different when encountered in the absence of occupational exposures. Work-related aspects of illness may go unrecognized for many reasons, including long latency periods between the exposure and development of some diseases and the failure of health care professionals to recognize or report work-related illnesses or obtain information about a patient's work history.

The Bureau of Labor Statistics (BLS) records information about nonfatal occupational illness in the Survey of Occupational Injuries and Illnesses (SOII) using data from logs maintained by employers. The illnesses reported in SOII are those most easily and directly related to workplace activity. Illnesses with workplace associations that are not immediately obvious are vastly undercounted in SOII. Other illness surveillance systems use different approaches to record and classify illnesses for targeting prevention efforts. Data are presented here from SOII and other systems, including the Sentinel Event Notification System for Occupational Risk (SENSOR), the Third National Health and Nutrition Examination Survey (NHANES III), the Coal Workers' X-Ray Surveillance Program (CWXSP), the Adult Blood Lead Epidemiology and Surveillance Program (ABLES), the National Surveillance System for Hospital Health Care Workers (NaSH), and various reporting systems for human immunodeficiency virus (HIV) and acquired immune deficiency syndrome (AIDS), viral hepatitis, and TB. Details about each of the surveillance systems and information contacts are presented in Appendix A.

NONFATAL ILLNESS

Incidence of Occupational Illness in Private Industry

New nonfatal occupational illness cases recorded in SOII totaled 429,800 in 1997—the third year of decline in reported illnesses after a high of more than 500,000 cases in 1994 (Figure 5–1). Disorders associated with repeated trauma accounted for most of the decrease from 1994 to 1997. Sixty percent of nonfatal occupational illnesses reported in 1997 occurred in manufacturing (Figure 5–2). The overall incidence rate that year was 49.8 illnesses per 10,000 full-time workers, with the highest rates reported by establishments with 1,000 or more workers (Figure 5–3). The highest rate by industry division occurred in manufacturing (Figure 5–4).

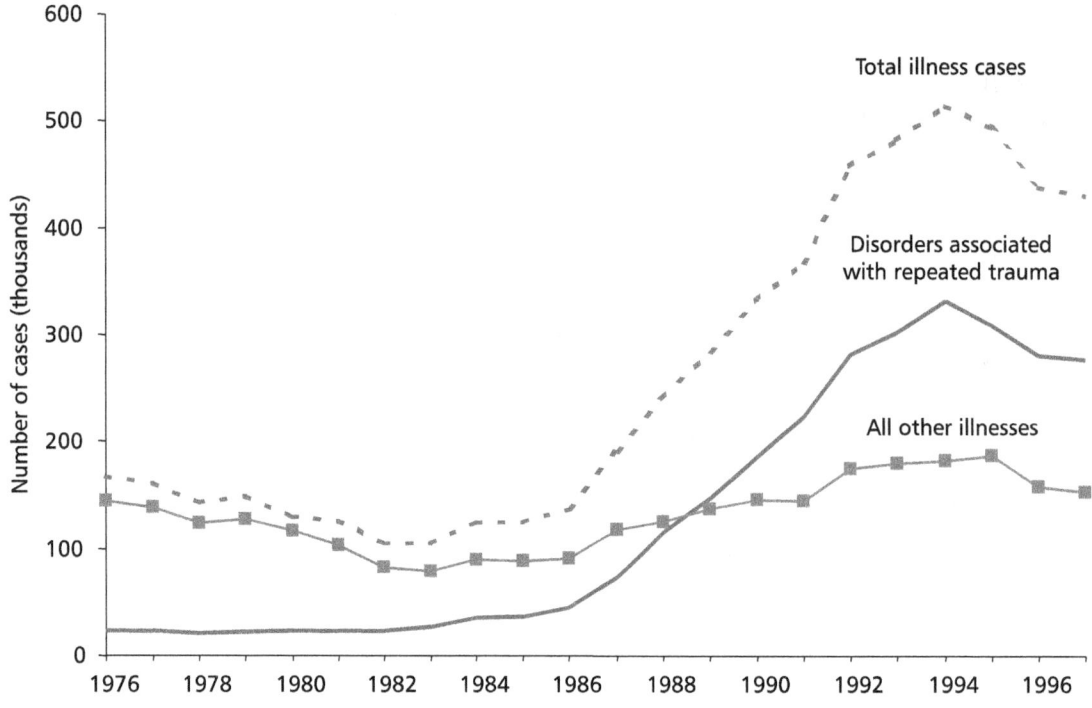

Figure 5–1. Incidence of nonfatal occupational illness cases in private industry, 1976–1997. (Source: SOII [1999].)

NONFATAL ILLNESS

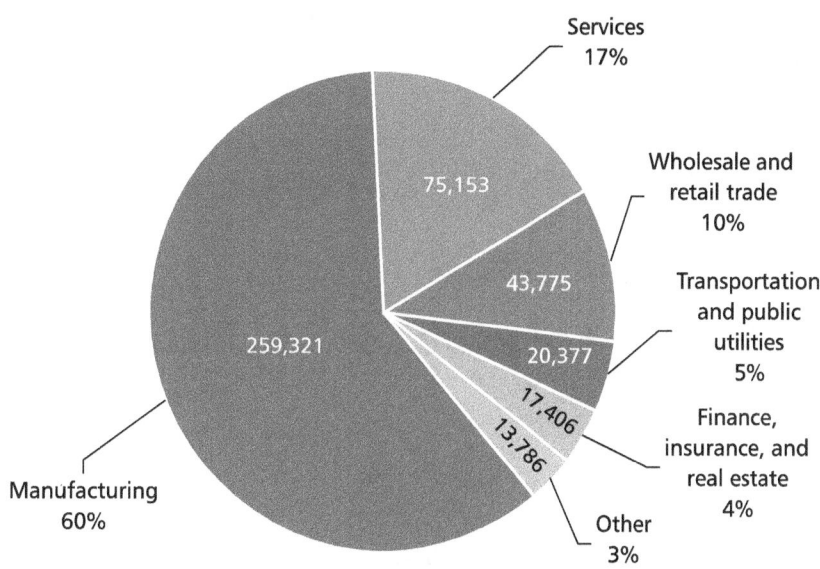

Figure 5–2. Number and distribution of nonfatal occupational illnesses in private industry by industry division, 1997. (Source: SOII [1999].)

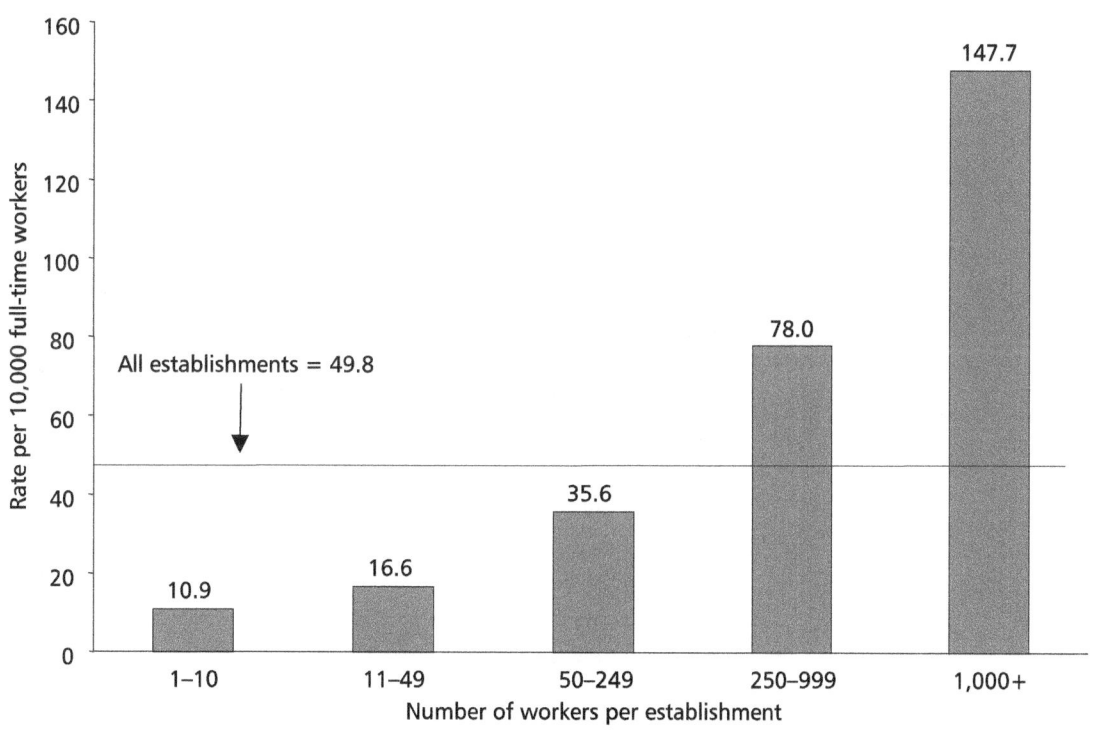

Figure 5–3. Incidence rates of nonfatal occupational illness in private industry by establishment employment size, 1997. (Source: SOII [1999].)

Nonfatal Illness

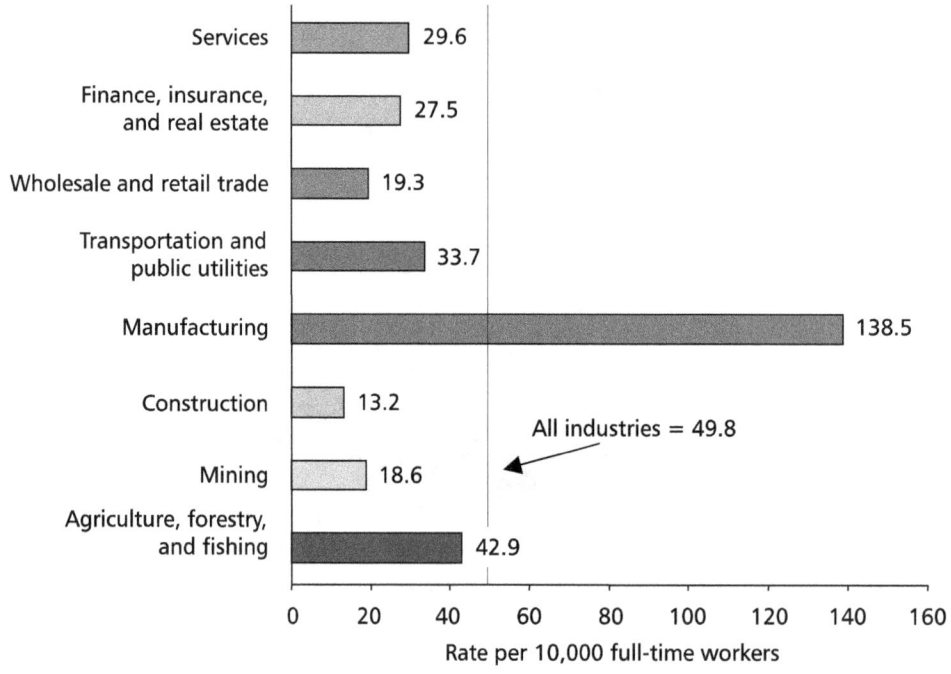

Figure 5–4. Incidence rates of nonfatal occupational illness in private industry by industry division, 1997. (Source: SOII [1999].)

Repeated Trauma Disorders

Repeated trauma disorders accounted for 64% (276,600 cases) of all nonfatal occupational illness cases recorded in SOII in 1997. Included in this category are carpal tunnel syndrome (CTS), tendinitis, and noise-induced hearing loss. Repeated trauma disorders accounted for most of the increases in nonfatal occupational illnesses recorded in SOII from 1976 through 1997 (Figure 5–1). Manufacturing accounted for 72% of the cases in private industry in 1997 (Figure 5–5). Industries associated with the highest rates of nonfatal occupational disorders involving repeated trauma were meat packing plants (1,192 cases per 10,000 workers), motor vehicles and car bodies (741 cases per 10,000 workers), and poultry slaughtering and processing (523 cases per 10,000 workers).

NONFATAL ILLNESS

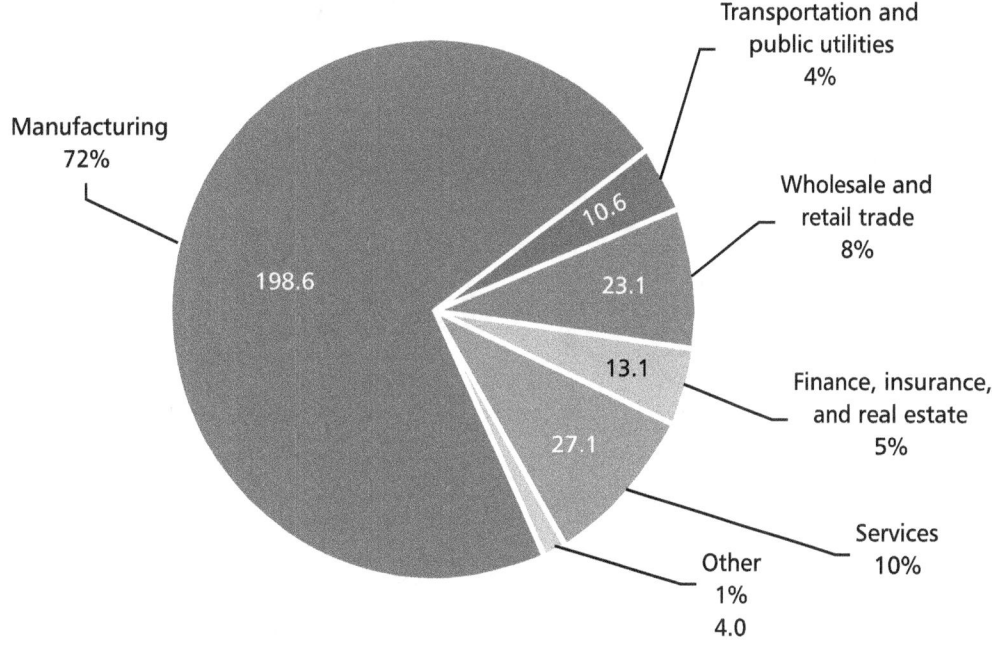

Figure 5–5. Number (thousands) and distribution of repeated trauma disorders in private industry by industry division, 1997. (Source: SOII [1999].)

Carpal Tunnel Syndrome

Cases Recorded by SOII

CTS accounted for more than 29,000 nonfatal occupational illness cases with days away from work recorded in SOII in 1997. Women accounted for 70% of these cases, and more than half of all CTS cases required 25 or more days away from work. Most CTS cases occurred in the manufacturing (42%) and service (21%) industries in 1997 (Figure 5–6) among operators, fabricators, and laborers (39%) and technical, sales, and administrative support personnel (30%) (Figure 5–7). The vast majority of SOII cases of CTS (98%) were attributed to job tasks requiring repetitive motion.

NONFATAL ILLNESS

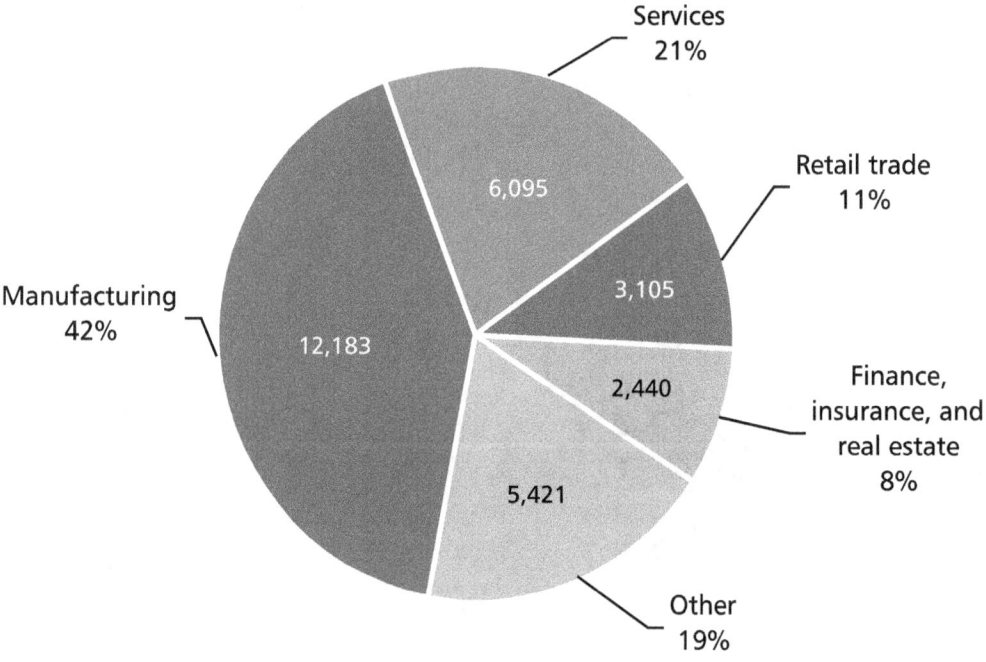

Figure 5–6. Number and distribution of CTS cases with days away from work in private industry by industry division, 1997. (Source: SOII [1999].)

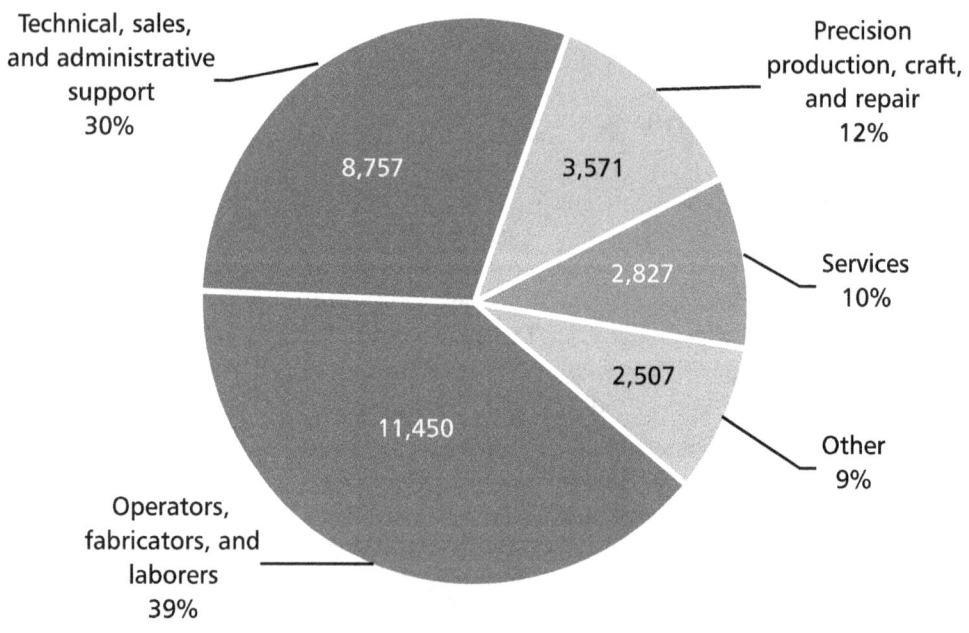

Figure 5–7. Number and distribution of CTS cases with days away from work in private industry by occupational group, 1997. (Source: SOII [1999].)

NONFATAL ILLNESS

Cases Identified by SENSOR

In collaboration with the National Institute for Occupational Safety and Health (NIOSH), the California Department of Health Services conducts a SENSOR program for CTS using first reports filed by physicians seeking reimbursement through the State workers' compensation system. The CTS case definition for SENSOR includes (1) symptoms such as pain, burning, or numbness in the hands or wrists, (2) objective evidence from a physical examination or electrodiagnostic tests, and (3) a history of work involving one of the known risk factors. Of the approximately 1,300 CTS cases identified by the California SENSOR program in 1998, the industries with the most cases were services (30%), manufacturing (17%), and wholesale trade (15%) (Figure 5–8). Most cases occurred among technical, sales, and administrative support personnel (44%) and managerial and professional specialty personnel (14%) (Figure 5–9). Of the cases in which an activity or exposure was associated with the injury, 49% reported using a computer (Figure 5–10).

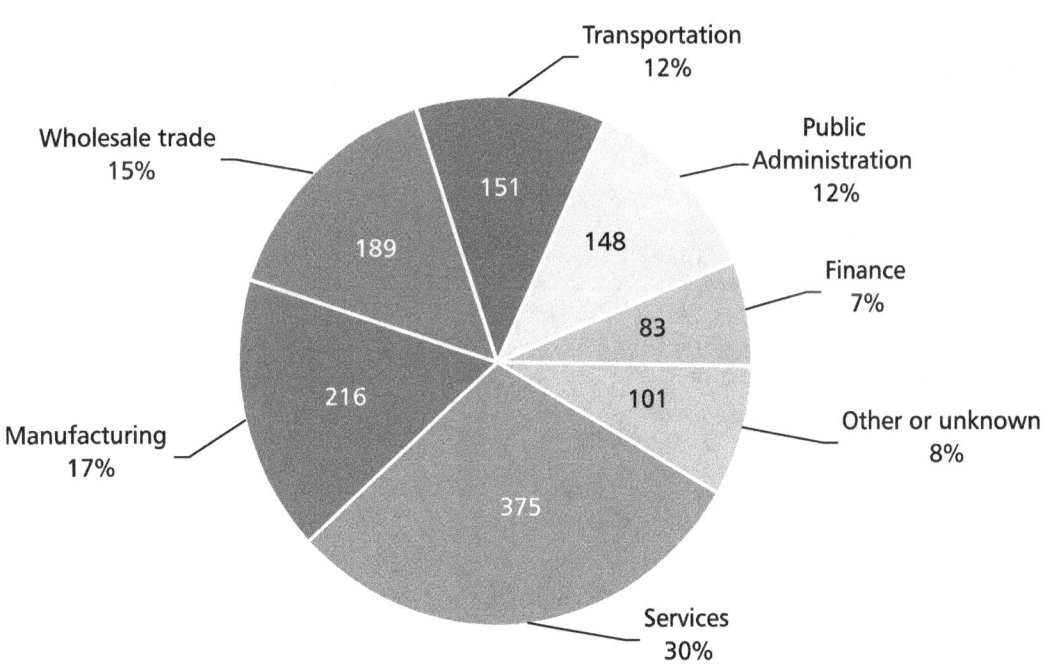

Figure 5–8. Number and distribution of CTS cases in California by industry group, 1998. (Source: SENSOR [California Department of Health Services 1999].)

NONFATAL ILLNESS

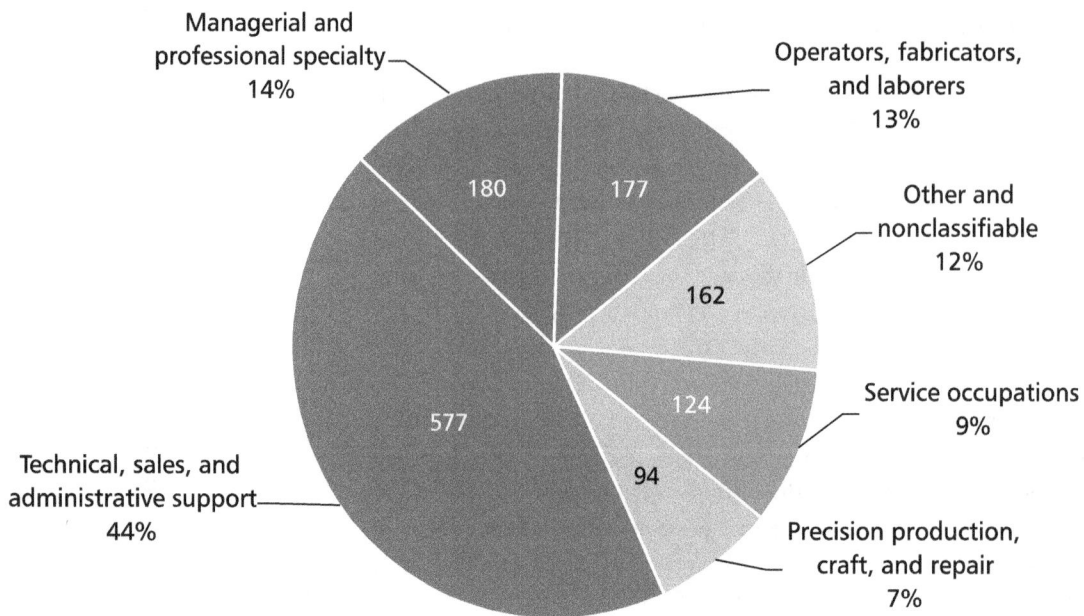

Figure 5–9. Number and distribution of CTS cases in California by occupational group, 1998. (Source: SENSOR. [California Department of Health Services 1999].)

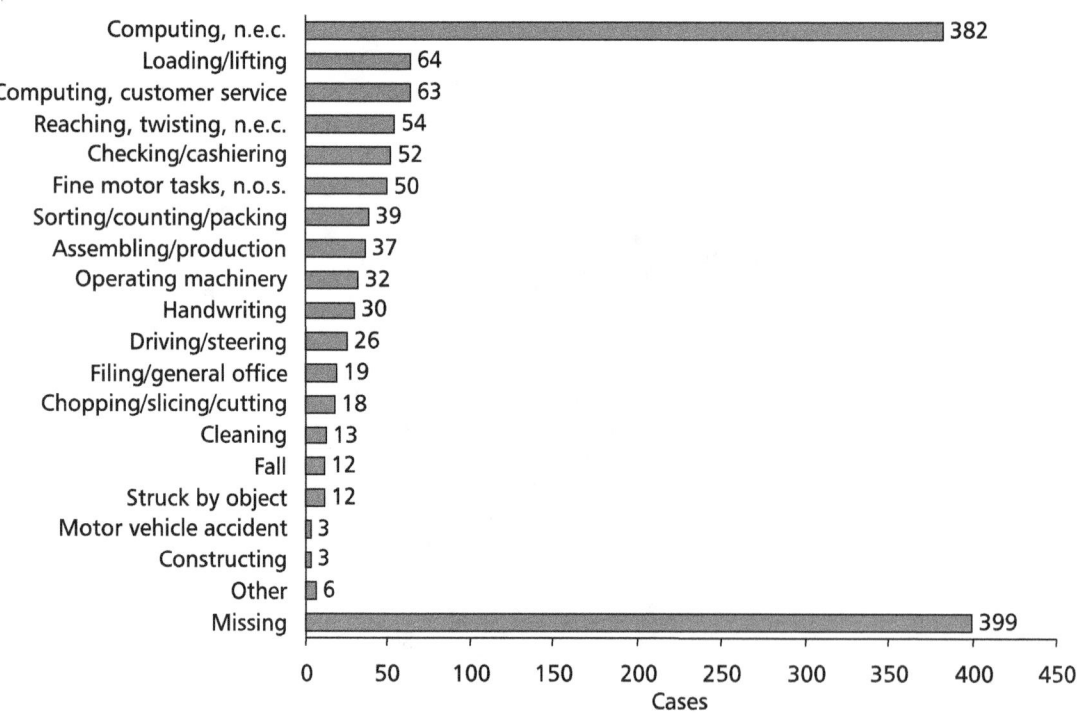

Figure 5–10. Number of CTS cases in California by type of activity or exposure, 1998. (Source: SENSOR [California Department of Health Services 1999].)

Nonfatal Illness

Tendinitis

Nearly 18,000 tendinitis cases recorded in SOII in 1997 required days away from work. Women accounted for more than 60% of those cases, and the upper extremities were affected in more than 70% of cases. Most cases occurred in the manufacturing (45%) and services (20%) industries (Figure 5–11) among operators, fabricators, and laborers (47%) and technical, sales, and administrative personnel (17%) (Figure 5–12). Worker motion or position was the event or exposure accounting for 73% of cases.

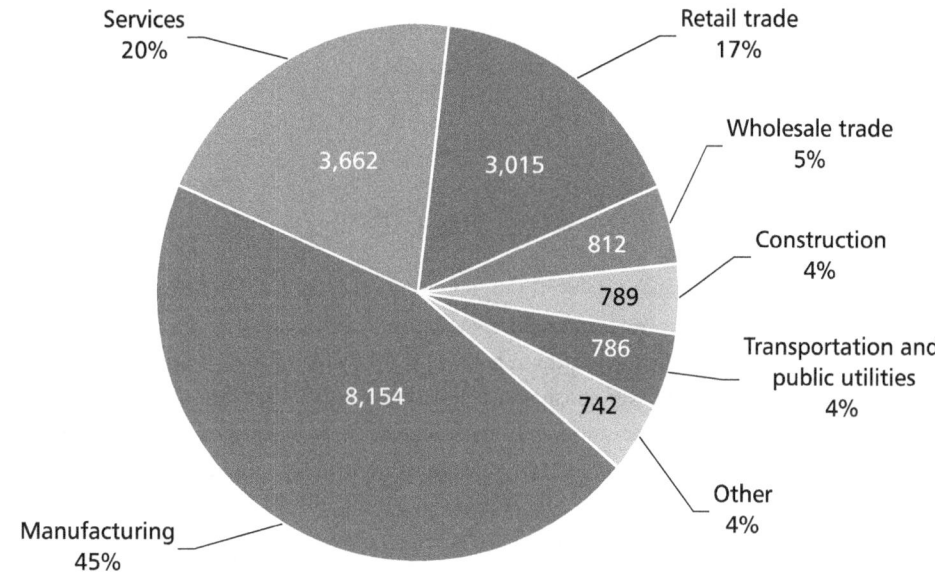

Figure 5–11. Number and distribution of tendinitis cases with days away from work in private industry by industry division, 1997. (Source: SOII [1999].)

NONFATAL ILLNESS

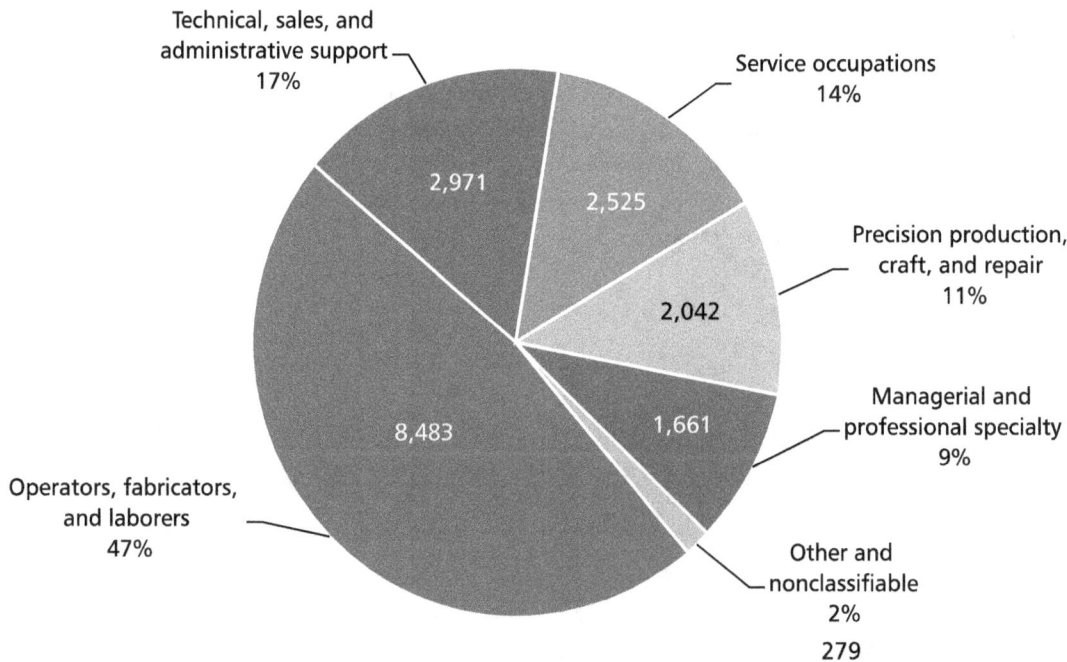

Figure 5–12. Number and distribution of tendinitis cases with days away from work in private industry by occupational group, 1997. (Source: SOII [1999].)

Noise-Induced Hearing Loss

A SENSOR program to protect workers from noise-induced hearing loss was initiated in Michigan in 1992. The case definition for occupational noise-induced hearing loss under the program requires audiometric findings consistent with noise-induced hearing loss and a history of noise exposure at work sufficient to cause hearing loss. This case definition includes (1) workers with standard threshold shifts reported by company hearing conservation programs and (2) workers with a permanent noise-induced hearing loss diagnosed by a clinician. From 1992 to 1998, there were 13,177 cases of noise-induced hearing loss reported by companies, audiologists, otolaryngologists, the Bureau of Workers' Compensation, and hospitals. Companies accounted for 85.2% of these cases (Figure 5–13). The SENSOR program interviews workers identified with permanent hearing loss by clinicians. In 1998, most of these cases were associated with manufacturing (Figure 5–14). Within the manufacturing sector, 60% of cases were associated with transportation manufacturing, which includes automobile manufacturing.

Nonfatal Illness

According to patient interviews, 25% to 76% of companies in major industry divisions did not test hearing at the time the worker was exposed to noise (Figure 5–15). Patients with hearing loss reported by companies (more than 85% of the reports) tended to be younger than patients whose hearing loss was reported by health professionals (Figure 5–16). Of the cases in which sex was listed, 89% were men.

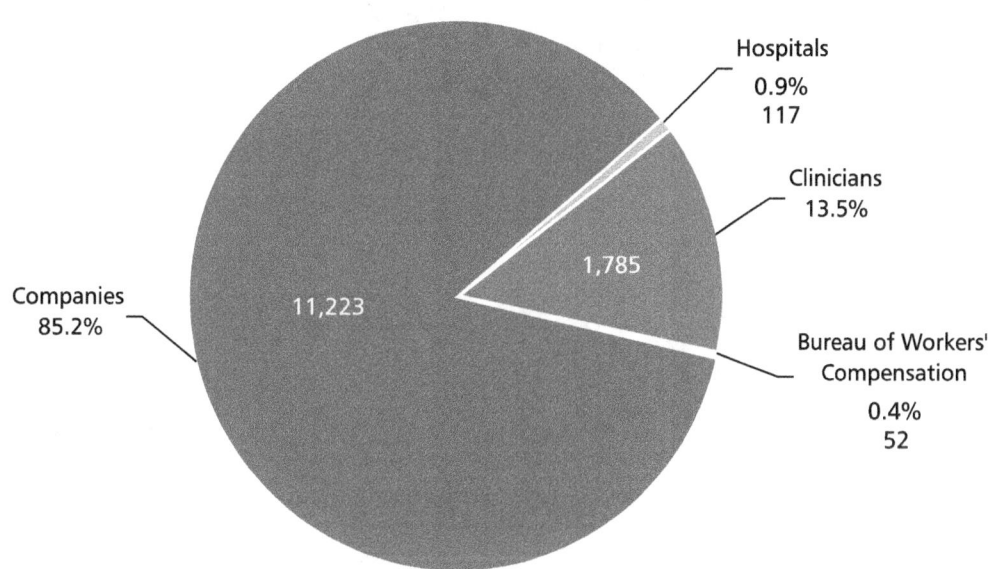

Figure 5–13. Number and distribution of noise-induced hearing loss cases in Michigan by source of reports, 1992–1998. Total number of cases was 13,177. (Source: SENSOR [Rosenman and Reilly 1999].)

NONFATAL ILLNESS

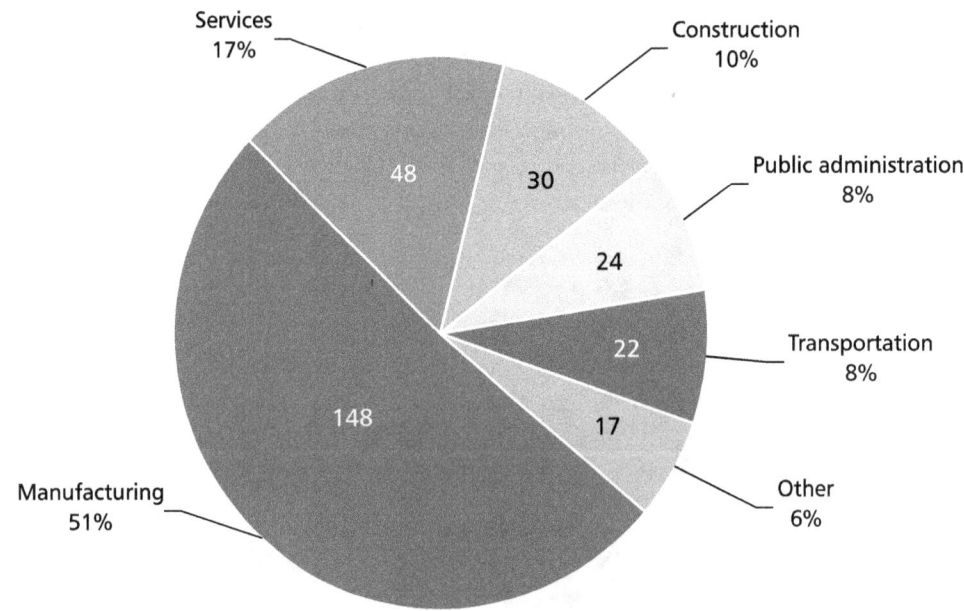

Figure 5–14. Number and distribution of permanent hearing loss cases reported by clinicians by industry division, 1998. (Source: SENSOR [Rosenman et al. 1999].)

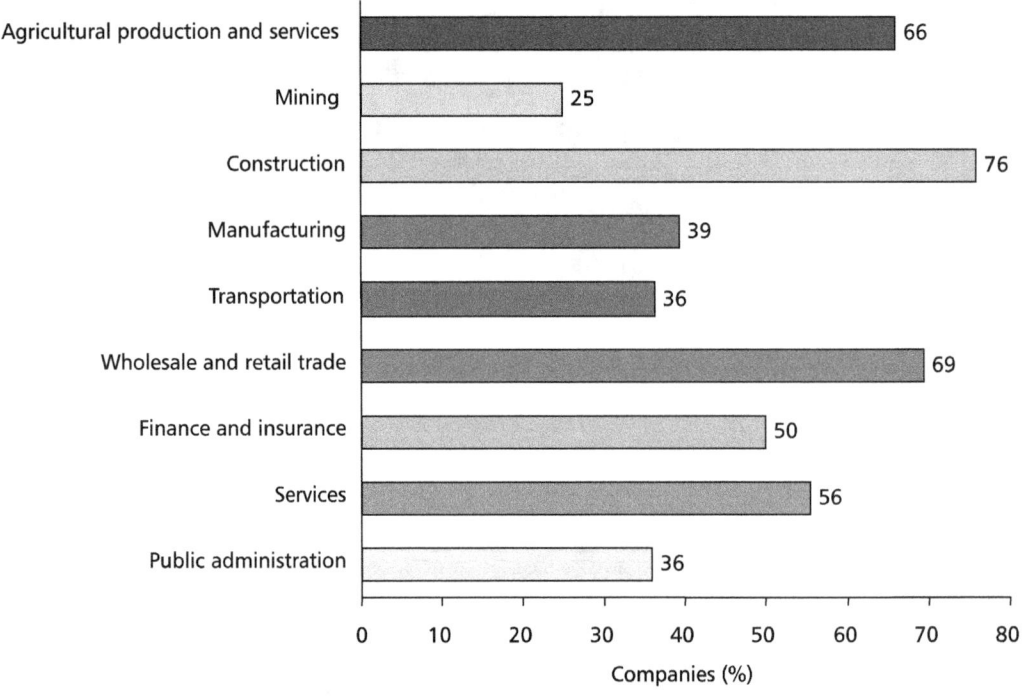

Figure 5–15. Percentage of companies within major industry divisions that did not test hearing at the time the worker was exposed to noise, as reported by patient interviews, 1992–1998. (Source: SENSOR [Rosenman et al. 1999].)

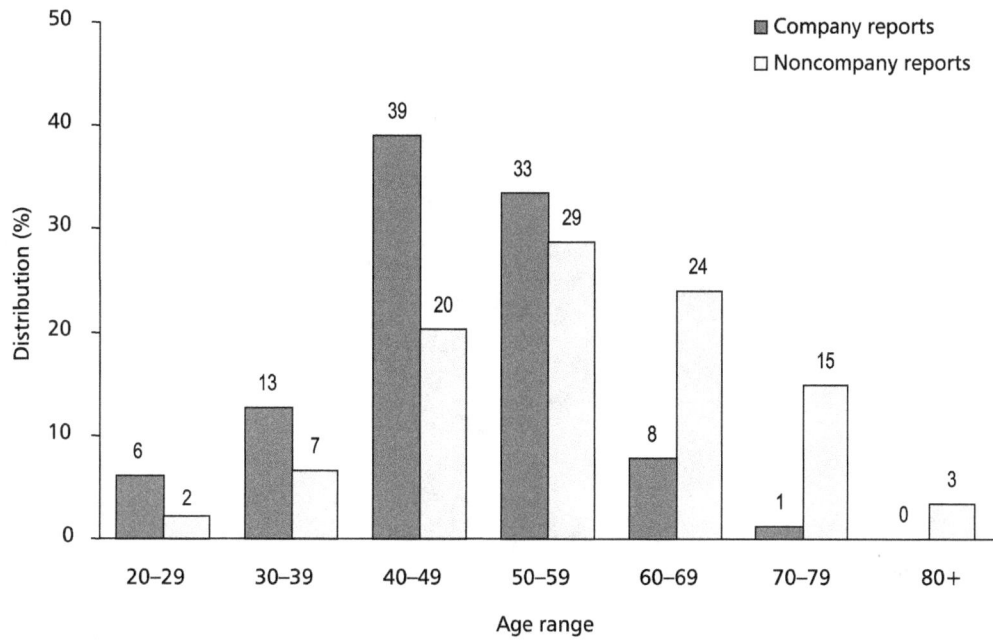

Figure 5–16. Distributions of noise-induced hearing loss cases by age range of patients and by company and noncompany reports, 1998. Age was unknown for 31 workers reported by company medical departments and 12 workers reported by noncompany hearing health professionals. (Source: SENSOR [Rosenman et al. 1999].)

Skin Diseases or Disorders

Skin diseases or disorders accounted for 13% (57,900) of all illness cases reported in SOII in 1997. These disorders include allergic and irritant dermatitis, skin cancer, and other conditions. Manufacturing accounted for 45% of the skin diseases or disorders in private industry in 1997 (Figure 5–17). The highest reported incidence rate was in the canned and cured fish and seafoods industry (181 cases per 10,000 workers). Other industries with the highest rates of occupational skin disease or disorder were meat packing plants (104 cases per 10,000 workers), ball and roller bearings (92 cases per 10,000 workers), and leather tanning and finishing (86 cases per 10,000 workers). Dermatitis, a subcategory of skin diseases and disorders, was associated with nearly 6,600 cases involving time away from work in 1997. A median number of 3 days away from work was associated with dermatitis. Exposures to chemicals and chemical products accounted for 53% of job-related dermatitis cases. The manufacturing and service industry divisions accounted for the most dermatitis cases with days away from work (29% each) (Figure 5–18). Occupational groups that experienced most dermatitis conditions were operators, fabricators, and laborers (36%) and precision production, craft, and repair personnel (18%) (Figure 5–19).

Nonfatal Illness

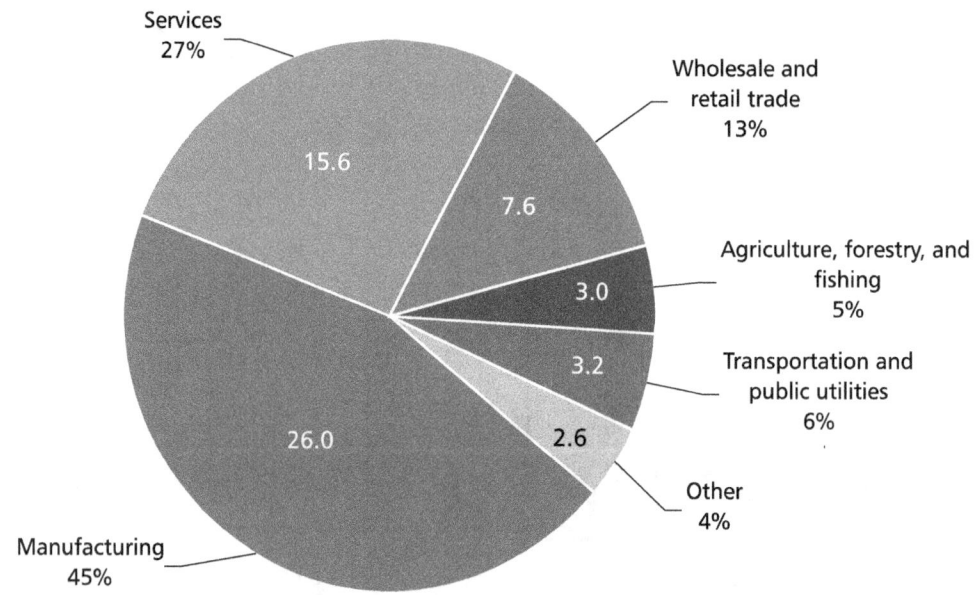

Figure 5–17. Number (thousands) and distribution of skin disease or disorder cases in private industry by industry division, 1997. (Source: SOII [1999].)

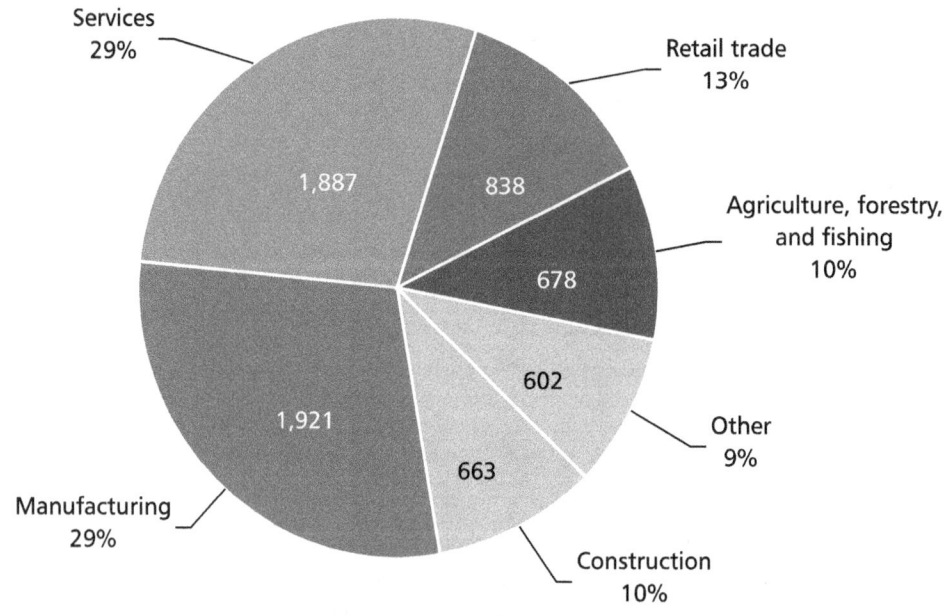

Figure 5–18. Number and distribution of dermatitis cases with days away from work in private industry by industry division, 1997. (Source: SOII [1999].)

Nonfatal Illness

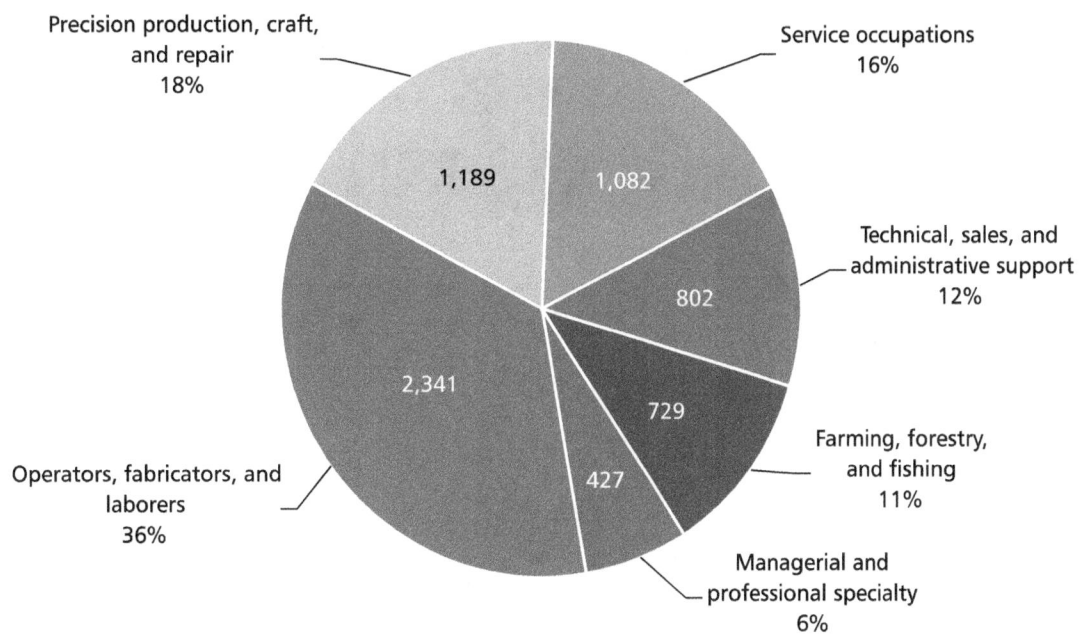

Figure 5-19. Number and distribution of dermatitis cases with days away from work in private industry by occupational group, 1997. (Source: SOII [1999].)

Respiratory Disorders

Dust Diseases of the Lungs

Dust diseases of the lungs accounted for less than 1% (2,900) of the nonfatal occupational illness cases recorded in SOII in 1997. These diseases include silicosis, asbestosis, and coal workers' pneumoconiosis (CWP). The most cases of occupational dust diseases of the lungs occurred in the manufacturing (33%) and service (27%) industries in 1997 (Figure 5-20). The highest dust disease incidence rates occurred in aluminum sheet, plate, and foil manufacturing (33 per 10,000 workers), anthracite mining (30 per 10,000 workers), and ship building and repairing (12 per 10,000 workers).

NONFATAL ILLNESS

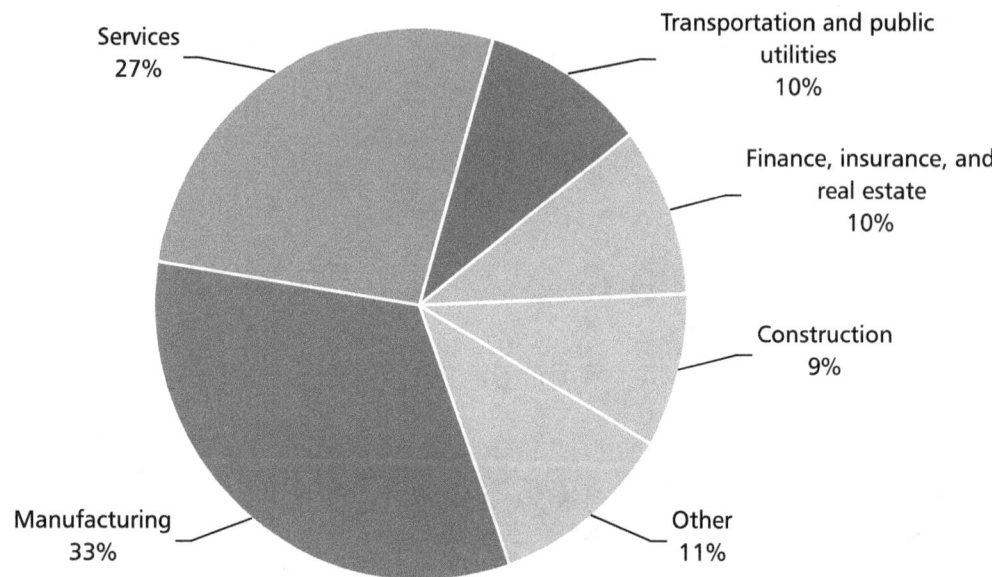

Figure 5–20. Distribution of occupational cases of dust diseases of the lungs in private industry, by industry division, 1997. Total number of cases was 2,900. (Source: SOII [1999].)

Coal Workers' Pneumoconiosis

The prevalence and severity of CWP are examined in Coal Workers' X-Ray Surveillance Program (CWXSP). CWP is defined as having X-ray evidence of lung abnormalities (grade 1/0 or higher) using the International Labour Organization (ILO) *Guidelines for the use of ILO International Classification of Radiographs of Pneumoconioses* [ILO 1980]. Among workers with 25 or more years of underground tenure, the prevalence of CWP category 1/0 or greater decreased from more than 28% during 1970–1973 to less than 10% during 1992–1995 (Figure 5–21). In the same tenure group, the prevalence of the more severe CWP category 2/1 or greater decreased from more than 10% during 1970–1973 to less than 2% during 1992–1995 (Figure 5–22). Decreases in prevalence are also apparent in groups with less tenure in underground mining (Figures 5–21 and 5–22).

NONFATAL ILLNESS

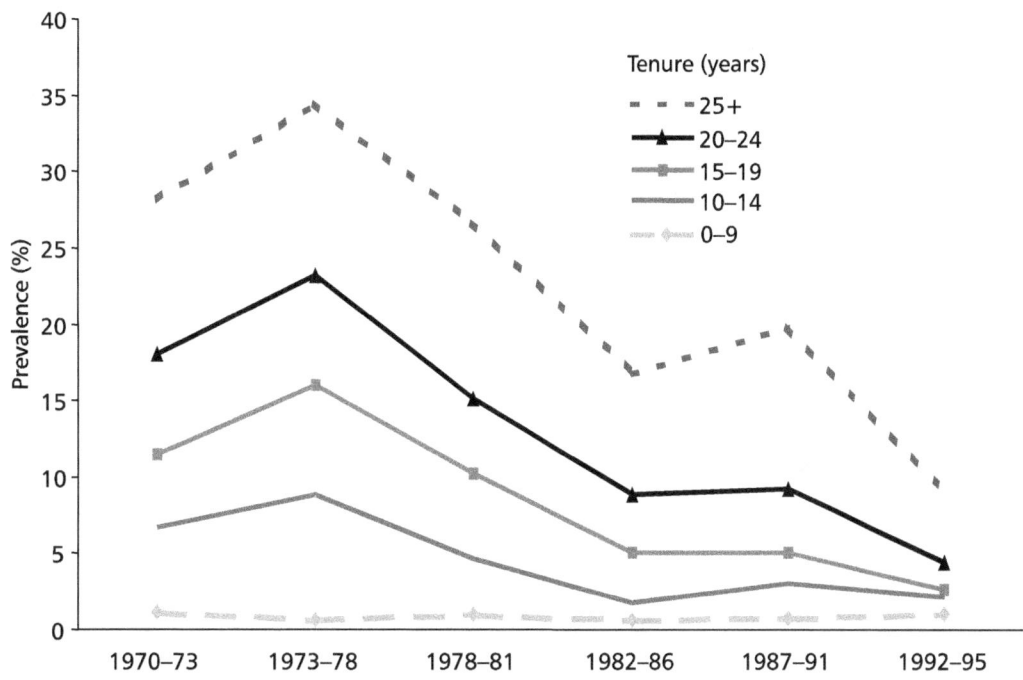

Figure 5–21. Prevalence of examined miners with CWP category 1/0 or greater by tenure in mining, 1970–1995. (Source: CWXSP [1999].)

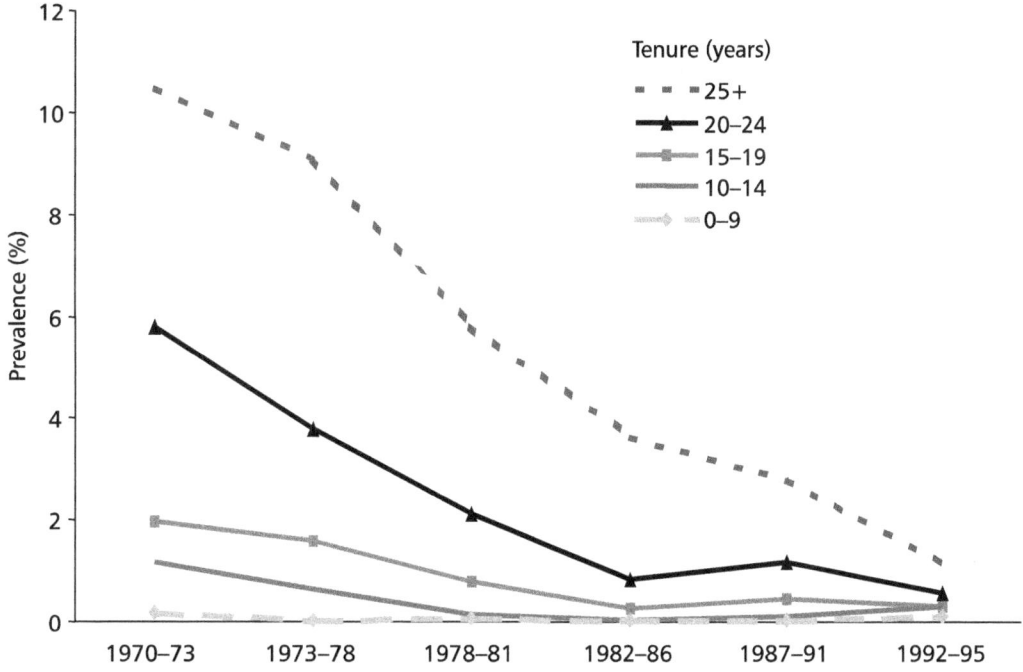

Figure 5–22. Prevalence of examined miners with CWP category 2/1 or greater by tenure in mining, 1970–1995. (Source: CWXSP [1999].)

Nonfatal Illness

Silicosis

Silicosis is a chronic inflammatory condition of the lung caused by the inhalation of silica particles; this condition is almost universally caused by occupational exposures. Prevalence of silicosis can be examined through the SENSOR program. For SENSOR purposes, silicosis cases require a history of occupational exposure to airborne silica dust and one or both of the following: (1) a chest radiograph (or other imaging technique) interpreted as consistent with silicosis and (2) pathologic findings characteristic of silicosis.

From 1993 to 1995, seven States participated in the SENSOR silicosis program. Together these States identified 604 cases of silicosis, mostly through hospital reports (64%), reports by health care professionals (11%), and death certificates (9%) (Figure 5–23). The cases originated mostly in manufacturing industries (75%), construction (9%), and mining (7%) (Figure 5–24). Operators, fabricators, and laborers represented the majority of cases (61%) (Figure 5–25).

Among silicosis patients who were interviewed, most had chronic disease with onset of symptoms 10 or more years after exposure. Exposure to high airborne concentrations of silica can cause disease within a few years, and acute silicosis (much less common) may result in death within months of intense occupational exposure. Although most of the interviewed workers had been occupationally exposed for more than 20 years, 8% had fewer than 10 years of exposure.

Nonfatal Illness

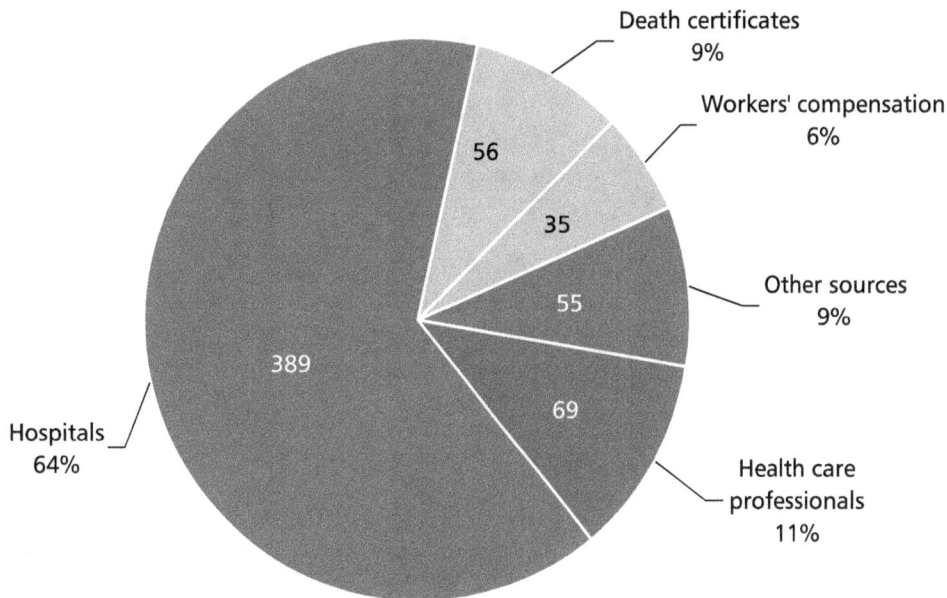

Figure 5–23. Number and distribution of silicosis cases in all seven reporting States by source of report, 1993–1995. (Source: SENSOR [NIOSH 1999].)

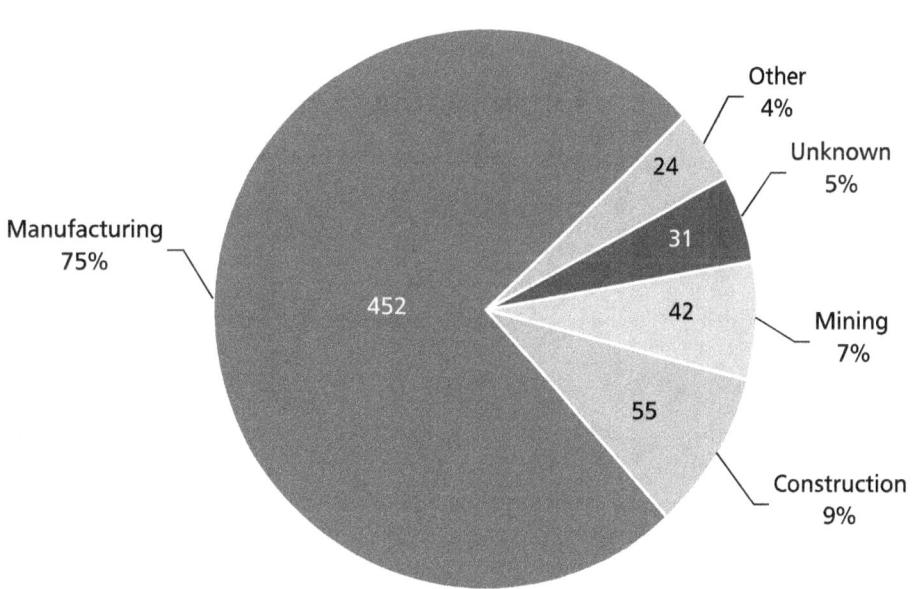

Figure 5–24. Number and distribution of silicosis cases in all seven reporting States by industry division, 1993–1995. (Source: SENSOR [NIOSH 1999].)

NONFATAL ILLNESS

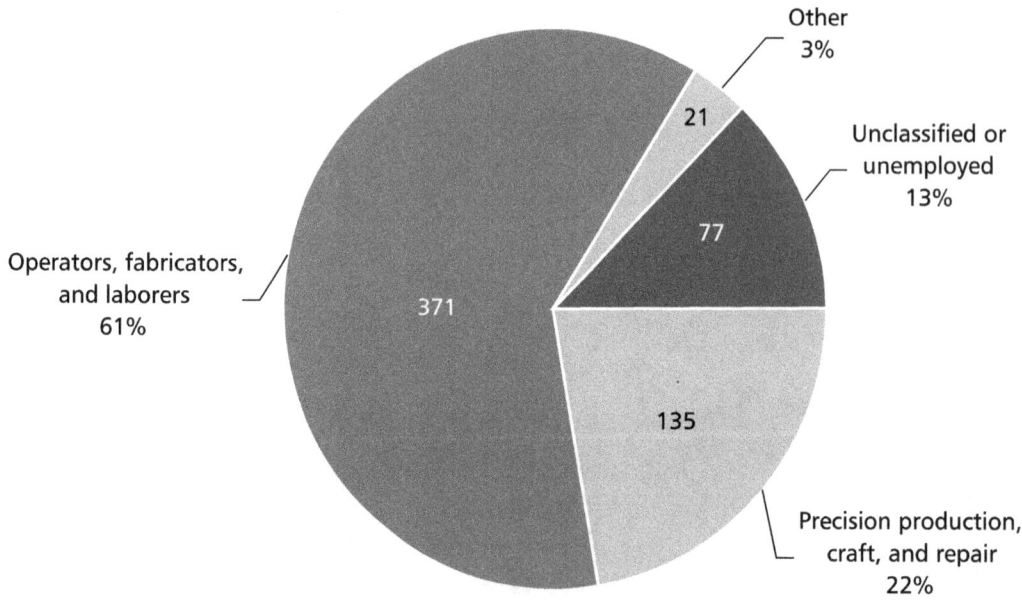

Figure 5–25. Number and distribution of silicosis cases in all seven reporting States by major occupational category, 1993–1995. (Source: SENSOR [NIOSH 1999].)

Respiratory Disorders Attributable to Toxic Agents

Respiratory disorders attributable to toxic agents in the work environment accounted for 5% (20,300) of the illness cases recorded in SOII in 1997. These disorders include allergic and irritant asthma, chronic bronchitis, and reactive airways dysfunction (an asthma-like syndrome). The industry divisions reporting the most cases in 1997 were manufacturing (37%) and services (34%) (Figure 5–26). SOII reported the highest industry incidence rates in leather tanning and finishing (77 per 10,000 workers), motorcycles, bicycles, and parts (50 per 10,000 workers), ammunition, except for small arms not elsewhere classified (n.e.c.) (36 per 10,000 workers), ship building and repairing (36 per 10,000 workers), and musical instruments (34 per 10,000 workers).

NONFATAL ILLNESS

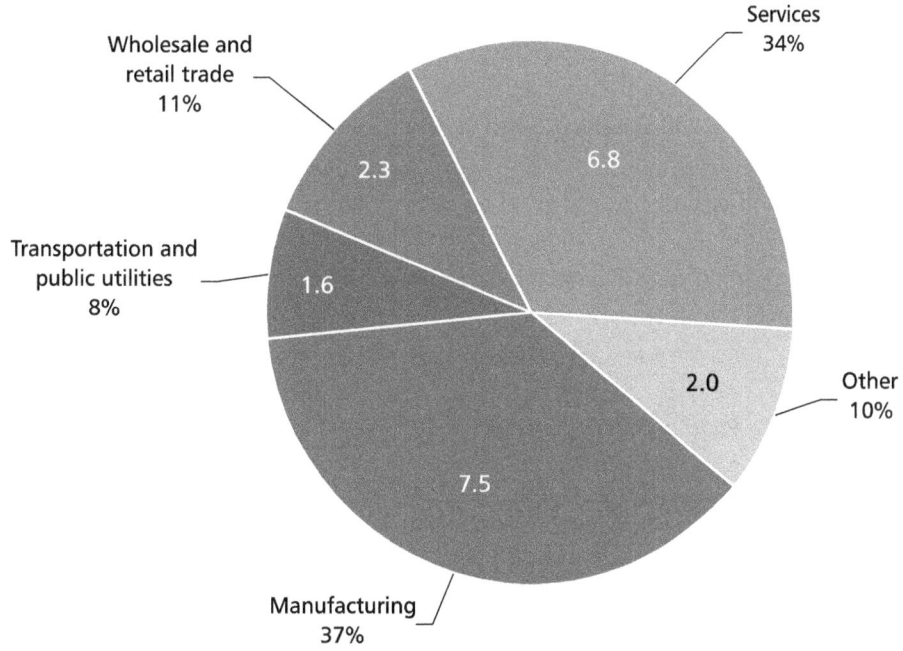

Figure 5–26. Number (thousands) and distribution of respiratory disorder cases attributed to toxic agents in private industry by industry division, 1997. (Source: SOII [1999].)

Asthma and Chronic Obstructive Pulmonary Disease

NHANES III

Workers' prevalence rates for asthma and chronic obstructive pulmonary disease (COPD) (such as chronic bronchitis and emphysema) are recorded in NHANES III (Figures 5–27 and 5–28). These conditions may be caused or exacerbated by workplace exposures, but no particular attribution to workplace factors is made in NHANES III. Variations in prevalence rates among workers in different industries (particularly among nonsmokers) may suggest an occupational association in some cases.

NONFATAL ILLNESS

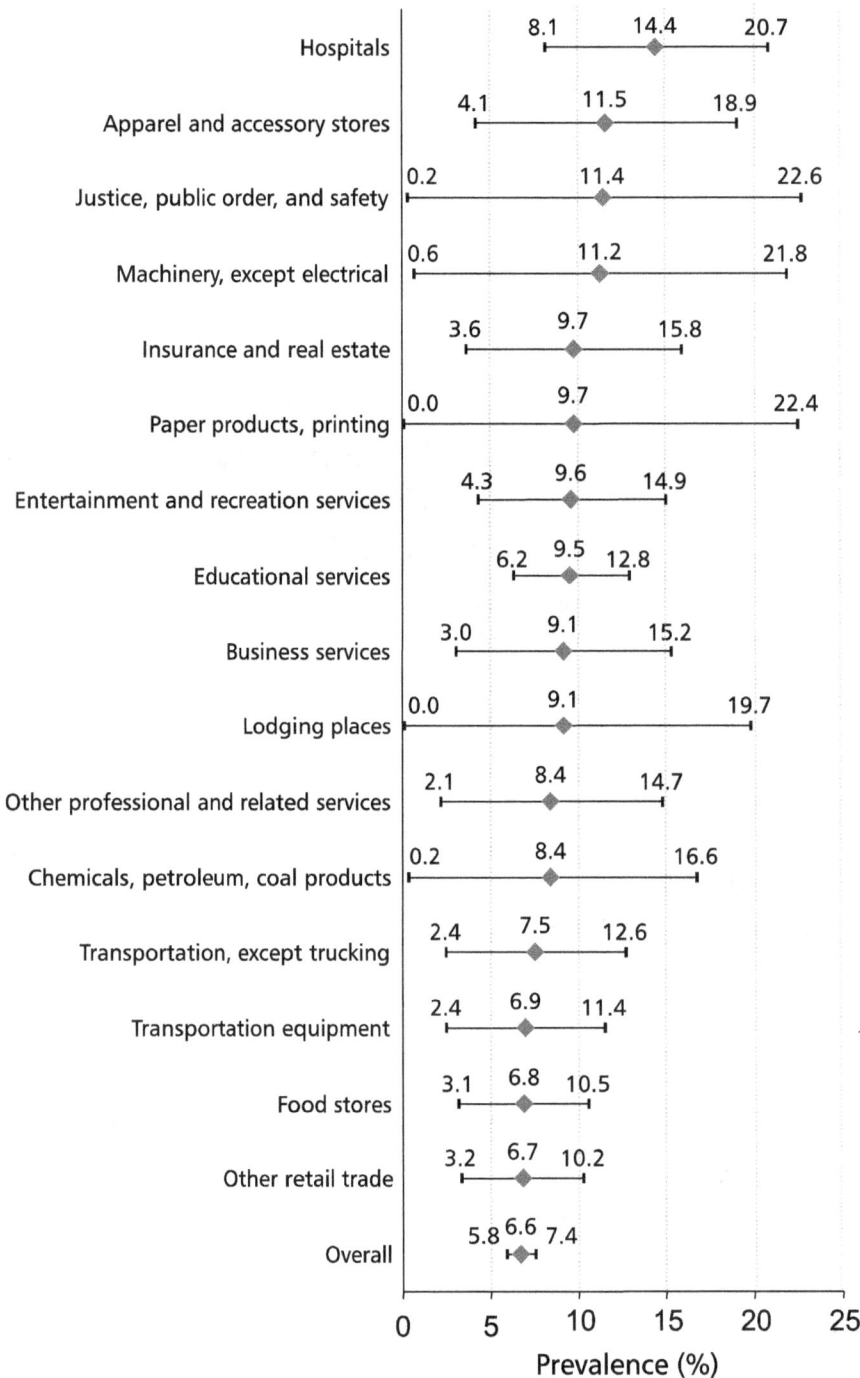

Figure 5–27. Estimated prevalence rates (and 95% confidence intervals [CIs]) for asthma among workers who are nonsmokers, by usual industry of workers' employment—U.S. residents aged 17 and older, 1988–1994. (Source: NHANES III [1999].)

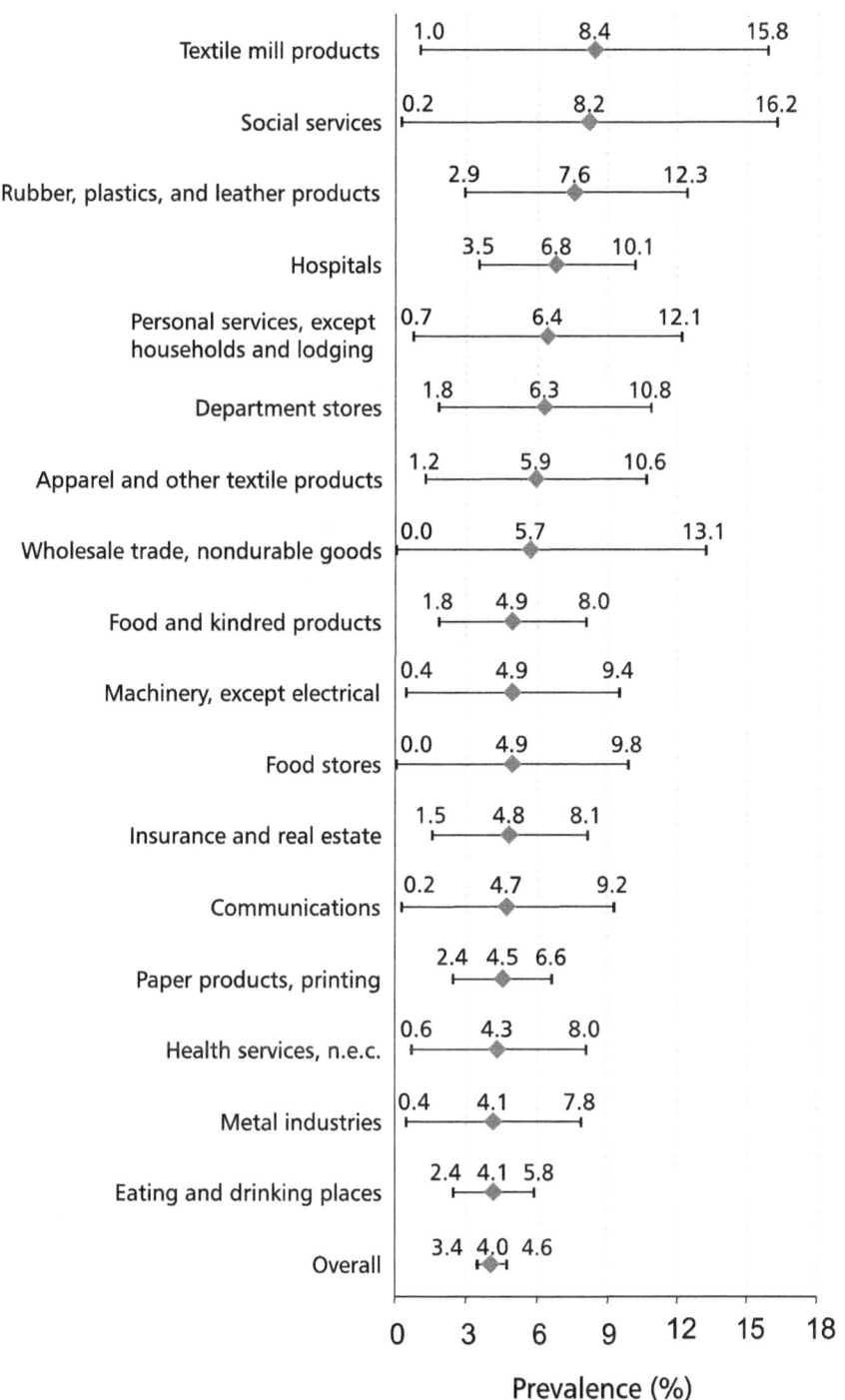

Figure 5-28. Estimated prevalence rates (and 95% CIs) for COPD among workers who are nonsmokers, by usual industry of workers' employment—U.S. residents aged 17 and older, 1988–1994. (Source: NHANES III [1999].)

NONFATAL ILLNESS

SENSOR

Under the SENSOR program, several State health departments have developed surveillance systems for work-related asthma (including occupational asthma, occupationally induced reactive airways dysfunction syndrome [RADS], and work-aggravated asthma). Occupational asthma is now the most common disease reported in occupational respiratory disease surveillance systems in several developed countries. However, most cases either are not recognized as work-related or are not reported as such. Population-based estimates suggest that about 20% of new-onset asthma in adults is work-related.

Four States—New Jersey, Michigan, Massachusetts, and California—had active SENSOR programs during the years for which data are included in this report (1993–1995). California relies on the first reports filed by physicians seeking reimbursement through the State workers' compensation system. The three remaining States rely primarily on more active physician reporting. In all four States, 90% of the 1,101 occupational asthma cases were identified through physician reports (Figure 5–29). Most cases occurred in manufacturing (42%) and services (31%) (Figure 5–30) among operators, fabricators, and laborers (32%) and technical, sales, and administrative support personnel (21%) (Figure 5–31). The categories of agents most frequently associated with occupational asthma cases were all isocyanates (toluene diisocyanate, methylene diisocyanate, and other diisocyanates) (9%), indoor environments (8%), and mineral and inorganic dusts not otherwise specified (n.o.s.) (7%) (Figure 5–32).

NONFATAL ILLNESS

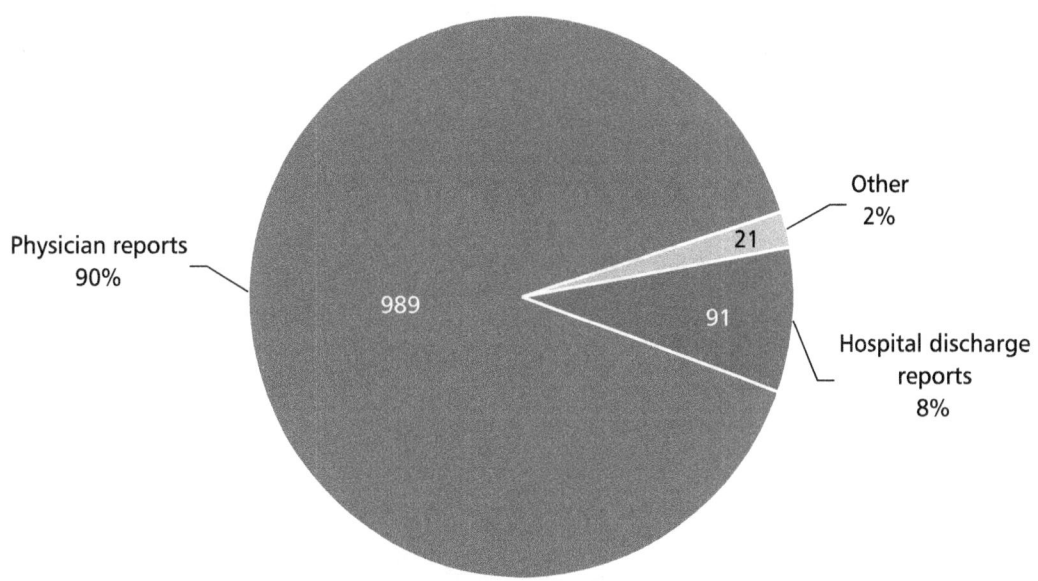

Figure 5–29. Number and distribution of occupational asthma cases for all four reporting States by source of report, 1993–1995. (Source: SENSOR [NIOSH 1999].)

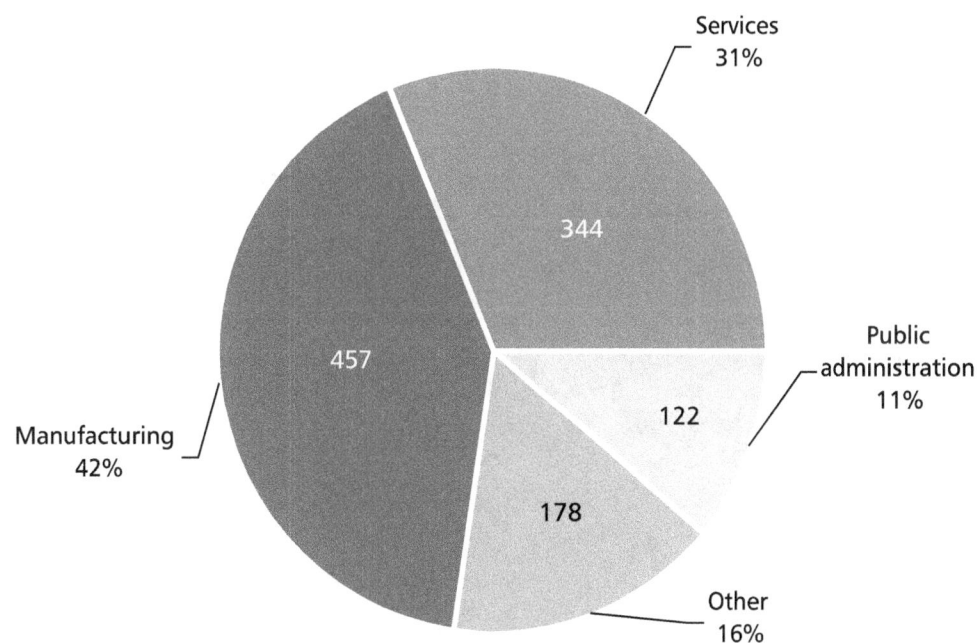

Figure 5–30. Number and distribution of occupational asthma cases for all four reporting States by industry division, 1993–1995. (Source: SENSOR [NIOSH 1999].)

Nonfatal Illness

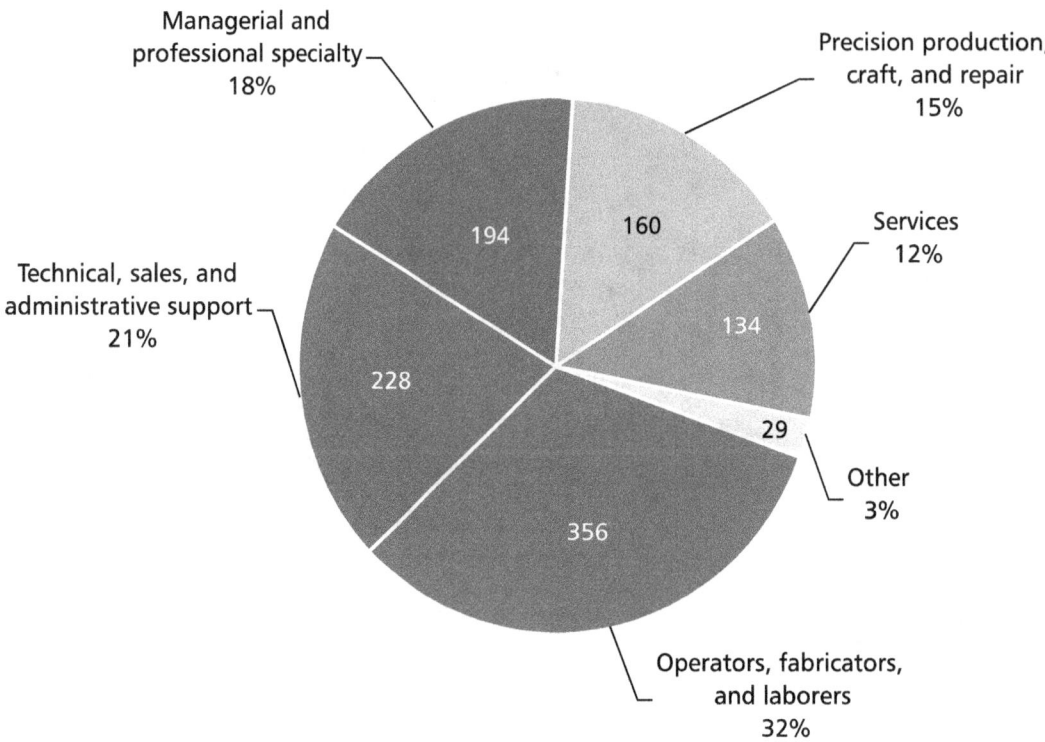

Figure 5–31. Number and distribution of occupational asthma cases for all four reporting States by occupation, 1993–1995. (Source: SENSOR [NIOSH 1999].)

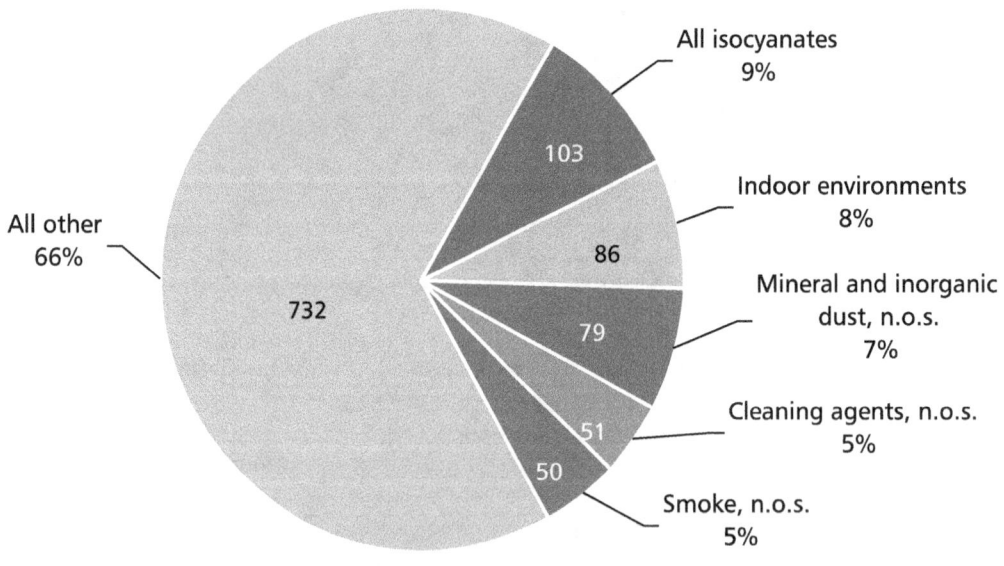

Figure 5–32. Number and distribution of occupational asthma cases for all four reporting States by most frequently associated agents, 1993–1995. (Source: SENSOR [NIOSH 1999].)

Nonfatal Illness

Poisoning and Toxicity

Poisoning

Poisoning represented 1% (5,100) of all nonfatal occupational illness cases recorded in SOII in 1997. Poisoning cases include exposures to heavy metals (including lead), toxic gases (such as carbon monoxide and hydrogen sulfide), organic solvents, pesticides, and other substances (such as formaldehyde). Manufacturing accounted for 55% of poisoning cases reported in private industry (Figure 5–33). The highest incidence rates occurred in the production of storage batteries (120 cases per 10,000 workers) and costume jewelry (78 cases per 10,000 workers), and in the secondary smelting and refining of nonferrous metals (62 cases per 10,000 workers).

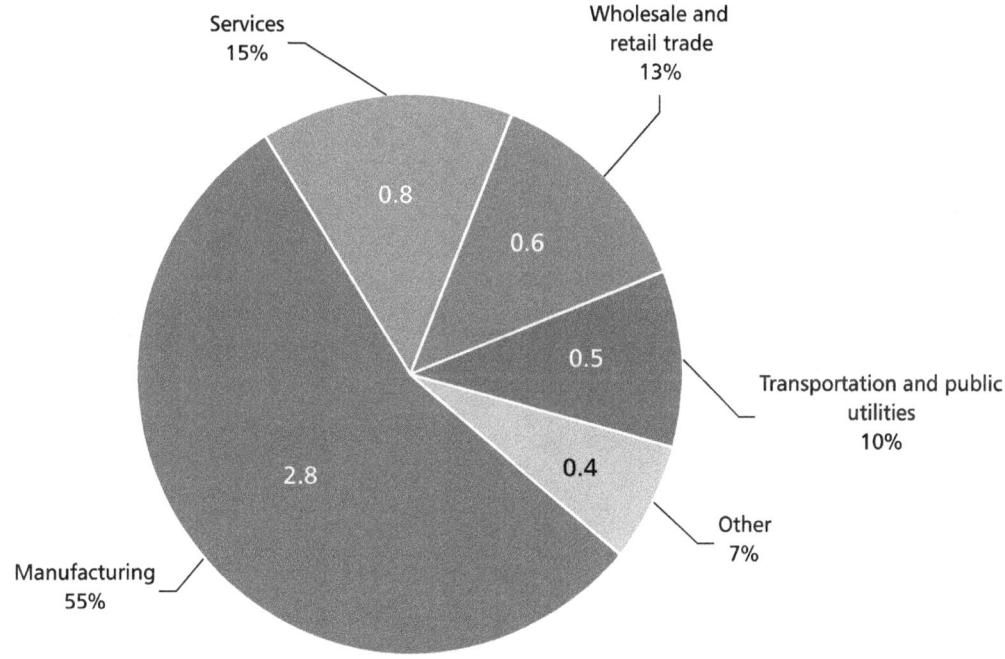

Figure 5–33. Number (thousands) and distribution of poisoning cases in private industry by major industry division, 1997. (Source: SOII [1999].)

Nonfatal Illness

Lead Toxicity

ABLES monitors elevated blood lead levels (BLLs) in adults (persons aged 16 and older). Twenty-seven States participated in this program in 1998 by collecting BLLs from local health departments, private health care professionals, and private and State reporting laboratories (Figure 5–34). During that year, a total of 10,501 adults in 25 of those States were reported to have BLLs of 25 µg/dL or greater. Prevalence rates for BLLs of 25 µg/dL or greater (based on all persons reported in a given year) do not reveal an obvious trend for the period 1993 through 1998, nor do the incidence rates (based on new cases reported in a given year) (Figure 5–35). However, prevalence and incidence rates for BLLs of 50 µg/dL or greater in 10 ABLES States decreased from 1993 to 1998 (Figure 5–36).

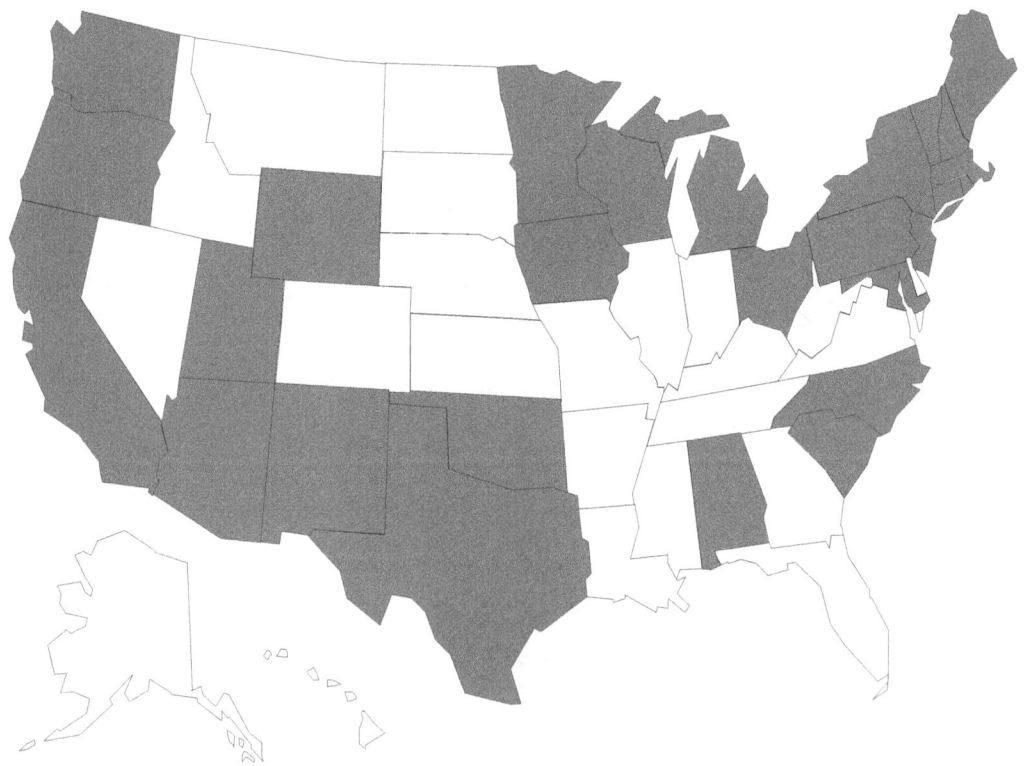

Figure 5–34. States (shaded) participating in the ABLES program in 1998. (Source: ABLES [1999].)

NONFATAL ILLNESS

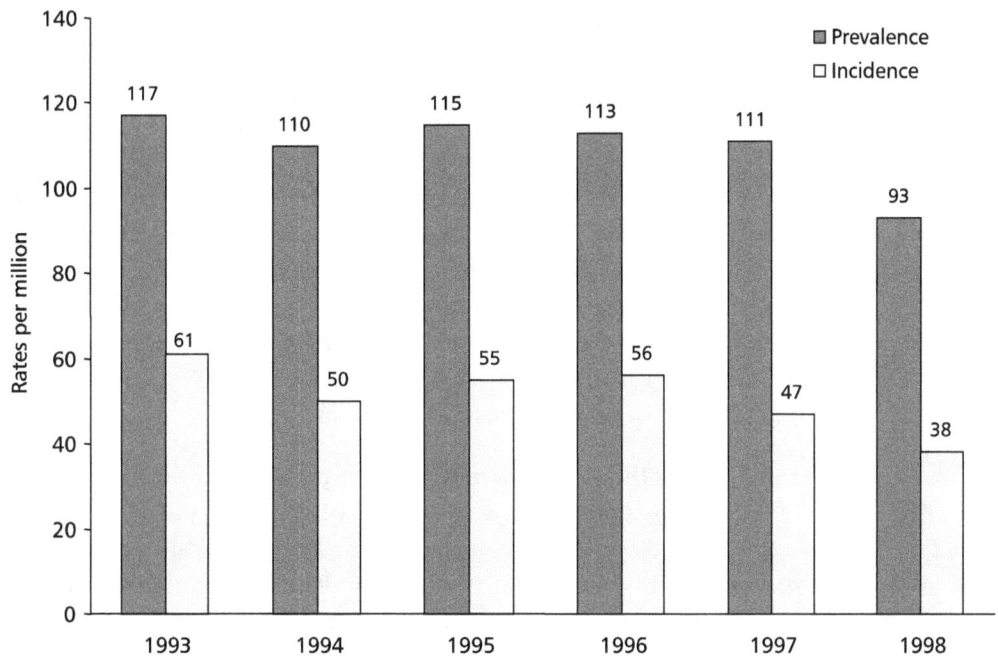

Figure 5–35. Prevalence and incidence rates of adults aged 16 to 64 with BLLs greater than 25 μg/dL, 1993–1998. (Source: ABLES [1999].)

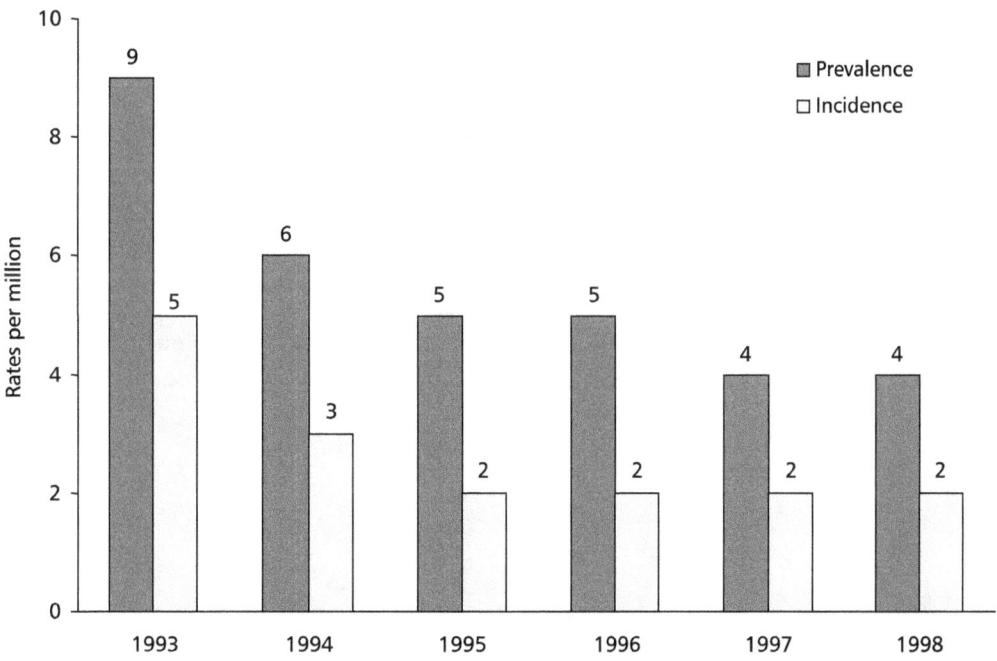

Figure 5–36. Prevalence and incidence rates for BLLs equal to or greater than 50 μg/dL in adults aged 16 to 64 from 10 States (California, Connecticut, Iowa, Maryland, Massachusetts, New Jersey, New York, Oregon, Texas, Utah), 1993–1998. (Source: ABLES [1999].)

Nonfatal Illness

Pesticide and Insecticide Toxicity

Several surveillance systems track acute occupational illness and injury related to pesticides. Two systems are national, and several additional systems cover individual States. The Toxic Exposure Surveillance System (TESS) is maintained by the American Association of Poison Control Centers. Between 1993 and 1996, about 81% of the U.S. population was covered by a participating poison control center. During those years, more than 6,300 pesticide poisonings that occurred in the workplace were documented in TESS. Most of the poisonings were associated with insecticides (Figure 5–37). Among those cases, 41% involved organophosphates, and 29% involved pyrethrins/pyrethroids.

SOII collects information about pesticide poisonings associated with lost workdays. Between 1992 and 1996, the annual number of nonfatal occupational illnesses and injuries related to pesticides ranged from 504 to 914 (Figure 5–38). Most of those illnesses were associated with exposure to insecticides. Because SOII records only cases that result in lost work time, illnesses may be more severe than those recorded by other surveillance systems.

Thirty-one States have reporting requirements for pesticide-related illness and injury, but only eight States conduct surveillance for this condition. In California, Florida, New York, Oregon, and Texas, surveillance activities for acute occupational illness and injury related to pesticides are conducted in a SENSOR program supported in part by the U.S. Environmental Protection Agency (EPA). Besides tabulating case reports, these systems perform in-depth investigations for case confirmation, conduct screening of other workers at a patient's worksite, and develop targeted interventions. Over a 5-year period (1992–1996), the annual number of cases in New York, Oregon, and Texas ranged from 72 to 170 (Figure 5–39). Most cases involved exposures to insecticides. In addition, 33% of the cases involved agricultural exposures, including pesticide mixing, loading, and application.

NONFATAL ILLNESS

Pesticide-related illness has been a reportable condition in California since 1971. The California Department of Pesticide Regulation (CDPR) has responsibility for collecting and evaluating these reports. Between 60% and 75% of cases are identified from workers' compensation reports. Most of the remainder are reported by physicians. The annual number of acute occupational illnesses and injuries related to pesticides in California ranged from 656 to 979 (Figure 5–40). Insecticides were responsible for the largest proportion of cases. Among insecticides, insecticide combinations and organophosphates were most commonly responsible (Figure 5–41). More than half of the reported cases occurred in agriculture (56%); services and public administration together contributed 28% (Figure 5–42).

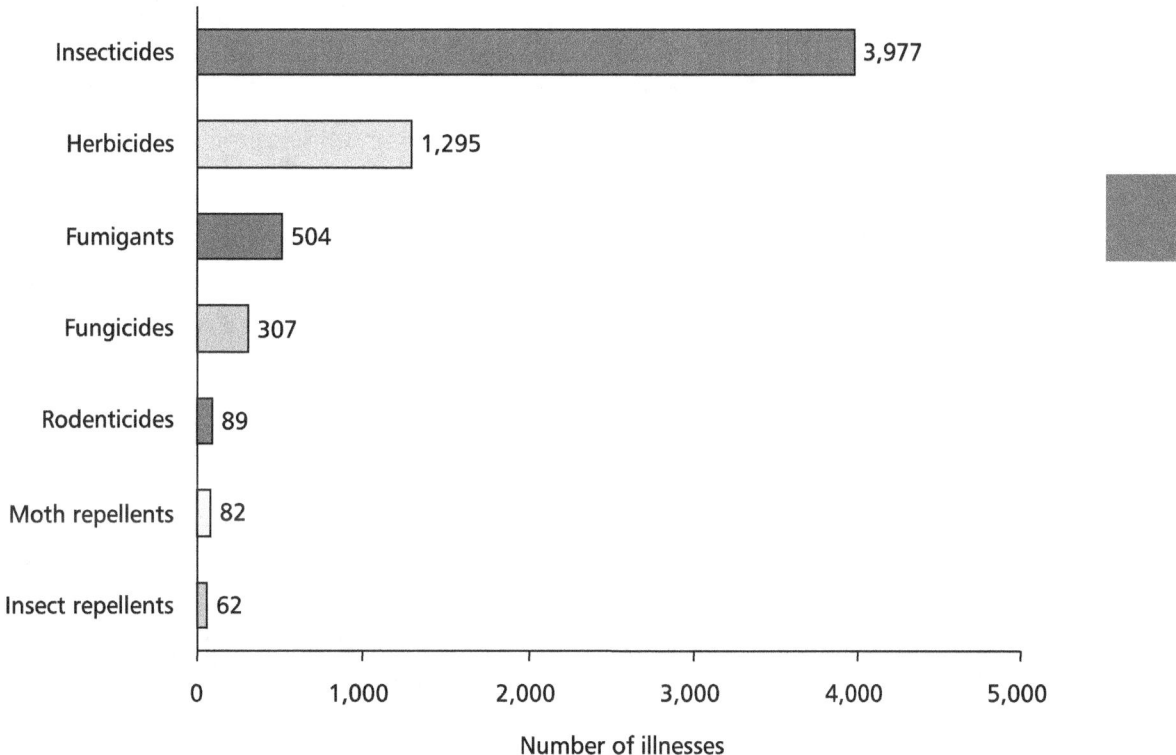

Figure 5–37. Number of acute occupational illnesses related to pesticides by pesticide category (excludes antimicrobials), 1993–1996. (Source: TESS [1998].)

NONFATAL ILLNESS

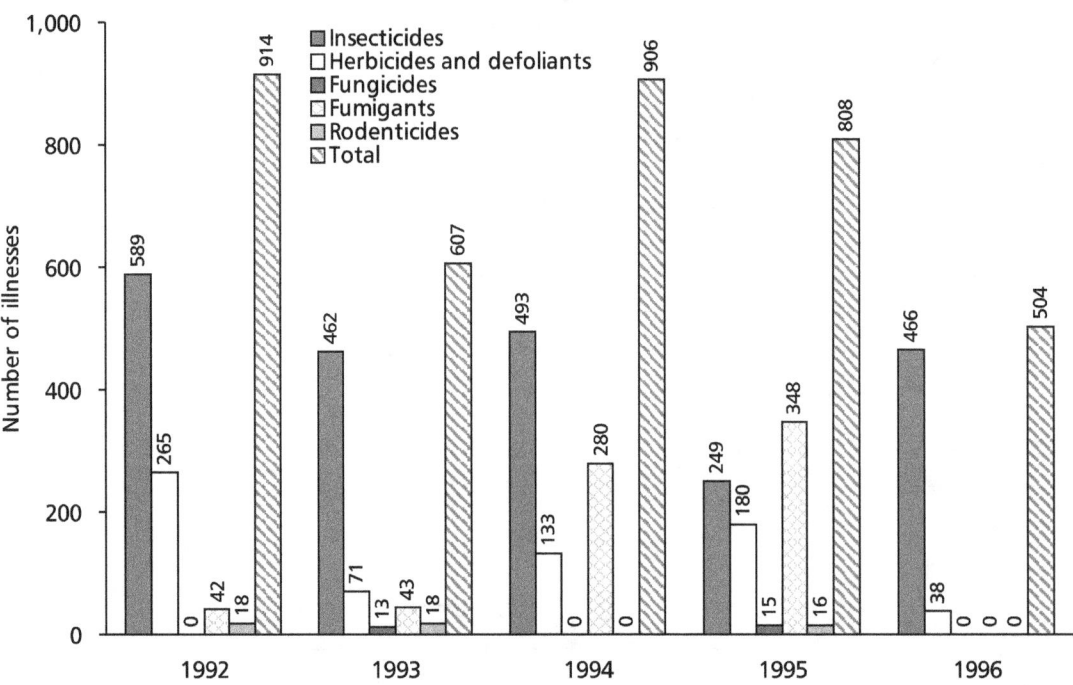

Figure 5-38. Number of occupational pesticide-related illnesses with days away from work in private industry by pesticide category, 1992–1996. (Source: SOII [1999].)

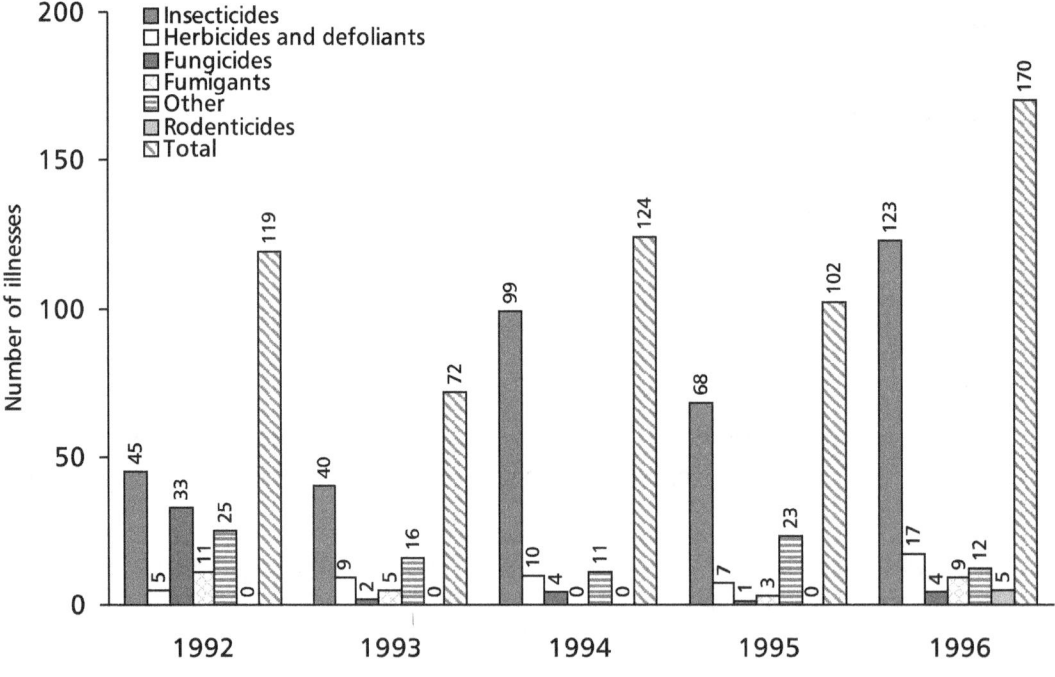

Figure 5-39. Number of occupational illnesses related to pesticides in New York, Oregon, and Texas by pesticide category, 1992–1996. (Source: SENSOR [New York State Department of Health 1999; Oregon Health Division 1999; PEST 1999].)

Nonfatal Illness

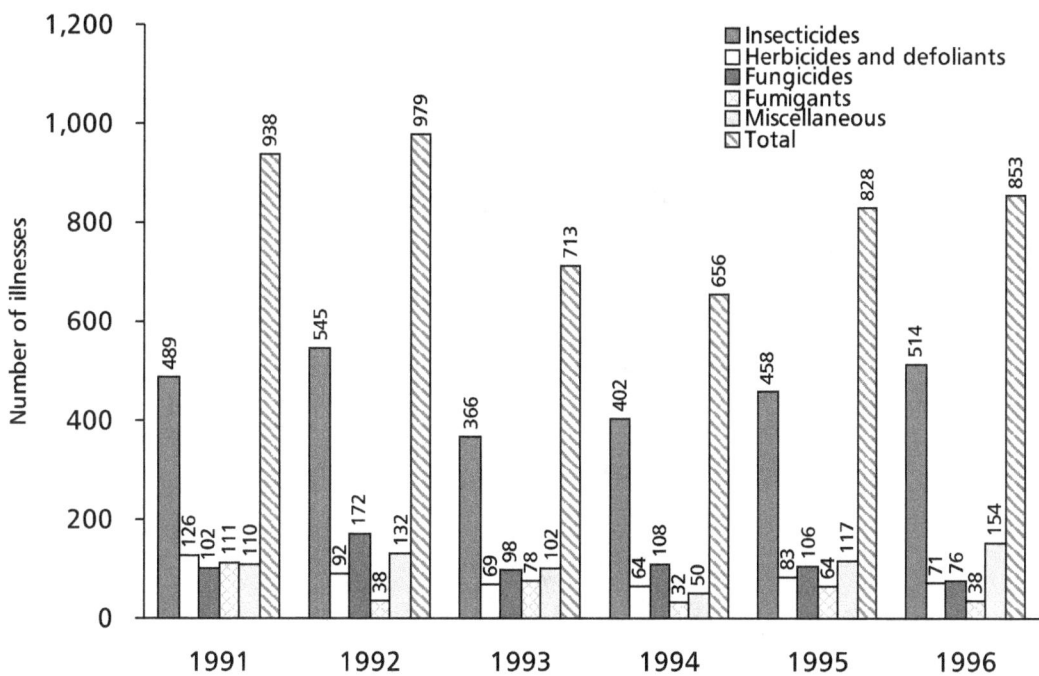

Figure 5–40. Number of occupational illnesses related to pesticides in California by pesticide category (excludes antimicrobials and unknown agents), 1991–1996. (Source: CDPR [1999].)

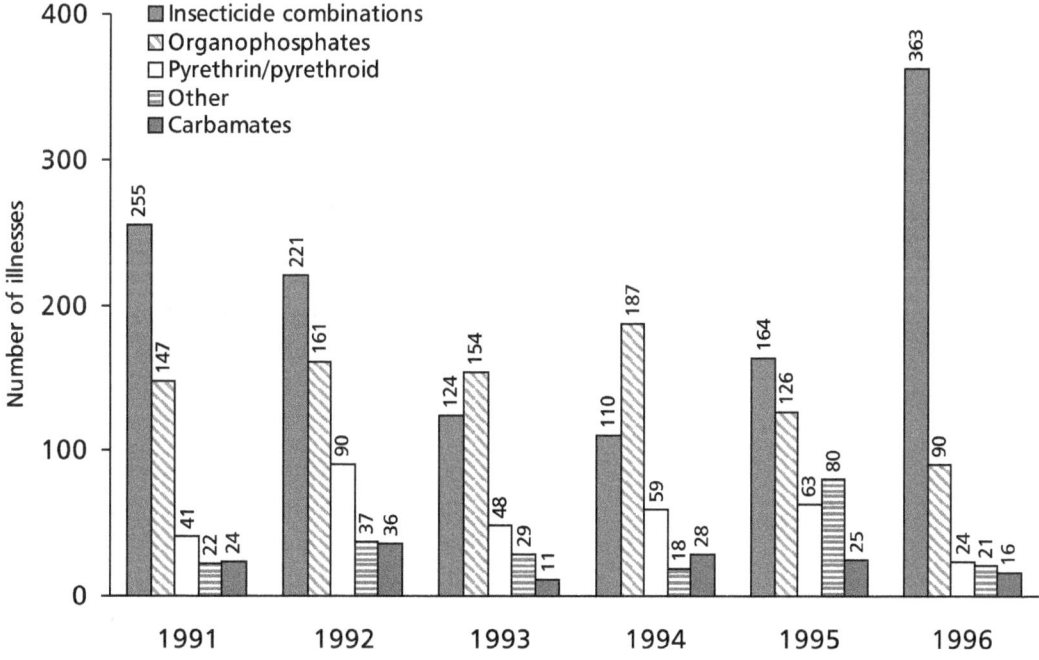

Figure 5–41. Number of occupational illnesses related to insecticides in California by insecticide category, 1991–1996. (Source: CDPR [1999].)

NONFATAL ILLNESS

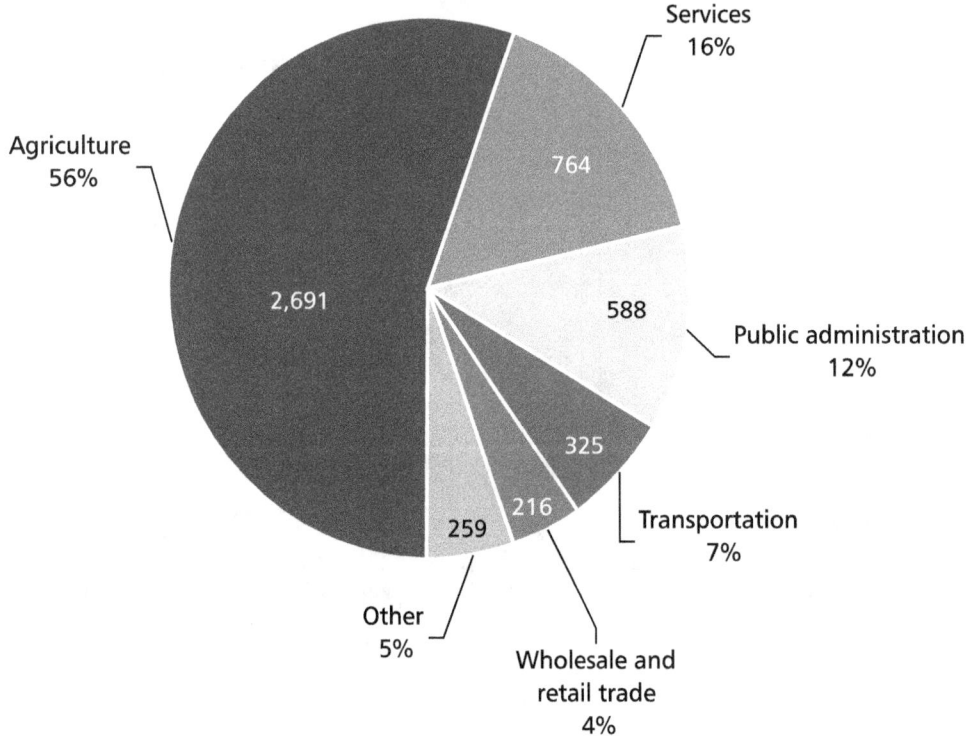

Figure 5–42. Number and distribution of occupational illnesses related to pesticides (excluding antimicrobials and unknown agents) in California, by industry division, 1991–1996. (Source: CDPR [1999].)

Infections in Health Care Workers

The 10 million health care workers in the United States constitute approximately 8% of the workforce. Health care workers can be exposed to a variety of occupational hazards, including repeated trauma, toxins, and a broad range of infectious agents. Surveillance data on infections in these workers are included in four Federal health databases:

- NaSH tracks exposures to and infections from several agents, including TB, vaccine-preventable diseases, and bloodborne pathogens.

- The Viral Hepatitis Surveillance Program (VHSP) and the Sentinel Counties Study of Acute Viral Hepatitis track hepatitis infection.

NONFATAL ILLNESS

- Cases of AIDS and HIV infection among health care workers are ascertained from several sources, including the HIV/AIDS Reporting System (HARS), which is maintained by CDC.

- *staffTRAK–TB* is used by health department TB control programs to monitor skin testing in employees of their clinics and affiliated institutions.

Between June 1995 and October 1999, 60 participating NaSH hospitals reported 6,983 cases of exposure to blood or body fluids. Most of these cases occurred in nurses (43%) and physicians (29%) (Figure 5–43). The largest number of exposures to blood or body fluids occurred in inpatient (30%) and operating/procedure room settings (29%) (Figure 5–44). The major route of exposure was percutaneous (puncture/cut injury) (Figure 5–45).

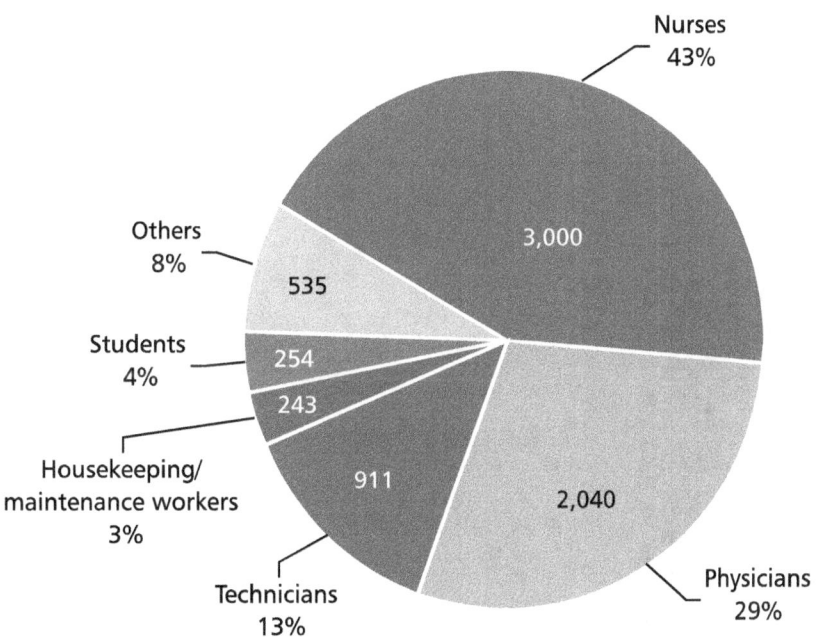

Figure 5–43. Number and distribution of reported health care worker exposures to blood or body fluids in 60 participating hospitals by occupational group, June 1995 to October 1999. (Source: NaSH [1999].)

149

Nonfatal Illness

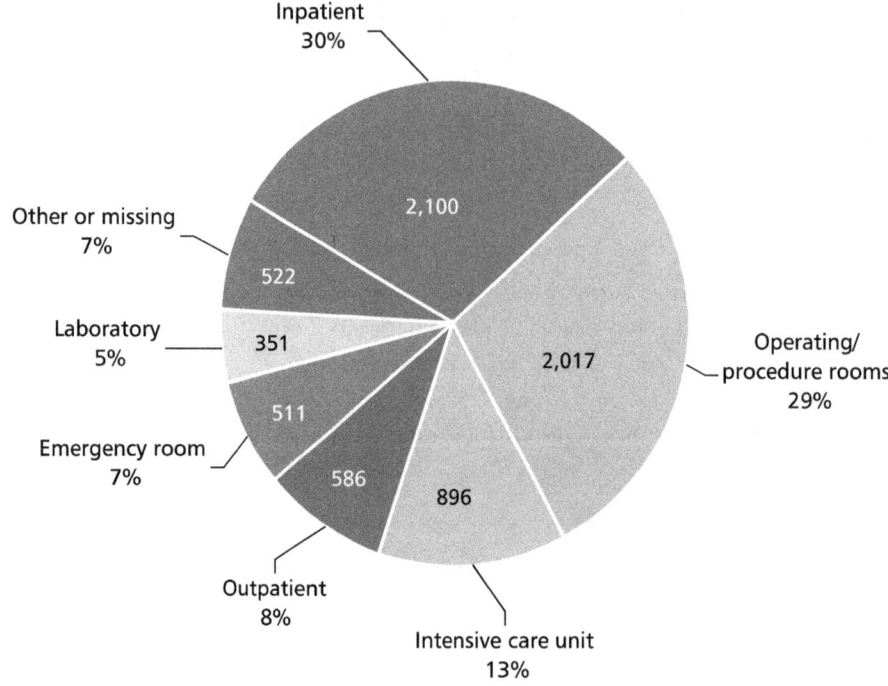

Figure 5–44. Number and distribution of reported health care worker exposures to blood or body fluids in 60 participating hospitals by work location, June 1995 to October 1999. (Source: NaSH [1999].)

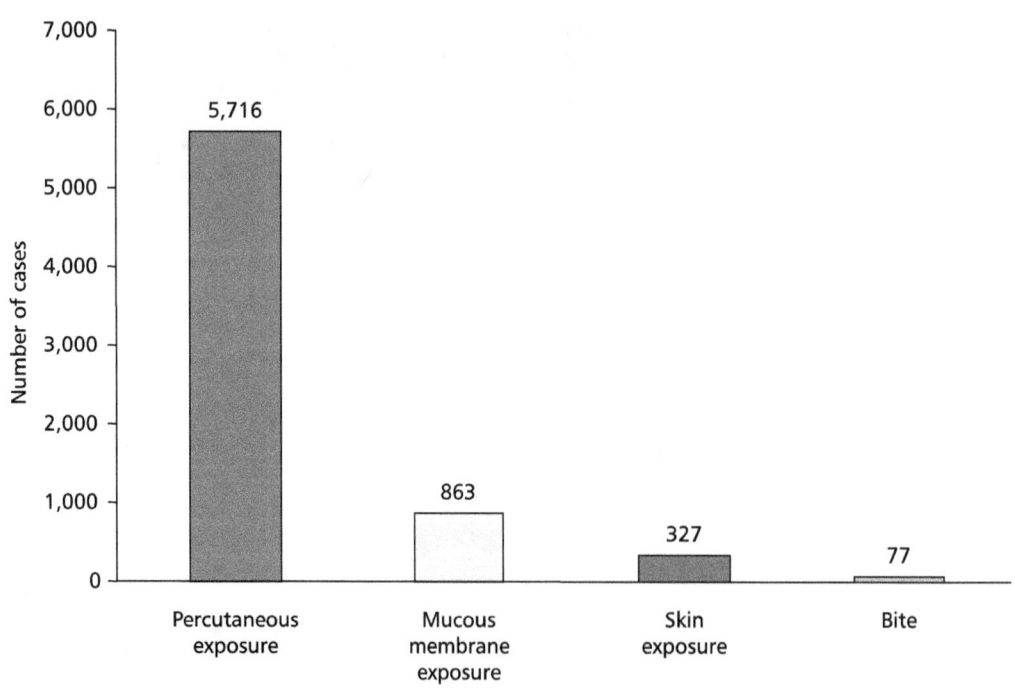

Figure 5–45. Number of reported health care worker exposures to blood or body fluids in 60 participating hospitals by exposure type, June 1995 to October 1999. (Source: NaSH [1999].)

Nonfatal Illness

Consequences of Bloodborne Exposures

Hepatitis B Virus

VHSP and the Sentinel Counties Study of Acute Viral Hepatitis indicate a 93% decline in hepatitis B viral infections in health care workers over a 10-year period—from approximately 12,000 cases in 1985 to 800 cases in 1995 (Figure 5–46). Infections also declined among the general population during this time, but not as dramatically. The greater decline among health care workers may be attributed to the adoption of universal precautions against exposure to body fluids and vaccinations against hepatitis B.

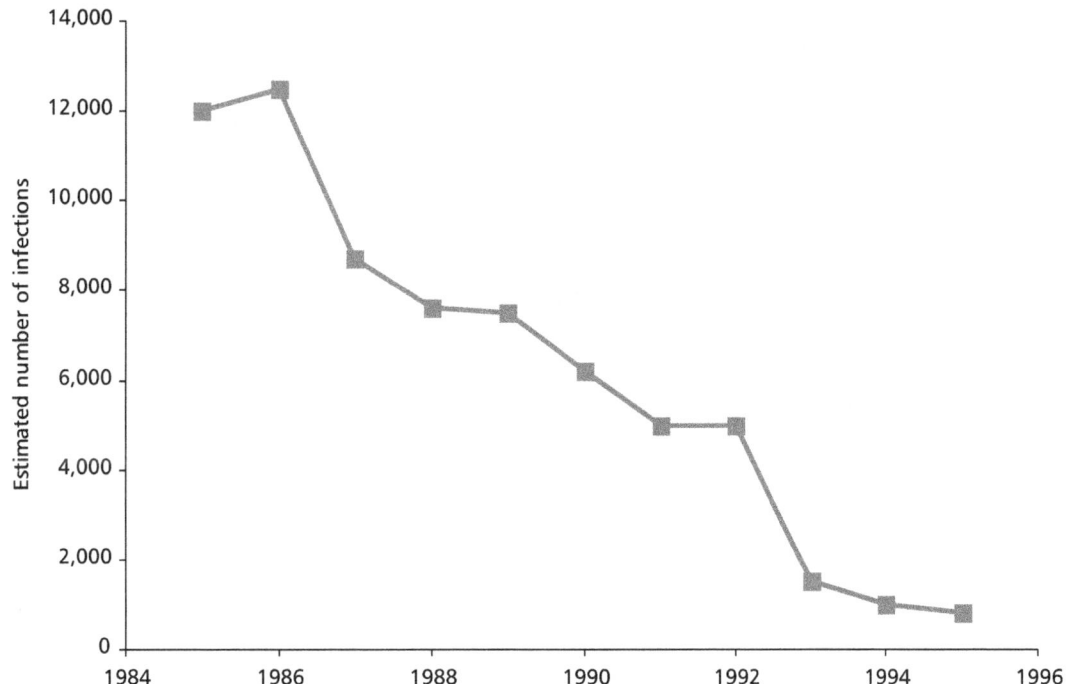

Figure 5–46. Estimated number of hepatitis B infections among U.S. health care workers, 1985–1995. (Source: VHSP [1999]; NCID [1999].)

Nonfatal Illness

Hepatitis C Virus

Hepatitis C virus infection is the most common chronic bloodborne infection in the United States. Although the prevalence of hepatitis C virus infection in health care workers is similar to that in the general population (1% to 2%), health care workers have an increased occupational risk from needlestick injuries. The number of health care workers who have acquired hepatitis C infections occupationally is not known. But approximately 2% to 4% of acute infections in the United States occurred among health care workers exposed to blood in the workplace. Most workers exposed to hepatitis C were physicians or nurses (Figure 5–47).

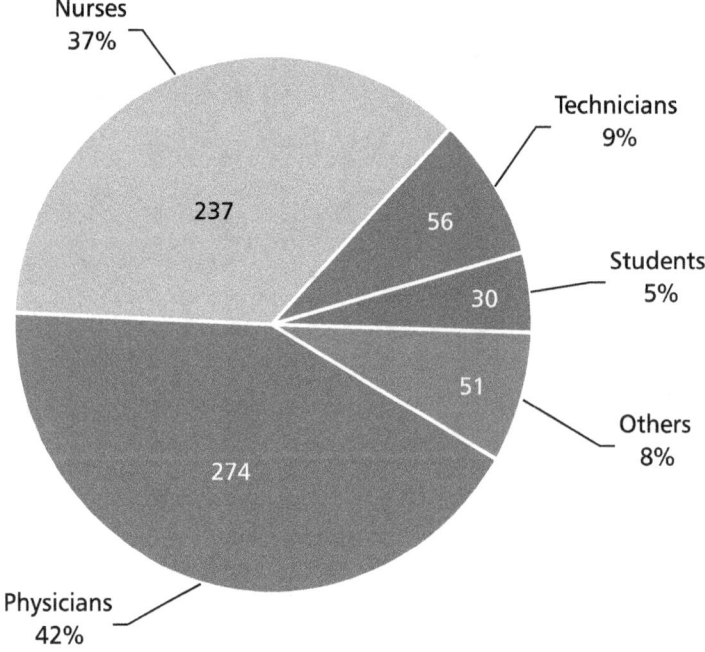

Figure 5–47. Number and distribution of health care workers exposed to hepatitis C virus by occupational group, June 1995 to October 1999. (Source: NaSH [1999].)

Nonfatal Illness

Human Immunodeficiency Virus

Fifty-five cases of documented and 136 cases of possible occupational HIV transmission were recorded in HARS through June 1999. Among the documented cases of HIV seroconversion following occupational exposure, 85% resulted from percutaneous exposure and 93% involved exposure to blood or visibly bloody fluid. Most documented cases of occupational HIV transmission occurred among nurses (42%) and laboratory workers (35%) (Figure 5–48).

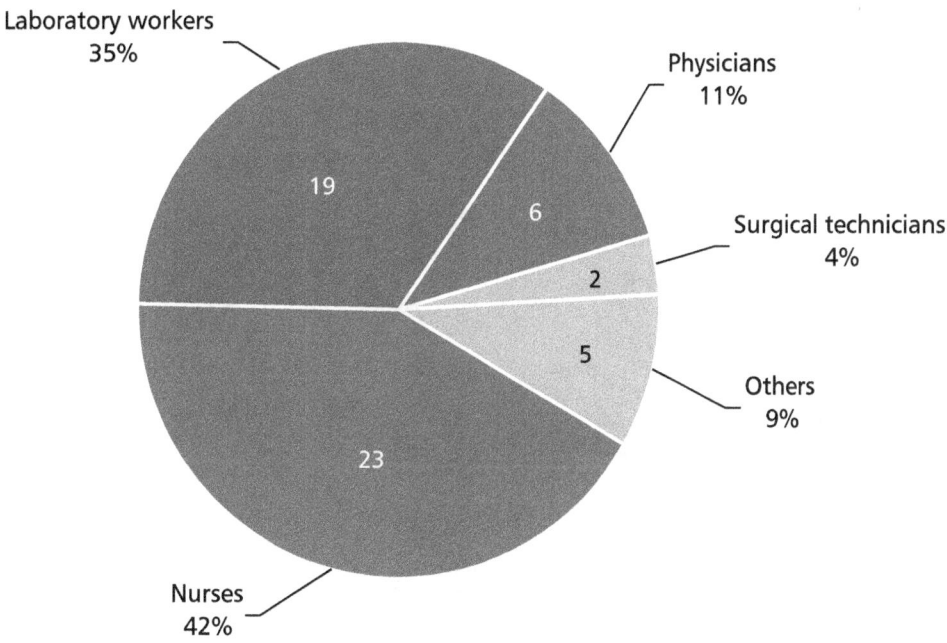

Figure 5–48. Number and distribution of health care worker cases with documented occupational transmission of HIV by occupation through June 1999. (Source: HARS [CDC 1999].)

NONFATAL ILLNESS

Tuberculosis (TB)

Health care workers have long been at risk of contracting TB. This risk increased in the 1980s with the resurgence of TB in the United States and the subsequent development of drug-resistant TB bacteria during the AIDS epidemic. From 1994 through 1998, there were 2,732 cases of TB in health care workers reported to the Centers for Disease Control and Prevention (CDC) through *staffTRAK–TB* from the 50 States, the District of Columbia, and Puerto Rico. Incidence rates in health care workers are shown in Figure 5–49 for each year from 1994 through 1998. These rates are not associated specifically with occupational exposure because that information is not available. Cases in health care workers constituted 3% of all TB cases.

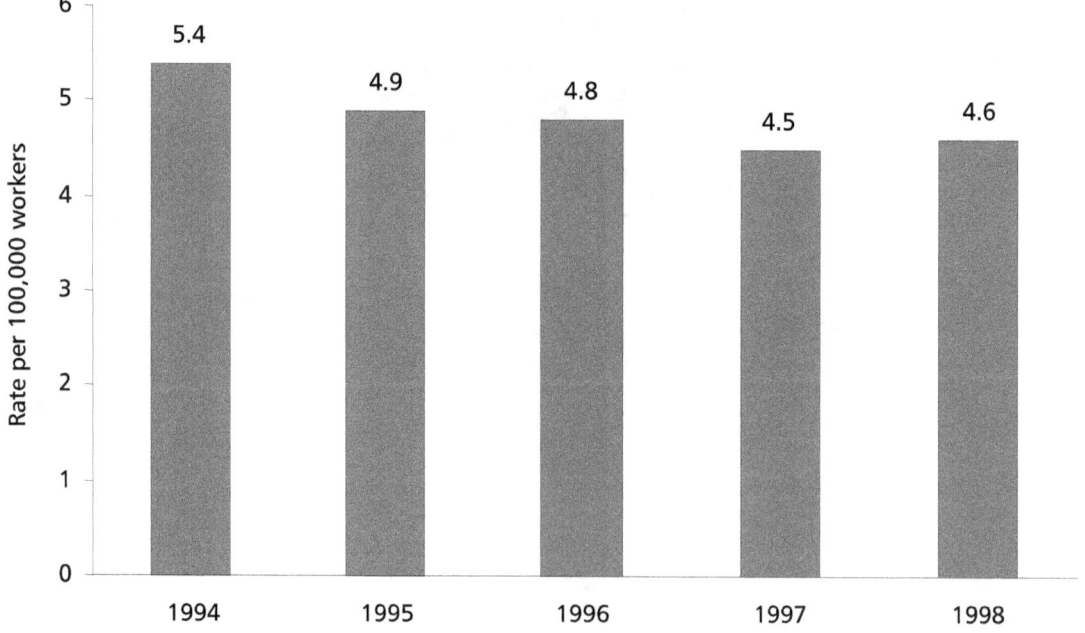

Figure 5–49. Incidence rates of TB in health care workers, 1994–1998. (Source: *staffTRAK–TB* [1999].)

Physical Agents

Disorders attributable to physical agents represented 4% (16,600) of all nonfatal occupational illness cases recorded in SOII in 1997. Disorders attributable to physical agents include heatstroke, sunstroke, heat exhaustion, and other effects of environmental heat; freezing and frostbite; effects of ionizing radiation (isotopes, X-rays, radium); and effects of nonionizing radiation (welding flash, ultraviolet rays, microwaves, and sunburn). Illnesses from toxic exposures are excluded. Among industry divisions, manufacturing accounted for 55% of the disorders attributable to physical agents in private industry in 1997 (Figure 5–50). Among individual industries, the highest illness rates occurred in metal sanitary ware (294 cases per 10,000 workers), primary aluminum (89 cases per 10,000 workers), ship building and repairing (79 cases per 10,000 workers), and plumbing and heating, except electric (73 cases per 10,000 workers).

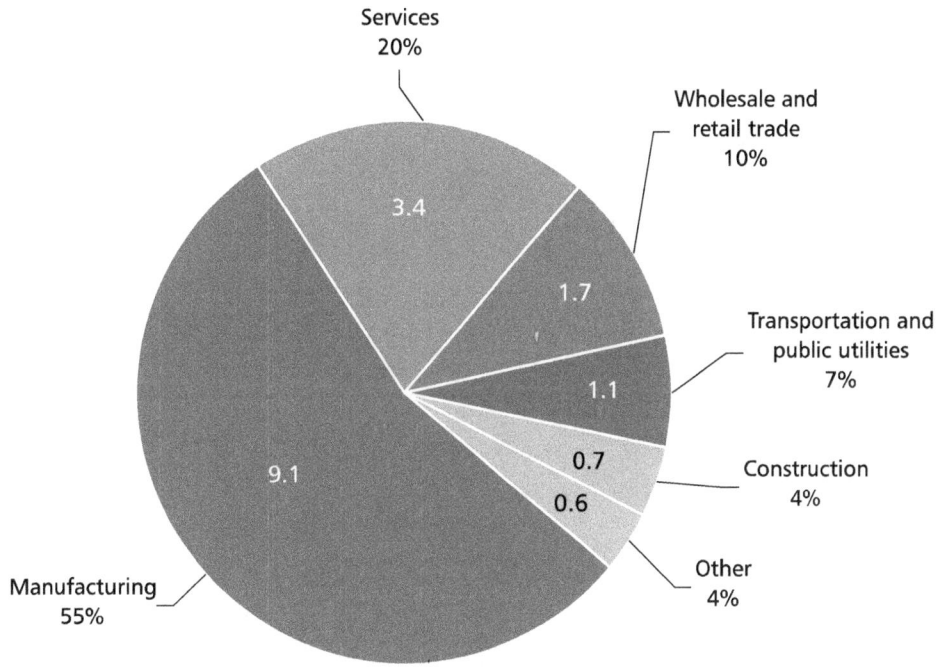

Figure 5–50. Number (thousands) and distribution of disorders attributable to physical agents in private industry by major industry division, 1997. (Source: SOII [1999].)

Nonfatal Illness

Anxiety, Stress, and Neurotic Disorders

Nearly 5,300 cases of anxiety, stress, or neurotic disorders with time away from work were recorded in SOII in 1997. These represent 1% of all reported nonfatal occupational illness cases. Women accounted for more than 60% of all occupational anxiety, stress, and neurotic disorder cases with time away from work. Half of all such disorder cases required 23 or more days away from work, and more than 40% of workers with these disorders required more than 31 days away from work. The industry divisions accounting for most cases were services (35%), wholesale and retail trade (20%), and manufacturing (20%) (Figure 5–51). The occupational groups most frequently experiencing these disorders were technical, sales, and administrative personnel (47%) and operators, fabricators, and laborers (18%) (Figure 5–52). The exposures most frequently associated with anxiety, stress, or neurotic disorders were harmful substances (30%) and assaults or violent acts (13%) (Figure 5–53).

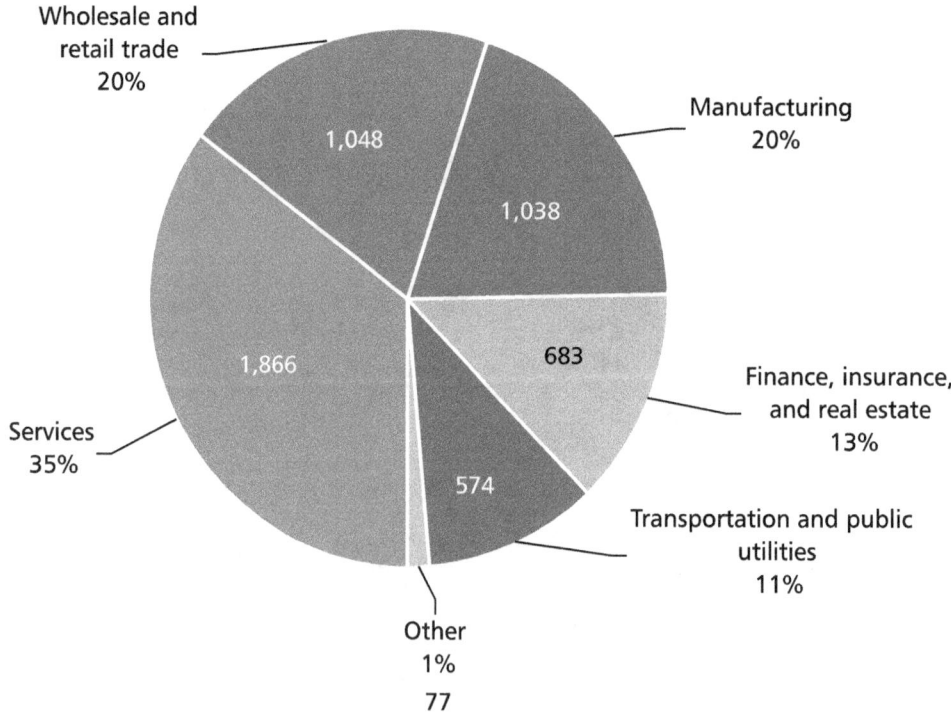

Figure 5–51. Number and distribution of anxiety, stress, and neurotic disorder cases with days away from work in private industry by industry division, 1997. (Source: SOII [1999].)

NONFATAL ILLNESS

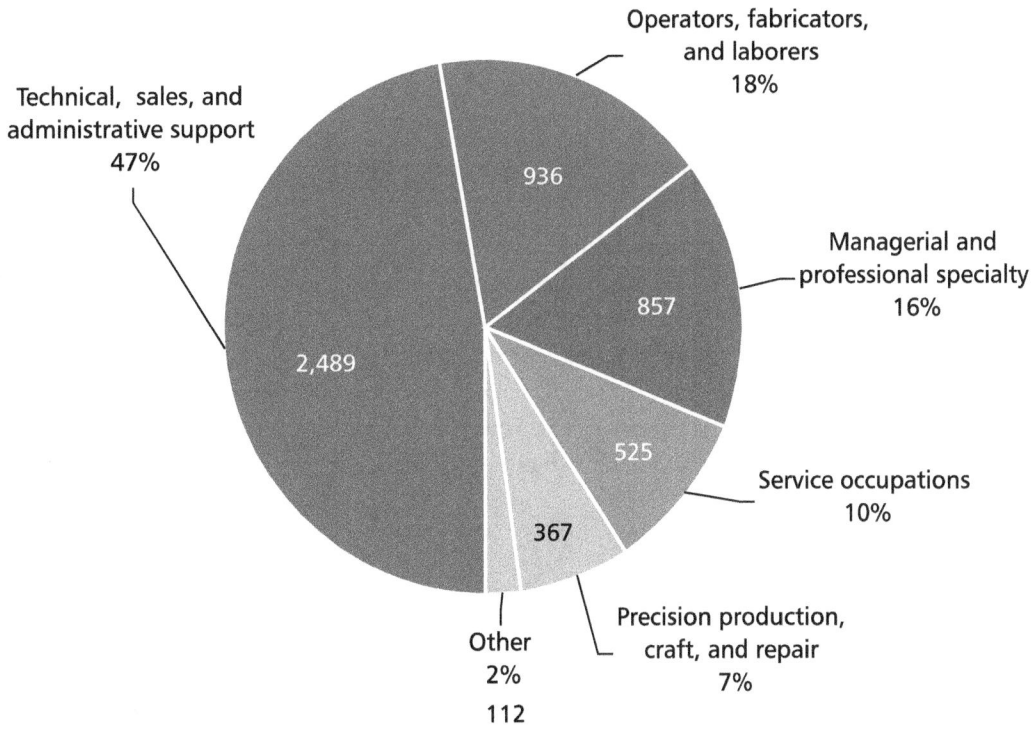

Figure 5-52. Number and distribution of anxiety, stress, and neurotic disorder cases with days away from work in private industry, by occupational group, 1997. (Source: SOII [1999].)

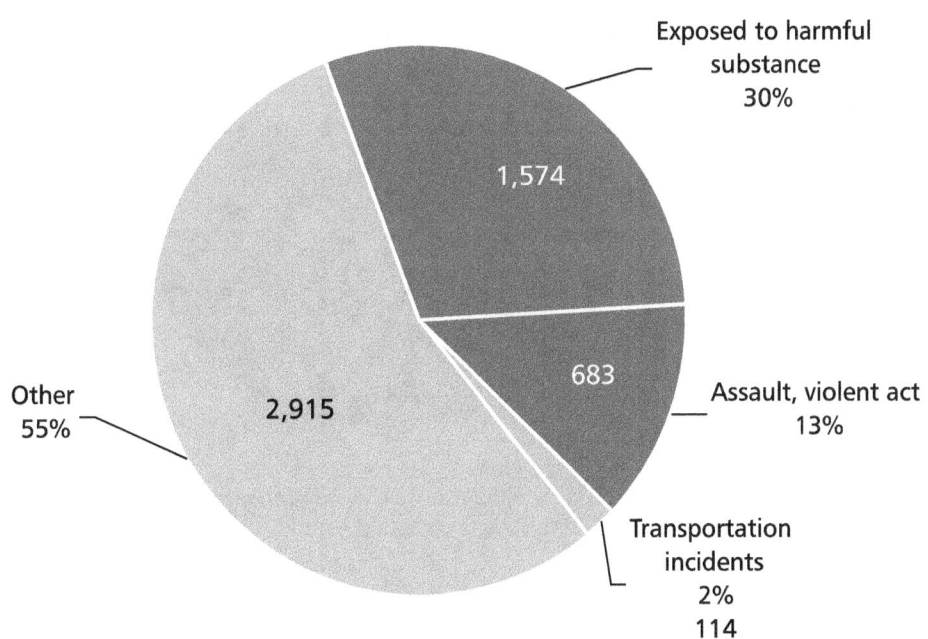

Figure 5-53. Number and distribution of anxiety, stress, and neurotic disorder cases with days away from work in private industry, by event or exposure, 1997. (Source: SOII [1999].)

Nonfatal Illness

All Other Nonfatal Occupational Illnesses

All other nonfatal occupational illnesses represented 12% (50,400) of all illness cases recorded in SOII in 1997. This category captures illnesses such as anthrax, brucellosis, hepatitis B and C, HIV disease, malignant and benign tumors, food poisoning, histoplasmosis, and coccidioidomycosis. The largest percentages of such cases in 1997 occurred in services (41%) and manufacturing (29%) (Figure 5–54). Industries reporting the highest incidence rates were luggage (163 cases per 10,000 workers), secondary smelting and refining of nonferrous materials (120 cases per 10,000 workers), prefabricated metal buildings (66 cases per 10,000 workers), and iron and steel forgings (61 cases per 10,000 workers).

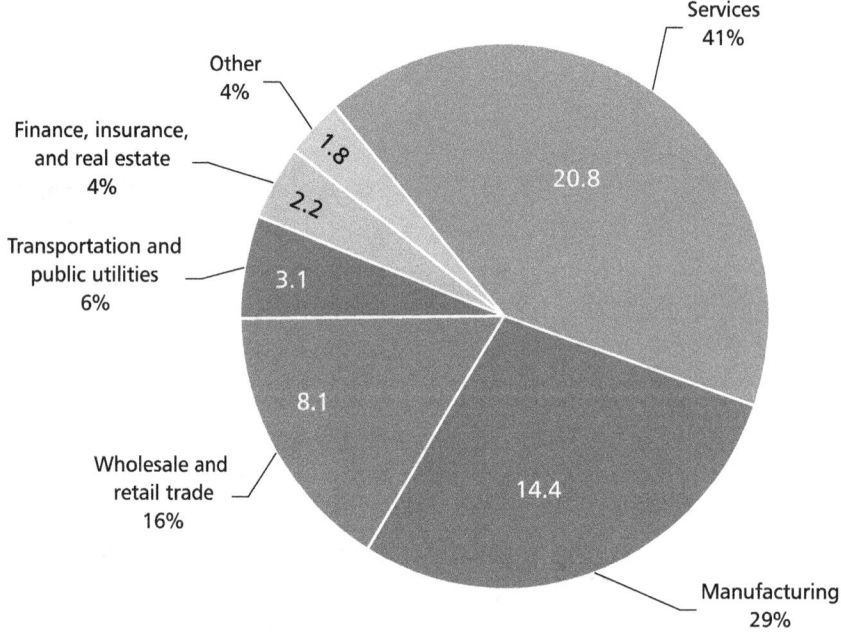

Figure 5–54. Number (thousands) and distribution of all other occupational illnesses in private industry by major industry division, 1997. (Source: SOII [1999].)

6 Focus on Mining

6 Focus on Mining

This section presents a detailed overview of mining and injuries. Historically, mining has been the industry sector with the highest fatal and nonfatal injury rates. Mining still has the highest fatal injury rate—more than five times the national average. Because of the accompanying social toll and public pressures for action, data on fatalities and injuries became available far earlier in the mining industry than in many others. State and Federal agencies began collecting data in the 1870s, and reliable information has been available for an entire century. Examination of the history of coal mining in this century shows important relationships between adverse outcomes (such as fatality rates) and regulatory actions such as enactment of Federal legislation and establishment of enforcement and consultation agencies.

Fatal Injuries

Historical Perspective

More than 103,000 workers died in the mining industry (including all commodities and work locations) during the 85-year period from 1911 to 1995 (Figure 6–1). From 1911 through 1915 alone, 16,646 fatalities occurred, with an annual average of 2,517 deaths in coal mining and 813 in metal and nonmetal mining. The corresponding annual average fatality rates during this 5-year period were 340 and 300 per 100,000 workers in coal mining and metal and nonmetal mining, respectively. The U.S. Bureau of Mines, established in 1910, focused on coal mine fires and explosions. In the first decade after its creation, disaster-related fatalities* in coal mines decreased substantially, with a 62% reduction in deaths from mine fires and a 45% reduction from explosions. Disaster-related fatalities in metal and nonmetal mining increased during the same period. The number and rate of fatalities decreased again during the Great Depression, a period accompanied by reductions in both the labor force and production, as measured by tonnage mined. Rates increased during the economic mobilization required during World War II. For the years 1941–1945, 694 million tons of coal were mined annually, compared with 520 million tons for the 5 preceding years—a 33% increase. Coal mining

---

*Five or more fatalities resulting from a single incident.

Focus on Mining

fatality rates increased during the decade preceding the Coal Mine Health and Safety Act of 1969 [Public Law 91–173]. Rates rose similarly for metal and nonmetal mining in the decade before the Federal Mine Safety and Health Act of 1977 [Public Law 95–164]. Fatality rates decreased following the passage of these two Federal mine acts.

Fatal Injuries during 1988–1997

Mining operations are located in every State as well as in Puerto Rico and the Virgin Islands. In 1997, 13,682 mining operations reported employment to the Mine Safety and Health Administration (MSHA)—2,609 were coal mining operations (average employment size of 35 full-time workers); 374 were metal mining operations (average employment size of 122 full-time workers); 786 were nonmetal mining operations (average employment size of 32 full-time workers); 3,712 were stone mining operations (average employment size of 19 full-time workers); and 6,201 were sand and gravel mines (average employment size of 5 full-time workers).

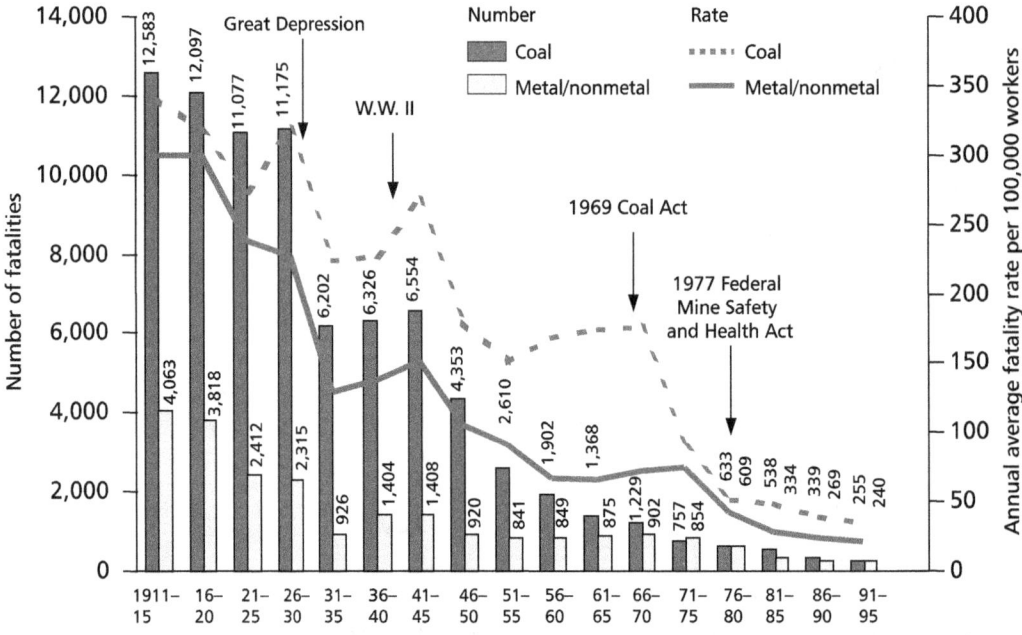

Figure 6–1. Number of fatalities (5-year aggregates) and annual average fatality rates in the mining industry by commodity, 1911–1995. Annual average fatality rates are calculated per 100,000 workers aggregated over 5-year periods. Metal and nonmetal includes metal, nonmetal, stone, and sand and gravel. (Source: MSHA [1999]; Adams and Wrenn [1941]; Adams and Kolhos [1941]; Reese et al. [1955]; MSHA [1984].)

Focus on Mining

At least one mine operator fatality occurred in each State but Delaware, Maine, and Rhode Island during the period 1988–1997 (Figure 6–2). The national annual average mine operator fatality rate during this period was 28.5 per 100,000 miners, which is more than five times the national annual average occupational fatality rate of 5.3 per 100,000 workers from 1980 through 1995 [NTOF 1999] (see Figure 2–1 and preceding text). Nine states had very high mine operator fatality rates that exceeded the national mining fatality rate by 50% or more. Nine other States and Puerto Rico had rates greater than the national rate.

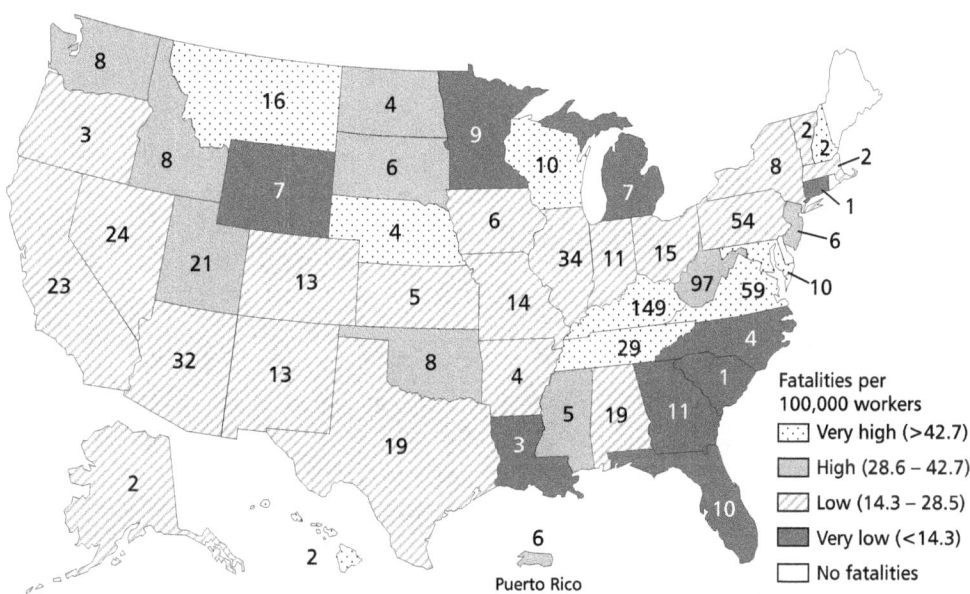

Figure 6–2. Number of mine operator fatalities (as numbers within State boundaries) and corresponding annual average fatality rates (shading within States) for each State and Puerto Rico, 1988–1997. (Source: MSHA [1999].)

Focus on Mining

During the 10-year period 1988–1997, the number of fatalities was highest in 1990 (n=122) and lowest in 1994 (n=82) (Figure 6–3). The overall downward trend since 1990 is attributable primarily to decreases in fatalities of mine operator workers in the coal and metal commodities. During 1988–1997, mine operator workers accounted for 806 (81.2%) of the total number of fatalities (n=993), with the remaining 187 (18.8%) being independent contractor workers. Coal operator workers accounted for 432 (43.5%) of the total, followed by mine operator workers in the commodities of stone (n=157, 15.8%), metal (n=98, 9.9%), sand and gravel (n=83, 8.4%), and nonmetal (n=36, 3.6%). Independent contractor workers in metal and nonmetal mining made up 11.3% (112) of the total, with 7.6% (75) attributed to independent contractors in coal.

Despite annual fluctuations, overall fatality rates have decreased for mine operator workers in the coal and metal commodities (Figure 6–4). Although fatality rates for independent contractor workers are the highest overall, reduced employment reporting requirements for this sector of workers compromise any direct comparison with rates for mine operator workers.

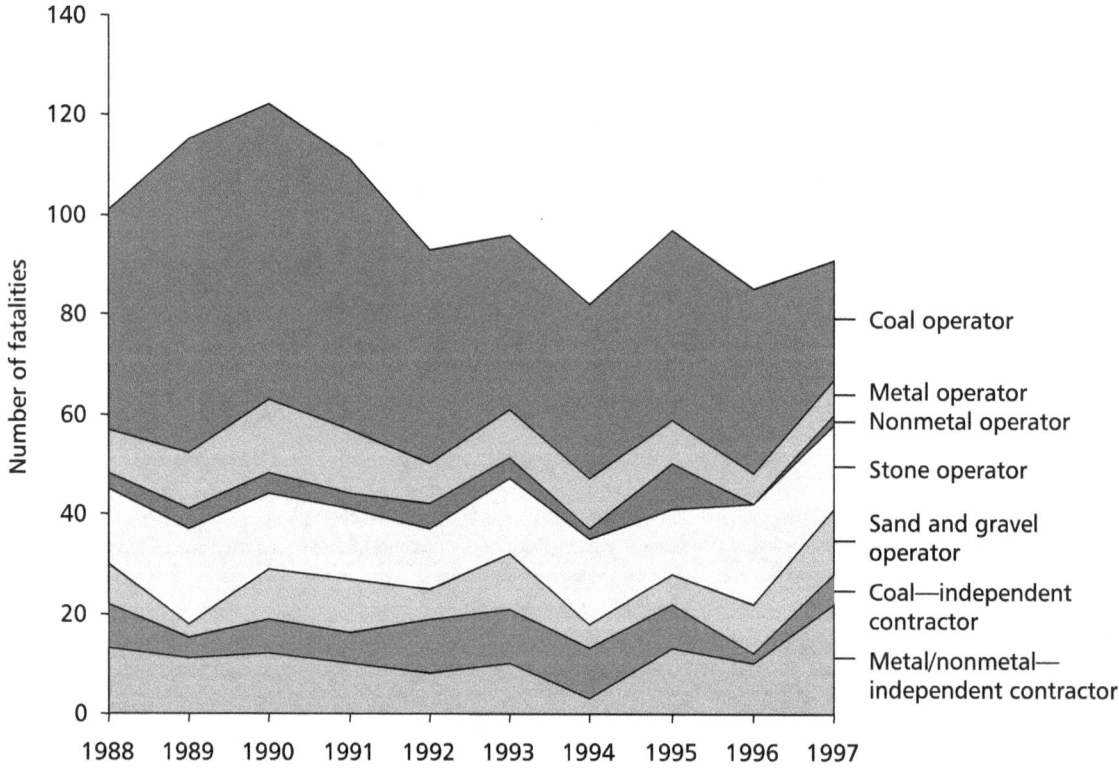

Figure 6–3. Number of fatalities, by type of employer (mine operator versus independent contractor) and commodity, 1988–1997. Metal and nonmetal includes metal, nonmetal, stone, and sand and gravel. (Source: MSHA [1999].)

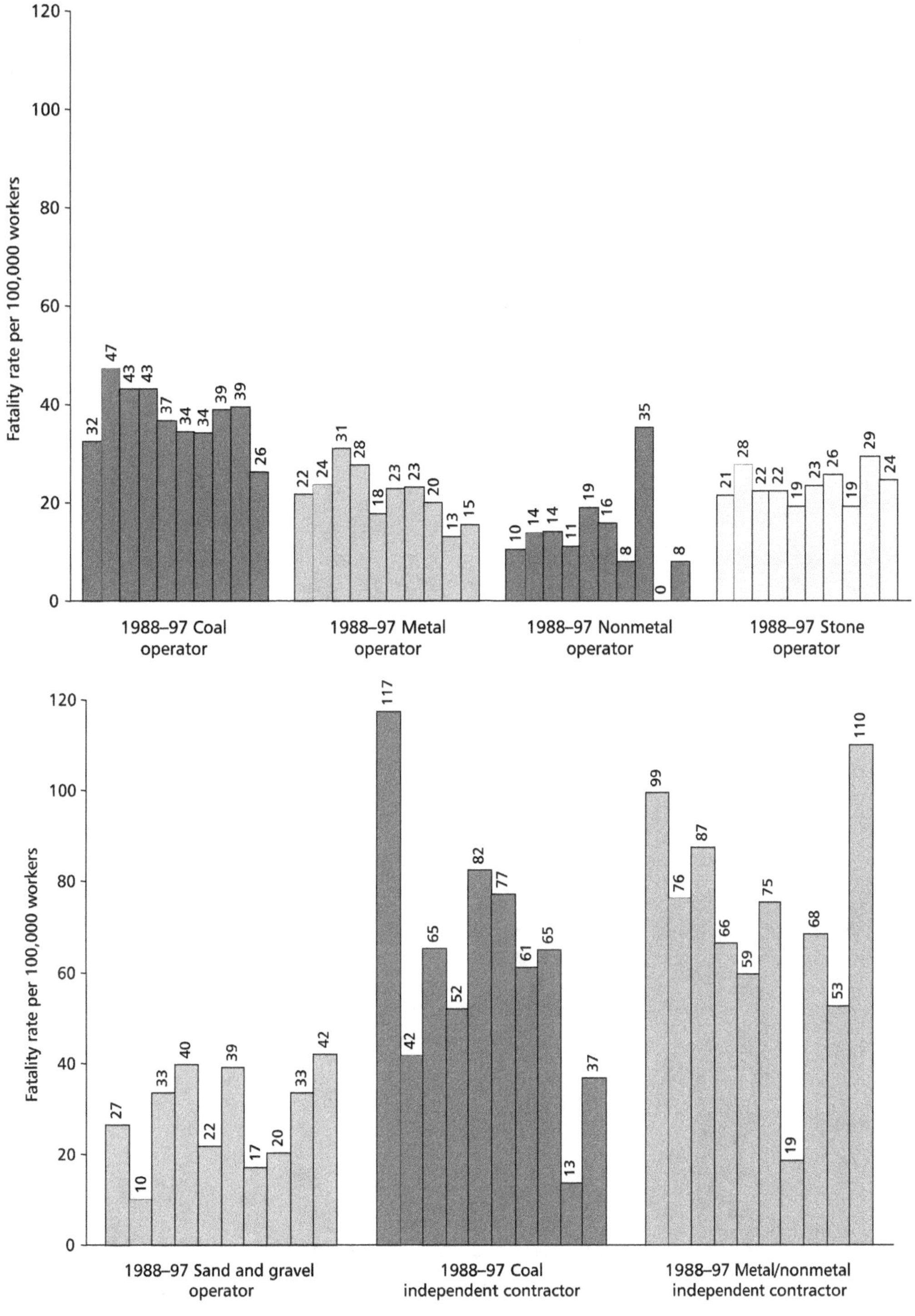

FOCUS ON MINING

Figure 6–4. Fatality rates by type of employer (mine operator versus independent contractor) and commodity, 1988–1997. Fatality rates are calculated per 100,000 full-time workers or 200 million employee hours. Metal and nonmetal includes metal, nonmetal, stone, and sand and gravel. (Source: MSHA [1999].)

Focus on Mining

Both the number and annual average rate of fatalities from 1988 through 1997 varied by work location and by type of employer and commodity (Table 6–1). Overall, underground work locations exhibited both the highest numbers and rates of fatalities, and preparation plants and mills exhibited the lowest fatality rates. Among mine operator workers, the combination of high numbers and rates of fatalities is most conspicuous for those working in underground coal mines, underground metal mines, stone surface mines (or quarries), and sand and gravel operations. Although independent contractor workers accounted for about one-fifth (19%) of all fatalities during this 10-year period, they accounted for almost one-third (30%) of the fatalities at the surface areas of underground mines, at surface mines, and at mills or preparation plants.

A trend toward decreasing fatality rates with increasing mine size is apparent in underground coal and metal mining operations (Figure 6–5). This trend is particularly significant for underground coal mining operations, which also accounted for 83% of all underground employee hours and 77% of all underground fatalities reported during the 10-year period 1988–1997. In addition, coal accounted for 90% of the employee hours worked underground in small mines (fewer than 50 workers), followed by stone mines (5.5%), metal mines (3%), and nonmetal mines (less than 1%).

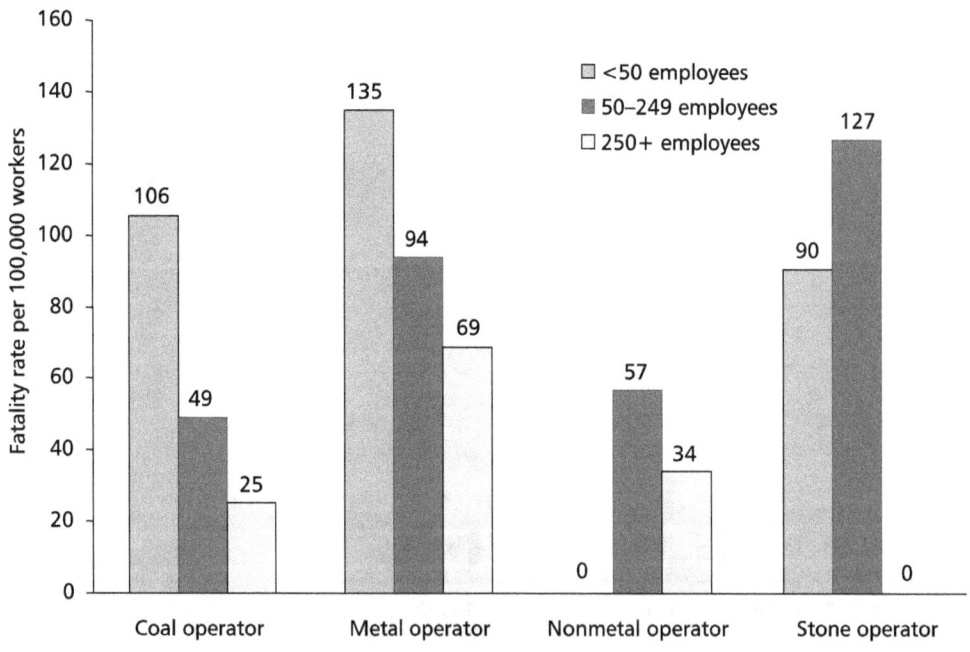

Figure 6–5. Fatality rates at underground mining operations, by commodity and employment size of operation, 1988–1997. Fatality rates are computed per 100,000 full-time workers or 200 million employee hours. (Note: There are no sand and gravel underground operations.) (Source: MSHA [1999].)

FOCUS ON MINING

Table 6–1. Number and annual average rate* of fatalities associated with various types of employers and commodities by work location, 1988–1997

Type of employer and commodity	All		Underground mines				Surface mines							
			Underground		Surface areas		Strip/open pit/quarry		Dredge		Other surface operations†		Mills/plants	
	Number	Rate	Number	Rate	Number	Rate	Number	Rate	Number	Rate	Number	Rate	Number	Rate
All	993	31.9	388	56.9	51	46.3	383	29.4	21	38.3	10	37.2	140	15.0
Mine operator:														
Coal	432	37.9	298	54.2	27	49.5	74	19.1	0	0.0	4	25.7	29	22.0
Metal	98	21.6	53	83.8	6	36.5	28	16.6	0	0.0	0	0.0	11	5.7
Nonmetal	36	13.4	14	40.0	0	0.0	11	18.5	0	0.0	NA‡	NA	11	6.6
Stone	157	23.5	15	93.5	3	77.1	95	33.7	0	0.0	0	0.0	44	12.1
Sand and gravel	83	28.2	NA	NA	NA	NA	65	26.3	18	38.3	NA	NA	NR§	NR
Independent contractor:														
Coal	75	59.1	2	17.6	12	52.3	40	62.5	0	0.0	5	253.2	16	60.3
Metal and nonmetal**	112	71.3	6	102.0	3	49.0	70	75.0	3	278.1	1	232.7	29	57.7

Source: MSHA [1999].
*Computed per 100,000 full-time workers or 200 million employee hours.
†Includes culm banks, auger mining, independent shops and yards, and surface mining n.e.c.
‡NA=Not applicable for this commodity.
§NR=Not reported separately. Sand and gravel operators report mill employment under strip or dredge operations.
**Includes metal, nonmetal, stone, and sand and gravel.

Focus on Mining

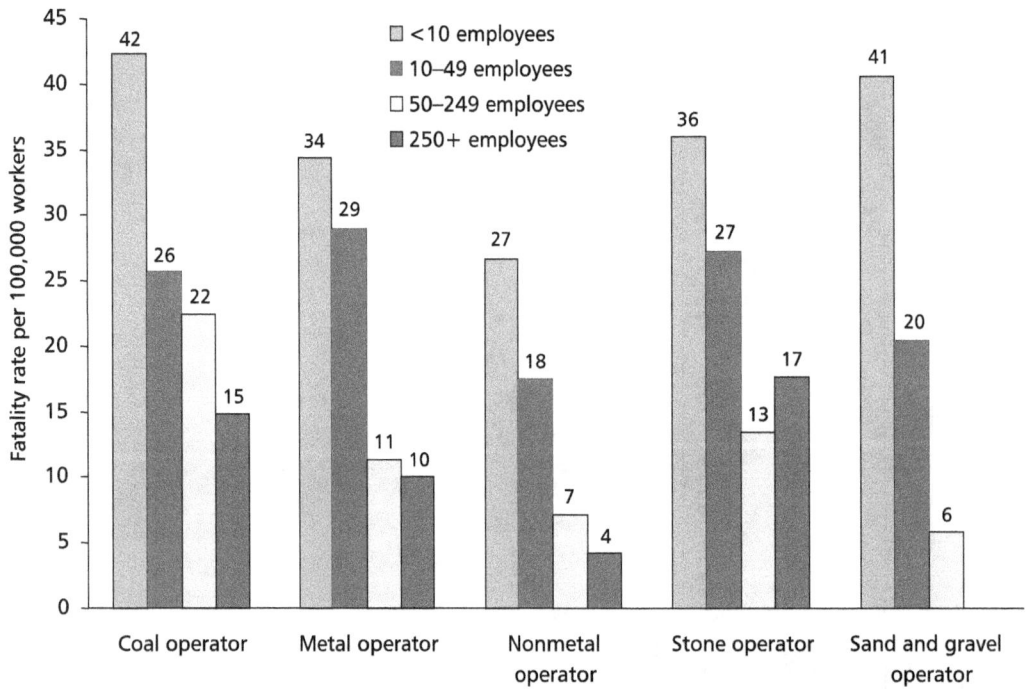

Figure 6–6. Fatality rates at surface mining operations, by commodity and employment size of operation, 1988–1997. Fatality rates are computed per 100,000 full-time workers or 200 million employee hours. (Source: MSHA [1999].)

Fatality rates from 1988 through 1997 at surface work locations were highest at the smallest mining operations (fewer than 10 workers) in every commodity (Figure 6–6). The proportion of employee hours worked at the surface locations of these mines was highest in sand and gravel (43%), followed by stone (8%), coal (6%), nonmetal (5%), and metal (less than 1%). Elevated rates persist for small operations with 10 to 49 workers, although these rates are only slightly elevated in coal. No sand and gravel operations employed more than 250 workers.

The major types of incidents associated with fatal injuries are shown in Table 6–2 by commodity and type of employer. Overall, powered haulage incidents accounted for the largest percentage of fatalities (30.8%), followed by fall of ground (18.9%) and machinery (16.3%). Some types of incidents, such as falls of ground, are substantially more frequent among mine operators than among independent contractors. The rate differences between type of employer and among commodities suggest that different strategies are needed to reduce fatality rates in various sectors of the mining industry.

Focus on Mining

Table 6–2. Number* and annual average rate† of fatalities associated with various types of employers and commodities, by type of incident,‡ 1988–1997

Type of employer and commodity	All		Type of incident							
			Powered haulage		Fall of ground (from in place)		Machinery		Electrical		Slip or fall of person	
	Number	Rate	Number	Rate	Number	Rate	Number	Rate	Number	Rate	Number	Rate
All	993	31.9	306	9.8	188	6.0	162	5.2	80	2.6	60	1.9
Mine operator:												
Coal	432	37.9	101	8.9	135	11.8	76	6.7	33	2.9	9	0.8
Metal	98	21.6	30	6.6	18	4.0	6	1.3	11	2.4	8	1.8
Nonmetal	36	13.4	11	4.1	11	4.1	4	1.5	5	1.9	2	0.7
Stone	157	23.5	64	9.6	16	2.4	18	2.7	12	1.8	13	1.9
Sand and gravel	83	28.2	38	12.9	1	0.3	16	5.4	5	1.7	7	2.4
Independent contractor:												
Coal	75	59.1	34	26.8	2	1.6	10	7.9	8	6.3	8	6.3
Metal and nonmetal§	112	71.3	28	17.8	5	3.2	32	20.4	6	3.8	13	8.3

See footnotes at end of table.

(Continued)

169

FOCUS ON MINING

Table 6–2 (Continued). Number* and annual average rate† of fatalities associated with various types of employers and commodities, by type of incident,‡ 1988–1997

Type of employer and commodity	Type of incident									
	Falling, rolling, or sliding rock or material		Ignition/ explosion of gas/dust		Explosives and breaking agents		Unknown or n.e.c.		Hand tools	
	Number	Rate	Number	Rate	Number	Rate	Number	Rate	Number	Rate
All	57	1.8	35	1.1	31	1.0	27	0.9	15	0.5
Mine operator:										
Coal	13	1.1	28	2.5	12	1.1	9	0.8	6	0.5
Metal	13	2.9	1	0.2	4	0.9	1	0.2	0	0.0
Nonmetal	0	0.0	0	0.0	0	0.0	1	0.4	1	0.4
Stone	16	2.4	2	0.3	7	1.0	3	0.4	3	0.4
Sand and gravel	4	1.4	1	0.3	0	0.0	6	2.0	1	0.3
Independent contractor:										
Coal	4	3.2	3	2.4	2	1.6	2	1.6	1	0.8
Metal and nonmetal§	7	4.5	0	0.0	6	3.8	5	3.2	3	1.9

(Continued)

See footnotes at end of table.

FOCUS ON MINING

Table 6–2 (Continued). Number* and annual average rate[†] of fatalities associated with various types of employers and commodities, by type of incident,[‡] 1988–1997

Type of employer and commodity	Exploding vessels under pressure		Handling materials		Type of incident					
					Fire		Hoisting		Inundation	
	Number	Rate	Number	Rate	Number	Rate	Number	Rate	Number	Rate
All	11	0.4	8	0.3	7	0.2	4	0.1	2	0.1
Mine operator:										
Coal	3	0.3	1	0.1	4	0.4	0	0.0	2	0.2
Metal	3	0.7	0	0.0	2	0.4	1	0.2	0	0.0
Nonmetal	1	0.4	0	0.0	0	0.0	0	0.0	0	0.0
Stone	1	0.1	1	0.1	1	0.1	0	0.0	0	0.0
Sand and gravel	1	0.3	3	1.0	0	0.0	0	0.0	0	0.0
Independent contractor:										
Coal	0	0.0	0	0.0	0	0.0	1	0.8	0	0.0
Metal and nonmetal[§]	2	1.3	3	1.9	0	0.0	2	1.3	0	0.0

Source: MSHA [1999].
*Note: See Appendix A, *Mining Injury and Employment Statistics*, for selection of fatalities.
[†]Computed per 100,000 full-time workers or 200 million employee hours.
[‡]MSHA's accident/injury/illness classification. See Appendix A, *Mining Injury and Employment Statistics*, for modifications.
[§]Includes metal, nonmetal, stone, and sand and gravel.

Focus on Mining

Lost-Workday Injuries

Lost-workday injury rates decreased between 1988 and 1997 in all five commodities for mine operator workers and independent contractor workers in metal and nonmetal mining (Figure 6–7). The lost-workday injury rates for independent contractor workers in coal follow a somewhat different trend, with a gradual increase from 1988 through 1992, then a slow decrease for the remainder of the period. Over the 10-year period, the highest lost-workday injury rate was observed in mine operator workers in coal (7.9 cases per 100 full-time workers), followed by mine operator workers in stone (4.7 cases per 100 full-time workers), sand and gravel (4.1 cases per 100 full-time workers), metal (3.9 cases per 100 full-time workers), and nonmetal mining (3.7 cases per 100 full-time workers). Among mine operator workers, the largest percentage of decrease in the lost-workday injury rate occurred in metal mining, dropping from a rate of 5.4 cases per 100 full-time workers in 1988 to 3.1 cases per 100 full-time workers in 1997 (a 43% decrease). Following metal mining, the decreases in lost-workday injury rates were as follows: coal (35%), nonmetal (31%), stone (30%), and sand and gravel mining (24%). A 46% decrease was observed for independent contractor workers in metal and nonmetal mining, and a 6% decrease was observed for independent contractor workers in coal mining.

The number and rate of lost-workday cases from 1988 to 1997 for various work locations are shown by type of employer and commodity in Table 6–3. Overall, underground work locations exhibited both the highest numbers and rates of lost-workday cases, whereas surface strip/open pit/quarry operations exhibited the lowest rates. Among mine operator workers, a combination of high numbers and rates of lost-workday cases are most conspicuous for workers in underground coal mines, stone surface mines (or quarries) and mills, and sand and gravel operations. Independent contractor workers accounted for less than 6% of all lost-workday cases but 19% of all fatalities (see Table 6–2).

The leading types of incidents associated with lost-workday cases (Table 6–4) are handling materials (34.4%), slip or fall of person (20.9%), powered haulage (10.9%), machinery (10.9%), and hand tools (9.5%). These incidents accounted for 86.6% of all cases between 1988 and 1997. Incidents involving handling materials accounted for more than one-third of all lost-workday cases, compared with fewer than 1% of the fatalities. Furthermore, these incidents account for 29.2% of the 6,840,987 lost workdays, followed by slip or fall of person (20.8%) and powered haulage (17.3%).

Focus on Mining

Sprains to the back region accounted for the largest proportion of lost workdays during the 10-year period (Figure 6–8). Sprains to the lower extremities (primarily the knee), amputations of the arms or hands (primarily the fingers), and fractures to the lower extremities also accounted for a substantial proportion of lost workdays.

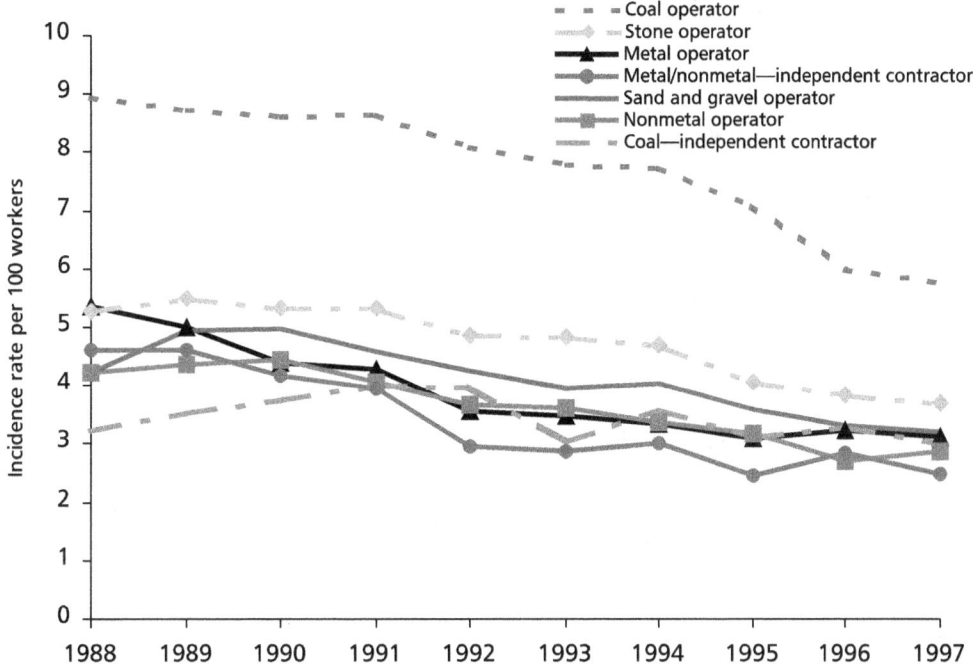

Figure 6–7. Incidence rates for lost-workday cases, by type of employer and commodity, 1988–1997. Incident rates are calculated per 100 full-time workers or 200,000 employee hours. Metal and nonmetal includes metal, nonmetal, stone, and sand and gravel. (Source: MSHA [1999].)

Focus on Mining

Table 6–3. Number and annual average rate* of lost-workday cases associated with various types of employers and commodities, by work location, 1988–1997

Type of employer and commodity	All		Work location									
			Underground mines				Surface mines					
			Underground		Surface areas		Strip/ open pit/ quarry		Dredge		Other surface operations[†]		Mills/plants	
	Number	Rate	Number	Rate	Number	Rate	Number	Rate	Number	Rate	Number	Rate	Number	Rate
All	170635	5.5	74264	10.9	6272	5.7	46396	3.6	2270	4.1	1250	4.7	40183	4.3
Mine operator:														
Coal	89895	7.9	65668	11.9	4348	8.0	12453	3.2	9	1.7	848	5.4	6569	5.0
Metal	17622	3.9	4534	7.2	641	3.9	5152	3.1	193	4.5	214	3.3	6888	3.5
Nonmetal	9855	3.7	1528	4.4	308	5.0	1623	2.7	31	5.6	NA[‡]	NA	6365	3.8
Stone	31642	4.7	748	4.7	270	6.9	13426	4.8	66	5.6	82	3.3	17050	4.7
Sand and gravel	12059	4.1	NA	NA	NA	NA	10106	4.1	1953	4.2	NA	NA	NR[§]	NR
Independent contractor:														
Coal	4363	3.4	1367	12.1	499	2.2	1385	2.2	2	1.0	96	4.9	1014	3.8
Metal and nonmetal**	5199	3.3	419	7.1	206	3.4	2251	2.4	16	1.5	10	2.3	2297	4.6

Source: MSHA [1999].
*Computed per 100 full-time workers or 200,000 employee hours.
[†]Includes culm banks, auger mining, independent shops and yards, and surface mining n.e.c.
[‡]NA=Not applicable for this commodity.
[§]NR=Not reported separately. Sand and gravel operators report mill employment under strip or dredge operations.
**Includes metal, nonmetal, stone, and sand and gravel.

Focus on Mining

Table 6–4. Number and annual average rate* of lost-workday cases and mean days lost associated with various types of employers and commodities, by type of incident,[†] 1988–1997

Type of employer and commodity	All			Handling materials			Type of incident — Slip or fall of person			Powered haulage			Machinery		
	Number	Rate	MDL[‡]	Number	Rate	MDL	Number	Rate	MDL	Number	Rate	MDL	Number	Rate	MDL
All	170,635	5.49	40.1	58,661	1.89	34.0	35,679	1.15	39.8	18,676	0.60	63.5	18,647	0.60	43.0
Mine operator:															
Coal	89,895	7.89	41.7	31,072	2.73	39.0	16,337	1.43	42.5	10,859	0.95	57.7	9,078	0.80	39.2
Metal	17,622	3.89	39.1	5,975	1.32	32.0	4,067	0.90	40.4	1,591	0.35	59.1	1,950	0.43	41.4
Nonmetal	9,855	3.67	35.7	3,910	1.46	27.7	2,294	0.85	33.3	912	0.34	77.9	939	0.35	37.9
Stone	31,642	4.74	33.6	11,108	1.66	26.3	7,588	1.14	33.1	2,825	0.42	63.4	3,752	0.56	36.1
Sand and gravel	12,059	4.10	39.9	3,760	1.28	23.7	3,077	1.05	34.6	1,512	0.51	90.7	1,334	0.45	60.6
Independent contractor:															
Coal	4,363	3.44	49.6	1,238	0.98	44.4	1,054	0.83	57.0	515	0.41	51.2	696	0.55	52.3
Metal and nonmetal[§]	5,199	3.31	55.8	1,598	1.02	30.3	1,262	0.80	54.6	462	0.29	113.2	898	0.57	85.2

See footnotes at end of table.

(Continued)

Focus on Mining

Table 6–4 (Continued). Number and annual average rate* of lost-workday cases and mean days lost associated with various types of employers and commodities, by type of incident,[†] 1988–1997

Type of employer and commodity	Hand tools			Fall of ground (from in place)			Type of incident — Stepping or kneeling on object			Unknown or n.e.c.			Electrical		
	Number	Rate	MDL[‡]	Number	Rate	MDL	Number	Rate	MDL	Number	Rate	MDL	Number	Rate	MDL
All	16,134	0.52	28.7	10,522	0.34	45.0	3,446	0.11	25.7	3,203	0.10	38.8	1,571	0.05	53.4
Mine operator:															
Coal	6,905	0.61	28.9	9,142	0.80	44.5	1,919	0.17	29.8	1,477	0.13	48.3	916	0.08	43.4
Metal	1,835	0.40	33.6	806	0.18	43.9	404	0.09	21.7	403	0.09	37.5	92	0.02	46.2
Nonmetal	931	0.35	26.4	166	0.06	45.5	178	0.07	20.7	234	0.09	49.7	74	0.03	23.2
Stone	4,039	0.60	26.5	123	0.02	79.9	589	0.09	19.7	664	0.10	25.8	251	0.04	46.4
Sand and gravel	1,554	0.53	26.4	6	0.00	41.5	193	0.07	18.5	256	0.09	22.6	135	0.05	64.2
Independent contractor:															
Coal	366	0.29	30.0	206	0.16	52.1	66	0.05	23.7	61	0.05	25.3	33	0.03	251.4
Metal and nonmetal[§]	504	0.32	35.2	73	0.05	45.4	97	0.06	21.1	108	0.07	15.8	70	0.04	136.1

(Continued)

See footnotes at end of table.

FOCUS ON MINING

Table 6–4 (Continued). Number and annual average rate* of lost-workday cases and mean days lost associated with various types of employers and commodities, by type of incident,[†] 1988–1997

Type of employer and commodity	Striking or bumping			Nonpowered haulage			Type of incident						Falling, rolling, or sliding rock or material		
							Exploding vessels under pressure			Fire					
	Number	Rate	MDL[‡]	Number	Rate	MDL	Number	Rate	MDL	Number	Rate	MDL	Number	Rate	MDL
All	1,504	0.05	32.8	722	0.02	44.0	398	0.01	43.6	388	0.01	33.1	363	0.01	86.6
Mine operator:															
Coal	1,130	0.10	35.3	308	0.03	57.5	181	0.02	27.0	144	0.01	27.0	154	0.01	80.1
Metal	91	0.02	28.7	126	0.03	38.1	53	0.01	36.0	39	0.01	29.5	46	0.01	32.7
Nonmetal	55	0.02	14.0	46	0.02	36.2	22	0.01	233.1	24	0.01	28.2	22	0.01	35.8
Stone	139	0.02	26.6	170	0.03	37.5	81	0.01	27.1	100	0.01	49.6	75	0.01	190.1
Sand and gravel	57	0.02	30.4	38	0.01	12.7	34	0.01	79.4	42	0.01	18.1	36	0.01	42.2
Independent contractor:															
Coal	18	0.01	27.4	12	0.01	26.7	14	0.01	10.3	24	0.02	28.2	9	0.01	33.9
Metal and nonmetal[§]	14	0.01	4.6	22	0.01	20.3	13	0.01	29.4	15	0.01	49.4	21	0.01	34.9

(Continued)

See footnotes at end of table.

Focus on Mining

Table 6–4 (Continued). Number and annual average rate* of lost-workday cases and mean days lost associated with various types of employers and commodities, by type of incident,[†] 1988–1997

Type of employer and commodity	Ignition/explosion of gas/dust			Explosives and breaking agents			Type of incident — Hoisting			Entrapment			Inundation		
	Number	Rate	MDL[‡]	Number	Rate	MDL	Number	Rate	MDL	Number	Rate	MDL	Number	Rate	MDL
All	307	0.01	45.3	219	0.01	139.1	160	0.01	99.5	23	0.00	46.9	12	0.00	114.7
Mine operator:															
Coal	131	0.01	45.3	101	0.01	25.6	33	0.00	64.2	4	0.00	115.0	4	0.00	72.0
Metal	34	0.01	22.4	48	0.01	349.6	55	0.01	54.6	3	0.00	32.0	4	0.00	228.3
Nonmetal	21	0.01	36.2	4	0.00	151.5	20	0.01	63.1	1	0.00	15.0	2	0.00	7.0
Stone	67	0.01	59.5	44	0.01	97.7	17	0.00	436.4	8	0.00	42.9	2	0.00	80.5
Sand and gravel	18	0.01	23.6	2	0.00	1.5	3	0.00	35.3	2	0.00	5.5	0	0.00	0.0
Independent contractor:															
Coal	22	0.02	53.5	12	0.01	102.9	16	0.01	55.7	1	0.00	58.0	0	0.00	0.0
Metal and nonmetal[§]	14	0.01	60.8	8	0.01	620.8	16	0.01	70.1	4	0.00	23.8	0	0.00	0.0

Source: MSHA [1999].
*Computed per 100 full-time workers or 200,000 employee hours.
[†]MSHA's accident/injury/illness classification. See Appendix A, *Mining Injury and Employment Statistics*, for modifications.
[‡]MDL = mean days lost. MDL is average number of days lost (including restricted workdays) per lost-workday case.
[§]Includes metal, nonmetal, stone, and sand and gravel.

Focus on Mining

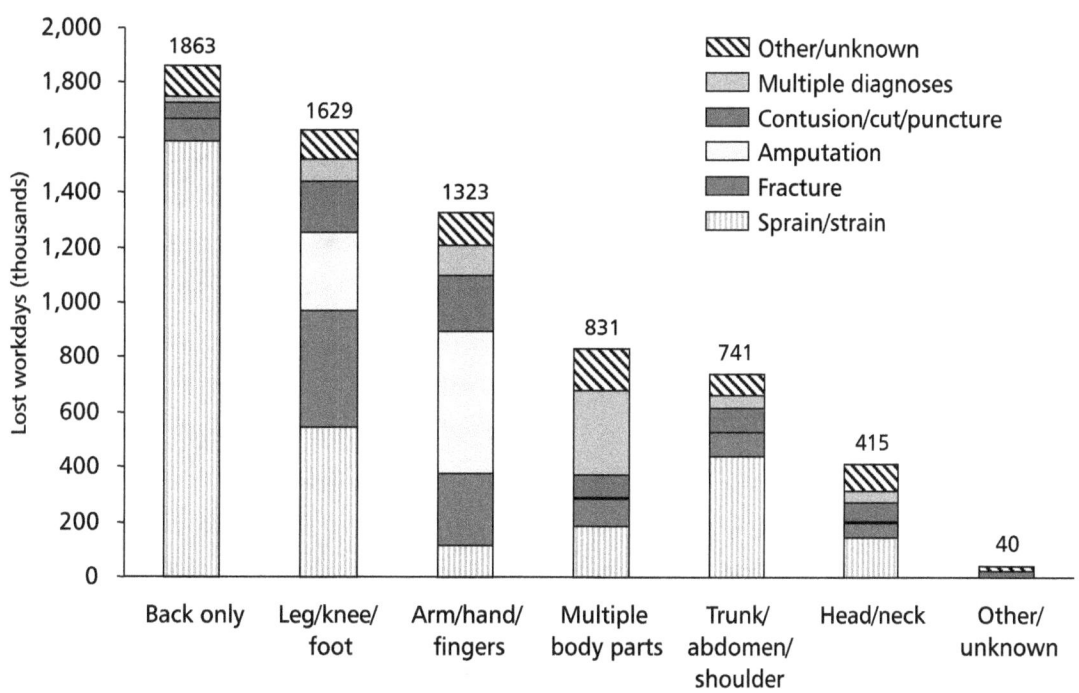

Figure 6–8. Lost workdays in mining, by part of body affected and nature of injury, 1988–1997. (Source: MSHA [1999].)

References

REFERENCES

ABLES [1999]. The Adult Blood-Lead Epidemiology and Surveillance Program (ABLES). Cincinnati, OH: U.S. Department of Health and Human Services, Public Health Service, Centers for Disease Control and Prevention, National Institute for Occupational Safety and Health. Database. [www.cdc.gov/niosh/ables.html].

Adams WW, Kolhos ME [1941]. Metal- and nonmetal-mine accidents in the United States during the calendar year 1939 (excluding coal mines). Washington, DC: U.S. Department of the Interior, Bureau of Mines, Bulletin 440.

Adams WW, Wrenn VE [1941]. Quarry accidents in the United States during the calendar year 1939. Washington, DC: U.S. Department of the Interior, Bureau of Mines, Bulletin 438.

BLS [1998]. Employment and wages annual averages, 1997. Washington, DC: U.S. Department of Labor, Bureau of Labor Statistics, Bulletin 2511, p. 535.

BLS [1999]. Current population survey. Washington, DC: U.S. Department of Labor, Bureau of Labor Statistics.

BLS [2000]. A special issue: charting the projections; 1998–2008. Occup Outlook Q 43(4):2–38. Winter 1999–2000.

Bureau of the Census [1992]. 1990 census of population and housing: alphabetical index of industries and occupations. Washington, DC: U.S. Government Printing Office, Publication CPH-R-3.

California Department of Health Services [1999]. Sentinel events notification system for occupational risks (SENSOR): occupational carpal tunnel syndrome—California year 2; second report, #60/CCU902990-13. Cincinnati, OH: U.S. Department of Health and Human Services, Public Health Service, Centers for Disease Control and Prevention, National Institute for Occupational Safety and Health.

CDC (Centers for Disease Control and Prevention) [1999]. HIV/AIDS Surveill Rep 11(1):26. [www.cdc.gov/hiv/stats/hasr1101.pdf].

References

CDPR [1999]. Pesticide illness surveillance program. Sacramento, CA: California Environmental Protection Agency, California Department of Pesticide Regulation, Worker Health and Safety Branch. Database.

CFOI [1999]. Census of fatal occupational injuries, 1992–1997. Washington, DC: U.S. Department of Labor, Bureau of Labor Statistics. Database. [www.bls.gov/iif/oshcfoi1.htm].

CWXSP [1999]. Coal workers' X-ray surveillance program 1970–1995. Morgantown, WV: U.S. Department of Health and Human Services, Public Health Service, Centers for Disease Control and Prevention, National Institute for Occupational Safety and Health. Database. [aspe.os.dhhs.gov/DATACNCL/datadir/cdc5.htm#cwxsp].

Fullerton HN Jr. [1999]. Labor force projections to 2008: steady growth and changing composition. Mon Labor Rev *122*(11):19–32.

ILO [1980]. Guidelines for the use of ILO international classification of radiographs of pneumoconioses. Rev. ed. Geneva, Switzerland: International Labour Office, Occupational Safety and Health Series No. 22 (Rev.).

MSHA [1984]. Summary of selected injury experience and worktime for the mining industry in the United States, 1931–77. Denver, CO: U.S. Department of Labor, Mine Safety and Health Administration, IR 1132.

MSHA [1999]. Quarterly employment and coal production: accidents/injuries/illnesses reported to MSHA under 30 CFR Part 50, 1986–1997. Denver, CO: U.S. Department of Labor, Mine Safety and Health Administration, Office of Injury and Employment Information.

NaSH [1999]. National Surveillance System for Hospital Health Care Workers, 1995–1999. Atlanta, GA: U.S. Department of Health and Human Services, Public Health Service, Centers for Disease Control and Prevention, National Center for Infectious Diseases. Database. [www.cdc.gov/ncidod/hip/SURVEILL/nash.htm].

NCHS [1999]. Mortality data, multiple-cause-of-death public-use data files, 1968–1996. Hyattsville, MD: U.S. Department of Health and Human Services, Public Health Service, Centers for Disease Control and Prevention, National Center for Health Statistics. Database. [www.cdc.gov/nchs/products/elec_prods/subject/mortmcd.htm].

References

NCID [1999]. Sentinel counties study of acute viral hepatitis, 1985–1995. Atlanta, GA: U.S. Department of Health and Human Services, Public Health Service, Centers for Disease Control and Prevention, National Center for Infectious Diseases. Database.

NEISS [1999]. National Electronic Injury Surveillance System. (Data collected by the Consumer Product Safety Commission; work-related case records maintained by the National Institute for Occupational Safety and Health.) Washington, DC: Consumer Product Safety Commission, Division of Hazard and Injury Data Systems; and Morgantown, WV: U.S. Department of Health and Human Services, Public Health Service, Centers for Disease Control and Prevention, National Institute for Occupational Safety and Health. Database.

New York State Department of Health [1999]. New York State pesticide poisoning registry, 1992–1996. Albany, NY: New York State Department of Health, Bureau of Occupational Health.

NHAMCS [1999]. National hospital ambulatory medical care survey, 1995–1997. Hyattsville, MD: U.S. Department of Health and Human Services, Public Health Service, Centers for Disease Control and Prevention, National Center for Health Statistics. Database. [www.cdc.gov/nchs/products/elec_prods/subject/nhamcs.htm].

NHANES III [1999]. Third National Health and Nutrition Examination Survey. Hyattsville, MD: U.S. Department of Health and Human Services, Public Health Service, Centers for Disease Control and Prevention, National Center for Health Statistics. Database. [www.cdc.gov/nchs/about/major/nhanes/datalink.htm].

NIOSH [1997]. Mortality by occupation, industry, and cause of death: 24 reporting States (1984–1988). Cincinnati, OH: U.S. Department of Health and Human Services, Public Health Service, Centers for Disease Control and Prevention, National Institute for Occupational Safety and Health, DHHS (NIOSH) Publication No. 97-114. [www.cdc.gov/niosh/97-114.html].

NIOSH [1999]. Work-related lung disease surveillance report 1999 (WoRLD). Cincinnati, OH: U.S. Department of Health and Human Services, Public Health Service, Centers for Disease Control and Prevention, National Institute for Occupational Safety and Health, DHHS (NIOSH) Publication No. 2000-105. [www.cdc.gov/niosh/docs/2000-105/pdfs/2000-105.pdf].

References

NOMS [1999]. National Occupational Mortality Surveillance System. Cincinnati, OH: U.S. Department of Health and Human Services, Public Health Service, Centers for Disease Control and Prevention, National Institute for Occupational Safety and Health. Database. [aspe.os.dhhs.gov/datacncl/datadir/cdc5.htm#noms].

NSSPM [1999]. National Surveillance System for Pneumoconiosis Mortality, 1968–1996. Morgantown, WV: U.S. Department of Health and Human Services, Public Health Service, Centers for Disease Control and Prevention, National Institute for Occupational Safety and Health. Database.

NTOF [1999]. National Traumatic Occupational Fatalities Surveillance System. Morgantown, WV: U.S. Department of Health and Human Services, Public Health Service, Centers for Disease Control and Prevention, National Institute for Occupational Safety and Health. Database.

Oregon Health Division [1999]. Pesticide Poisoning Prevention Program, 1992–1996. Portland, OR: Oregon Health Division; Center for Disease Prevention and Epidemiology; Environmental and Occupational Epidemiology Section. Database.

PEST [1999]. Pesticide Exposure Surveillance in Texas Program, 1992–1996. Austin, TX: Texas Department of Health, Bureau of Epidemiology, Division of Environmental Epidemiology and Toxicology. Database.

Ramazzini B [1713]. Diseases of workers. *De Morbis Artificum.* (translation by WC Wright, 1964). New York: Hafner Publishing Co.

Reese ST, Wrenn VE, Reid EJ [1955]. Injury experience in coal mining, 1952: analysis of mine safety factors, related employment, and production data. Washington, DC: U.S. Department of the Interior, Bureau of Mines, Bulletin 559.

Rosenman KD, Reilly MJ [1999]. Unpublished data, 1992–1998. East Lansing, MI: Michigan State University, Department of Medicine.

Rosenman KD, Reilly MJ, Deliefde B, Kalinowski DJ [1999]. 1998 annual report on occupational noise-induced hearing loss in Michigan. East Lansing, MI: Michigan State University, Department of Medicine; and Michigan Department of Consumer and Industry Services, Occupational Health Division.

SOII [1999]. Survey of occupational injuries and illnesses. Washington, DC: U.S. Department of Labor, Bureau of Labor Statistics. Database. [www.bls.gov/iif/].

staffTRAK–TB [1999]. Surveillance for tuberculosis infection in health care workers, 1994–1998. Atlanta, GA: U.S. Department of Health and Human Services, Public Health Service, Centers for Disease Control and Prevention, National Center for HIV, STD, and TB Prevention. Database.

TESS [1998]. AAPCC Toxic Exposure Surveillance System. Washington, DC: American Association of Poison Control Centers. Database. [www.aapcc.org/poison1.htm].

VHSP [1999]. Viral Hepatitis Surveillance Program. Atlanta, GA: U.S. Department of Health and Human Services, Public Health Service, Centers for Disease Control and Prevention, National Center for Infectious Diseases. Database. [www.cdc.gov/ncidod/osr/site/surv_resources/surv_sys.htm].

Washburn AE, LeBlanc PR, Fahy RF [1999]. Firefighter fatalities: National Fire Protection Association. NFPA Journal, July/August, pp. 56–70.

WHO [1977]. Manual of the international statistical classification of diseases, injuries, and causes of death, based on the recommendations of the Ninth Revision Conference, 1975. Geneva, Switzerland: World Health Organization.

Appendix A

Appendix A
Surveillance System Descriptions

Overview

The data described in this document represent compilations from several surveillance systems administered by various government agencies. In general, numerators and denominators are determined differently from one system to the next. This appendix provides a description of each of the surveillance systems used as data sources for much of the information contained in this chartbook. The appendix also provides contact information for acquiring additional details about the systems or using those systems for surveillance or research. Table A–1 describes selected surveillance systems.

Bureau of Labor Statistics (BLS) of the U.S. Department of Labor

BLS, under the U.S. Department of Labor, was established in 1884 and is charged with collecting annual data on occupational safety and health. BLS publishes either repeating annual bulletins or one-time bulletins addressing specific topics. Each BLS office is responsible for a program that gathers information about the American worker. The Office of Employment and Unemployment Statistics administers the Current Population Survey (CPS), and the Office of Safety, Health, and Working Conditions administers the Survey of Occupational Injuries and Illnesses (SOII) and the Census of Fatal Occupational Injuries (CFOI).

Current Population Survey (CPS)

BLS compiles statistics on the employment status and related data using CPS. This survey is published in a series of reports including, of most importance, the annual *Employment and Earnings* report. This survey provides denominators for many of the injury and illness incidence rates presented in this chartbook.

Appendix A

Table A–1. Descriptions of selected surveillance systems

Name of system	Agency	Employment and occupation coding (exposure surrogates)	Sample derivation/ population definition	Primary interest of data system	Fatal or nonfatal outcome	Injury or illness
ABLES*	NIOSH	None presently; NAICS and Bureau of the Census occupation codes planned	Adults (16 yr) with blood lead levels 25 µg/dL from 28 States	Facilitate State lead poisoning intervention; track trend and magnitude of adult lead exposures	Nonfatal	Illness
CFOI	BLS	BLS Occupational Injury and Illness Classification System, industry and occupations	All injury deaths identified through death certificates, workers' compensation reports, medical examiner reports, OSHA reports, and news media reports, verified to exclude duplicate counting	Counts of fatalities by various characteristics: worker, employer, and incident	Fatal	Injury
CWXSP	NIOSH	None	Working coal miners employed in underground coal mines	Degree of radiographic opacity	Nonfatal	Illness
HARS	All States report to CDC	Surveillance Branch, Division of HIV/AIDS Prevention (requests occupation and industry information for the health care setting only)	All cases nationwide	Monitoring the HIV epidemic	Nonfatal	Illness
MSHA Mine/ Contractor Address/ Employment and Accident/Injury/ Illness Database	MSHA	MSHA classifications for industry (mineral commodity) and occupation	Population of miners whose employment/injuries/ illnesses are required to be reported under 30 CFR Part 50	Counts and rates of fatalities, injuries, and illnesses occurring on mine property	Fatal and nonfatal	Injury and illness

See footnote at end of table.

(Continued)

APPENDIX A

Table A–1 (Continued). Descriptions of selected surveillance systems

Name of system	Agency	Employment and occupation coding (exposure surrogates)	Sample derivation/ population definition	Primary interest of data system	Fatal or nonfatal outcome	Injury or illness
Multiple-Cause-of-Death Data	NCHS	Coding available for selected States. Coding according to Bureau of the Census occupation and industry codes	Death records include codes for up to 20 conditions cited on death certificates, information from all States in the United States	Cause of death (underlying and contributory), occupation and industry codes where available	Fatal and nonfatal	Injury and illness
NaSH	NCID	NCID-generated coding	Convenience sample and sample drawn from NNIS	Incidence and trends of occupationally acquired infections in health care settings	Nonfatal	Illness
NEISS	CPSC for NIOSH (and others)	Bureau of the Census classification	Hospitals (65 of 91 stratified nationally on size) collect occupational identifiers	Injury and illness cases identified as work-related in emergency departments of participating hospitals	Nonfatal	Injury and illness
NHAMCS	NCHS	None	Representative sample of hospital emergency room visits	Type of injury or incident, body part, cause of incident; demographics of individual; work-relatedness by treating professional	Nonfatal	Injury
NHANES	NCHS	Bureau of the Census occupation and industry regrouped by NCHS	Cross-sectional household survey interviews, representative sample of U.S. civilian population	Characterizes health and nutritional status of U.S. civilian noninstitutionalized population	Not	Illness

(Continued)

See footnote at end of table.

Appendix A

Table A–1 (Continued). Descriptions of selected surveillance systems

Name of system	Agency	Employment and occupation coding (exposure surrogates)	Sample derivation/ population definition	Primary interest of data system	Fatal or nonfatal outcome	Injury or illness
NOMS	NIOSH using NCHS data	Bureau of the Census classification	Death certificates from NCHS	Cause of death, occupation and industry where available	Fatal	Illness
NSSPM	NIOSH using NCHS data	Bureau of the Census classification	Death certificates from NCHS	Cause of death (underlying and contributory), occupation and industry where available	Fatal	Illness
NTOF	NIOSH	Industry and occupation coding using 1980 and 1990 Bureau of the Census data	Death certificates from 52 U.S. vital statistics reporting units in 50 States for workers aged 16 or older	Represents the minimum number of work-related deaths in the United States for a given period	Fatal	Injury
Sentinel Counties Study of Acute Hepatitis	NCID	None	6 counties in the United States and convenience sample	Provides source data of viral hepatitis infection in the United States	Nonfatal	Illness
SENSOR	NIOSH	Varies by participating State and SENSOR condition	Case-based reporting from a variety of sources including physicians, agencies, workers' compensation, etc. Catchment area varies from geographic area (counties) to entire State	Case-based surveillance directly linked to intervention activities to maximize prevention	Fatal and nonfatal	Injury and illness

(Continued)

See footnote at end of table.

APPENDIX A

Table A–1 (Continued). Descriptions of selected surveillance systems

Name of system	Agency	Employment and occupation coding (exposure surrogates)	Sample derivation/ population definition	Primary interest of data system	Fatal or nonfatal outcome	Injury or illness
SOII	BLS	Industry coding using the SIC system, occupation coding using 1990 Bureau of the Census system	Stratified random sample of all private industry employers of one or more workers	The number of work-related injuries and illnesses reported by employers on the OSHA 200 Log	Nonfatal	Injury and illness
staffTRAK–TB	National Center for HIV, STD, and TB Prevention	1990 Bureau of the Census for population and housing and 1992 Bureau of the Census industry and occupation coding; also includes CDC NNIS coding for occupations	Demonstration project—participating health departments	Tuberculin skin testing targeting health departments	Nonfatal	Illness
VHSP	NCID	NCID-generated coding	All acute cases reported to local health departments	Identifies risk factors for infection	Nonfatal	Illness

*Abbreviations: ABLES = Adult Blood Lead Epidemiology and Surveillance Program; AIDS = acquired immune deficiency syndrome; BLS = Bureau of Labor Statistics; CDC = Centers for Disease Control and Prevention; CFOI = Census of Fatal Occupational Injuries; CPSC = Consumer Product Safety Commission; CWXSP = Coal Workers' X-Ray Surveillance Program; HARS = HIV/AIDS Reporting System; HIV = human immunodeficiency virus; MSHA = Mine Safety and Health Administration; NAICS = North American Industrial Classification System; NaSH = National Surveillance System for Hospital Health Care Workers; NCHS = National Center for Health Statistics; NCID = National Center for Infectious Diseases; NEISS = National Electronic Injury Surveillance System; NHAMCS = National Hospital Ambulatory Medical Care Survey; NHANES = National Health and Nutrition Examination Survey; NIOSH = National Institute for Occupational Safety and Health; NNIS = National Nosocomial Infection Surveillance System; NOMS =National Occupational Mortality Surveillance System; NSSPM = National Surveillance System for Pneumoconiosis Mortality; NTOF = National Traumatic Occupational Fatalities Surveillance System; OSHA = Occupational Safety and Health Administration; SENSOR = Sentinel Event Notification System for Occupational Risk; SIC = standard industrial classification; SOII = Survey of Occupational Injuries and Illnesses; staffTRAK–TB = Surveillance for Tuberculosis Infection in Health Care Workers; STD = sexually transmitted disease; TB = tuberculosis; VHSP = Viral Hepatitis Surveillance Program.

Appendix A

CPS is a monthly survey of households conducted for BLS by the Bureau of the Census through a scientifically selected sample that represents the civilian noninstitutional population. Respondents are interviewed to obtain information about the employment status of each member of the household aged 15 years and older, although data are routinely published on those aged 16 and older. The inquiry relates to activity or status during the calendar week (Sunday through Saturday) that includes the 12th day of the month. This is known as the "reference week." Actual field interviewing is conducted in the following week, referred to as the "survey week." The concepts and definitions underlying labor force data have been modified but not substantially altered since the inception of the survey in 1940.

Each month, about 50,000 occupied units are eligible for interview. The sample provides estimates for the Nation as a whole and provides data for model-based estimates for individual States and other geographic areas. Some 3,200 of these households are contacted but not interviewed because the occupants are not at home after repeated calls or are unavailable for other reasons. This figure represents a noninterview rate for the survey that ranges between 6% and 7%. In addition to the 50,000 occupied units, 9,000 sample units in an average month are visited but found to be vacant or otherwise not eligible for enumeration.

Part of the sample is changed each month. Three-fourths of the sample is common from one month to the next, and one-half is common with the same month a year earlier. Since 1953 (when the current 4–8–4 rotation system was adopted), households are interviewed for 4 consecutive months, leave the sample for 8 months, and then return to the sample for the same 4 months of the following year. Estimates obtained from CPS include employment, unemployment, earnings, hours worked, and other indicators. They are available for various demographic characteristics including age, sex, race, marital status, and education. They are also available by occupation, industry, and class of worker. Supplemental questions are also often added to the regular CPS questionnaire to produce estimates on other topics including school enrollment, income, previous work experience, health, employee benefits, and work schedules.

This information for the employed applies to the job held in the reference week. Persons with two or more jobs are classified in the job at which they worked the greatest number of hours. The unemployed are classified according to their last job. The occupational and industrial classification of CPS data is based on the coding systems used in the 1990 census until recently. Over the next years, BLS will begin using the newly developed North American Industry Classification System (NAICS) (www.census.gov/epcd/www/naics.html).

Appendix A

For further information contact
Office of Employment and Unemployment Statistics
Bureau of Labor Statistics
2 Massachusetts Avenue, NE
Washington, DC 20212
Telephone: 202–691–6400
(www.bls.gov)

Survey of Occupational Injuries and Illnesses (SOII)

The annual SOII is a surveillance system in which employer reports are collected by BLS from private industry establishments. About 165,000 establishments were included in the 1997 survey. A two-part survey is conducted and provides estimates for the United States and separately for participating States. Part 1, which has been collected since 1972, provides estimates of the number and incidence of injuries and illnesses by Standard Industrial Classification (SIC). Part 2, which was added to the survey in 1992, provides estimates of demographic characteristics of workers with injuries and illnesses involving time away from work. Part 2 also provides data on the circumstances of the injuries and illnesses with time away from work.

The survey sample is selected using stratified random sampling from all private industry employers of one or more workers. MSHA and the Department of Transportation's Federal Railroad Administration provide comparable occupational injury and illness data for coal, metal mining, nonmetal mining, and for railroad activities. The survey also gathers information on the average number of workers employed and the total hours worked at each establishment during the year. The survey collects the number of work-related injuries and illnesses that the employer has recorded on the Annual Log and Summary of Occupational Injuries and Illnesses (OSHA No. 200) kept by each establishment [29 CFR* 1904; OSHA 2000a]. Injury is a single reporting category. Illnesses are divided into seven broad categories: skin diseases or disorders, dust diseases of the lungs, respiratory conditions due to toxic agents, poisoning, disorders due to physical agents, disorders associated with repeated trauma, and all other occupational illnesses. These data are called the industry summary data (Part 1 of the survey). Since 1972, this information has been used to identify industries with high rates of injuries and illnesses.

Code of Federal Regulations. See CFR in references.

Appendix A

In 1992, a second part was added that collects descriptive information about a sample of the cases that resulted in at least 1 day away from work. Establishments take this information from workers' compensation reports, insurance forms, or other supplementary records (for example, the OSHA Form 101—Supplementary Record of Occupational Injury and Illness [29 CFR 1904; OSHA 2000b]). These data are called the case and demographic data. The descriptive information includes the personal characteristics of the injured or ill worker: industry, occupation, race/ethnicity, sex, age, and length of service with the employer. The injury or illness is characterized with information on the nature of the injury or illness; the part of body affected; the event or exposure leading to the injury or illness; and the source (the object, substance, bodily motion) that directly produced the injury or illness. The number of days away from work is collected as a surrogate for the severity of the case.

Industry is coded using the 1987 SIC Manual [OMB 1987], and occupation is coded using the 1990 Bureau of the Census classification system. For manufacturing industries, information is available at the 4-digit SIC level. For all other industries, the most detailed level is 3-digit. The nature of the injury or illness, the part of body affected, the source of injury or illness, and the event are coded according to the Occupational Injury and Illness Classification System developed by BLS [1992]. This system provides coding at four levels, from 1-digit to 4-digit, although not all categories can be expanded to the 4-digit level. NAICS will be used to code industry beginning with the 2003 survey (see Appendix B).

The complex statistical design of the annual survey required BLS to design special computer software to calculate estimates and variances. Data are not released when estimates do not meet publication guidelines as determined by BLS. The self-employed; farms with fewer than 11 employees; private households; and Federal, State, and local government agencies are excluded from the survey.

From 1972 through 1997, BLS disseminated this summary data either in a BLS publication or on the Internet. Information is available about the number and rates of all injuries and the seven illness categories by industry. The categories of all injuries and all illnesses are divided into cases without lost workdays and lost-workday cases. Lost-workday cases are further divided into cases with restricted work activity only and cases with days away from work. Selected information is available by number of employees. Information about workers with injuries and illnesses requiring recuperation away from work and the characteristics of their injuries and illnesses is available for 1992 through 1997.

Appendix A

For further information contact

Office of Safety, Health, and Working Conditions
Bureau of Labor Statistics
2 Massachusetts Avenue, NE
Washington, DC 20212
Telephone: 202–691–6170
(www.bls.gov/iif/)

Census of Fatal Occupational Injuries (CFOI)

CFOI is a Federal and State cooperative program that each year accesses multiple data sources to compile a complete roster of occupational fatal injuries. Since 1992, the fatality census has been conducted in all 50 States and the District of Columbia. CFOI includes data for all fatal work injuries—those that are covered by the Occupational Safety and Health Administration (OSHA) or other Federal or State agencies and those that are outside the scope of regulatory coverage.

A fatality is included in the census if the decedent was working for pay, compensation, or profit at the time of the event; engaged in a legal work activity; or present at the site of the incident as a requirement of his or her job. These criteria are generally broader than those used by Federal and State agencies. Thus any comparison between BLS census counts and those released by other agencies should take into account the different coverage requirements and definitions. Fatalities that occur during a person's commute to or from work are excluded from the BLS census.

An injury is defined as any intentional or unintentional wound or damage to the body resulting from (1) acute exposure to energy such as heat or electricity or kinetic energy from a crash, or (2) the absence of such essentials as heat or oxygen caused by an event, incident, or series of events within a single workday or shift. Included are open wounds, intracranial and internal injuries, heatstroke, hypothermia, asphyxiation, acute poisonings resulting from a short-term exposure limited to the worker's shift, suicides and homicides, and work injuries listed as underlying or contributory causes of death. Occupational fatal illnesses are not reported in the BLS census.

Data for CFOI are compiled from various Federal, State, and local administrative sources (including death certificates, workers' compensation reports and claims, reports to various regulatory agencies, medical examiner reports, and police reports) as well as news reports. Multiple sources are accessed because studies have shown that no single source captures all

APPENDIX A

occupational fatalities. Source documents are matched so that each fatality is counted only once. To ensure that a fatality occurred while the decedent was at work, information is verified from two or more independent source documents, or from a source document and a followup questionnaire.

Approximately 30 data elements are collected, coded, and tabulated, including information about the worker, the fatal incident, and the machinery or equipment involved. Industry and occupation describe the job the worker held at the time of the fatal incident. Industry is classified according to the 1987 SIC system. Occupation is coded according to the Bureau of the Census occupational classification system. Industry data are typically reported separately for the public and private sectors. The BLS Occupational Injury and Illness Classification System is used to classify the nature of the injury, part of body affected, primary and secondary sources of injury, and the exposure leading to the fatality. Other data elements include the worker's age, sex, and race; the time of day that the fatal event occurred; the activity the worker was performing when injured; and the location where the event occurred.

States may identify additional fatal work injuries after data collection closeout for a reference year. Other fatalities excluded from the published count because of insufficient information to determine work relationship may be subsequently verified as work related. States have 1 year to update their initial published counts. This procedure ensures that fatality data are disseminated as quickly as possible, and that no legitimate case is excluded from the counts.

Approximately 8 months after the end of the reference year, BLS publishes summary data in a national news release. Articles and detailed tables containing national and State data are published regularly in the BLS quarterly publication *Compensation and Working Conditions* and occasionally in the *Monthly Labor Review*. Other products of the CFOI program include profiles of occupations, industries, or types of events; a yearly compendium (*Fatal Workplace Injuries in [year]: A Collection of Data and Analysis*); and a data file for researchers that is available through a letter of agreement to protect confidentiality of workers and companies. Most of the published reports are available on the BLS Web site (www.bls.gov/iif/oshcfoi1.htm). States also produce news releases and reports on State-specific hazards. A list of participating agencies and their phone numbers is available in the 1998 news release on the BLS CFOI Internet site.

Appendix A

For further information contact
Office of Safety, Health, and Working Conditions
Bureau of Labor Statistics
2 Massachusetts Avenue, NE
Washington, DC 20212
Telephone: 202–691–6175
(www.bls.gov/iif/oshcfoi1.htm)

Centers For Disease Control and Prevention (CDC), U.S. Department of Health and Human Services

CDC is one of eight public health agencies in the U.S. Department of Health and Human Services with the mission of promoting health and quality of life by preventing and controlling disease, injury, and disability. CDC added prevention to its mission after an increasing percentage of its resources were spent on preventing diseases rather than controlling existing ones (www.cdc.gov/aboutcdc.htm#mission). CDC consists of 12 centers, institutes, and offices in 10 locations and additional employees in State health departments, quarantine offices, and other countries.

National Institute for Occupational Safety and Health (NIOSH)

NIOSH, an institute within CDC, was established by the Occupational Safety and Health Act of 1970. NIOSH maintains a range of surveillance systems and produces a wide variety of reports. These range from the approximately biennial work-related lung disease (WoRLD) reports to one-time reports covering a group of diseases or conditions over several years, such as *Fatal Injuries to Workers in the United States, 1980–1989: a Decade of Surveillance* [NIOSH 1993], *Mortality by Occupation, Industry, and Cause of Death: 1984–1988* [NIOSH 1997], and the *Atlas of Respiratory Disease Mortality, United States: 1982–1993* [Kim 1998]. Reports describing point prevalence of a range of diseases or conditions in certain industries (such as health hazard evaluation surveys) or project reports are generally accessible through the NIOSH Web site (www.cdc.gov/niosh/pubs.html).

Appendix A

National Electronic Injury Surveillance System (NEISS)

The Consumer Product Safety Commission (CPSC) developed NEISS to monitor injuries involving consumer products and to serve as a source for followup investigation of selected product-related injuries [McDonald 1994]. Data are collected at 101 hospitals selected from a stratified probability sample of all hospitals in the United States and its territories. The sampling frame was stratified by hospital size (determined by the annual total of emergency department visits) and geographic region, and the final sample of 101 hospitals was then selected. NIOSH entered into an interagency agreement to collect work-related injury data in 67 of the 101 hospitals. Each injury case in the sample was assigned a statistical weight based on the inverse of the hospital's probability of selection, and this weight was used to calculate national estimates. Confidence intervals (CIs) were calculated using methods described in detail elsewhere [Layne and Landen 1997].

A work-related case was defined as any injury sustained during (1) work for compensation, (2) volunteer work for an organized group, or (3) a work task on a farm. The *Operational Guidelines for Determination of Injury at Work* were provided to hospital coders to identify work-related injuries [NIOSH 1993]. Unlike CPSC consumer product data, the work-related data collected for NIOSH included all cases regardless of whether a consumer product was involved in the injury event.

Estimates of numbers of workers, used to calculate injury rates, were derived from the monthly Current Population Survey (CPS) of BLS, a national, population-based household survey that includes approximately 50,000 households each month [BLS 1997]. For this report, injury and illness rates were estimated as the number of cases per 100 full-time workers (2,000 working hr/full-time worker). Workers less than 16 years of age were excluded from this analysis.

For further information contact
Surveillance and Field Investigations Branch
Division of Safety Research
National Institute for Occupational Safety and Health
1095 Willowdale Road, MS 1812
Morgantown, WV 26505
Telephone: 304–285–5980

APPENDIX A

National Occupational Mortality Surveillance System (NOMS)

NOMS was developed to provide information about work-related deaths by industry and occupation using the United States Standard Certificate of Death. Information about the occupation of the decedent has been recorded since 1900, but this information has not been readily accessible until recently.

Over the last decade, NIOSH, the National Center for Health Statistics (NCHS), the National Cancer Institute (NCI), and Bureau of the Census have collaborated to (1) improve the quality of the occupational data collected on death certificates, (2) develop routine standardized coding of this information by State health departments, and (3) partially reimburse selected States for producing these data. The first report using these data was a Monthly Vital Statistics Report Supplement, based on the 1984 data from 12 States [Rosenberg et al. 1993]. The data included cause-specific estimates of relative risk for broad occupation and industry categories for both male and female workers.

The United States Standard Certificate of Death requests information about the usual occupation and kind of business or industry for each decedent. Beginning in 1983, an increasing number of State health departments have coded this information using standardized coding procedures. Twenty-four State health departments included the data in the coded death certificate information provided to NCHS for 1 or more years from 1984 through 1988.

The information about occupation and industry was coded according to the 1980 Bureau of the Census classification [Bureau of the Census 1982]. The underlying cause of death was coded according to the International Classification of Diseases, Ninth Revision (ICD–9) [WHO 1977]. Data included in this chartbook are based on an analysis of 185 selected causes of death for male workers and 188 selected causes for female workers. The analysis includes deaths that occurred in the 24-State reporting area among residents of one of the 24 States. The criteria for inclusion in the analysis differed for male and female workers. For male workers, all white and black decedents aged 20 and older were included. For female workers, decedents reported in the occupation category of "Housewives, homemakers" were not included in either the occupation or industry analysis. Therefore, all white and black female decedents aged 20 and older with an occupation code other than "Housewives, homemakers" were included.

Appendix A

Age-standardized proportionate mortality ratios (PMRs) for the four race-sex groups reported in this chartbook were calculated using a computer program developed at NIOSH [Dubrow et al. 1993]. For each race-sex group, the program calculates PMRs by comparing the proportion of deaths from a specific cause within an occupation or industry group with the proportion of deaths from that cause for all occupations or industries. Age stratification was done by 5-year age groups. The program provides 95% confidence limits for the PMRs. The limitations of PMR studies include potential inaccuracy of cause of death and imprecise exposure classification based on usual occupation [Breslow and Day 1987]. In addition, information is lacking about length of employment and possible confounders such as smoking, alcohol, or socioeconomic status. Although PMR studies are useful for hypothesis generation [Checkoway et al. 1989], lack of population data precludes obtaining death rates [Rothman 1986]. The PMR indicates only whether the age-standardized proportion of deaths from a specific cause appears to be higher or lower than the expected proportion for an occupation or industry. Also, the PMR for one cause of death may be relatively high if proportions of other causes of death are relatively low [DeCouflé et al. 1980]. Thus the PMR for each cause of death depends on the PMRs computed for the other causes in an occupation or industry analysis. This can be especially important if the occupation under study has relatively high or low mortality due to some common cause of death. If the PMR is low, the PMR for other causes may be artificially inflated. The PMR will be a poor estimate of the risk of death if the population-based standardized mortality ratio (SMR) for all causes for an occupation or industry group is greatly above or below 100.

Prior publications from this system include *Mortality by Occupation, Industry, and Cause of Death, 24 Reporting States (1984–1988)* [NIOSH 1997], which includes a data diskette for those years. More recent versions are being mounted on a Web-accessible version at the NCI.

For further information contact
Surveillance Branch
Division of Surveillance, Hazard Evaluations, and Field Studies
National Institute for Occupational Safety and Health
4676 Columbia Parkway, MS R18
Cincinnati, OH 45226
Telephone: 513–841–4219

APPENDIX A

National Surveillance System for Pneumoconiosis Mortality (NSSPM)

NSSPM is a pneumoconiosis mortality surveillance system developed and maintained by the Division of Respiratory Disease Studies, NIOSH. The system provides statistics for the surveillance of occupational respiratory diseases in an easily accessible, user-friendly format. The data are a subset of national mortality data obtained annually from NCHS since 1968. Currently, NSSPM contains death certificate information for 1968–1996 for all U.S. decedents aged 15 and older identified with any type of pneumoconiosis listed as either an underlying or contributing cause of death. Additional information includes age, race, sex, and State and county of residence at the time of death. Usual occupation and industry of each decedent have been available for several States since 1985.

Types of pneumoconioses included in the NSSPM are based on International Classification of Diseases coding categories (ICD–8 [WHO 1967] from 1968–1979, and ICD–9 [WHO 1977] from 1979–1996): asbestosis, coal workers' pneumoconiosis (CWP), silicosis, byssinosis, other/unspecified pneumoconioses, and all pneumoconioses aggregated.

NSSPM is designed to generate a variety of summary statistics, tables, charts, and maps. Examples of the types of statistics this system generates are counts of deaths, crude and age-adjusted rates, and years of potential life lost by year, age group, race, sex, and usual occupation or industry at the national, State, and county levels. Data from additional sources, such as population statistics, comparative standard population, and life table values are incorporated into the system.

For further information contact
Public Health Surveillance Team
Surveillance Branch
Division of Respiratory Disease Studies
National Institute for Occupational Safety and Health
1095 Willowdale Road
Morgantown, WV 26505–2888
Telephone: 304–285–6115

Coal Workers' X-Ray Surveillance Program (CWXSP)

The United States Federal Coal Mine Health and Safety Act of 1969 mandated several programs as part of a broad lung disease prevention effort. One program, CWXSP, makes available radiographic chest exami-

Appendix A

nations to working coal miners at all underground coal mines in the United States. Examinations are available to miners on a voluntary basis at least once every 5 years. If the results indicate radiographic evidence of coal workers' pneumoconiosis (CWP), the miner is given the right [under 30 CFR Part 90] to transfer to an area of the mine where exposures are continuously at or below 1 mg/m^3 respirable coal dust. This program provides secondary (medical monitoring) occupational disease prevention efforts for the mining industry.

The operation of CWXSP is guided by regulations published in Title 42 of the Code of Federal regulations [42 CFR 37]. In brief, the regulations specify the following:

- Every new miner must be examined within 6 months of employment and again 3 years later.
- If the 3-year X-ray shows signs of CWP, a third X-ray must be taken 2 years later.
- After these initial examinations, miners are eligible to have a voluntary chest X-ray once every 5 years.
- At NIOSH-certified facilities, miners can receive radiographic examinations that consist of chest radiographs and supporting demographic and work history information.
- Chest radiographs must be classified by at least two NIOSH-certified readers according to a standardized system for classifying radiographs of the pneumoconioses [ILO 1980].
- A final determination value is based on these classifications.
- A final determination value of small opacity profusion category 1/0 or higher is accepted as evidence of CWP.
- A miner with evidence of CWP may choose to transfer to a low-dust work environment.

For further information contact

Coal Workers' X-Ray Surveillance Program Activity
Surveillance Branch
Division of Respiratory Disease Studies
National Institute for Occupational Safety and Health
1095 Willowdale Road
Morgantown, WV 26505–2888
Telephone: 304–285–5724

APPENDIX A

National Traumatic Occupational Fatalities Surveillance System (NTOF)

NTOF is composed of information obtained from death certificates from the 52 U.S. vital statistics reporting units in the 50 States, New York City, and the District of Columbia for workers aged 16 and older for whom an external cause of death [WHO 1977] was noted, and for whom the certifier entered a positive response to the *Injury at Work?* item.

The industry and occupation on death certificates are defined as the "usual" industry and occupation of the victim. Industry and occupation narratives were coded according to the 1980 and 1990 Bureau of the Census classification schemes [Bureau of the Census 1982, 1992]. These data are reported by the major industry and occupation divisions as defined by the Bureau of the Census.

Limitations of death certificates used to ascertain work-related fatality information have previously been described [Russell and Conroy 1991; Stout and Bell 1991; NIOSH 1993]. Incomplete or unclear information on the death certificate and the lack of a national standard for completing the *Injury at Work?* item on the death certificate during this period are particular problems.

For much of the period this system has been used, no standardized guidelines were available for completing the *Injury at Work?* item on the death certificate. This item was subject to certifier interpretation. Although the lack of standardized reporting of this item may result in both false positives and false negatives, the numbers reported here are apt to represent the minimum number of occupational deaths that occurred in the United States during the period.

Death certificates ask for the "usual" occupation and industry of the person who died, which may not necessarily reflect the occupation or industry engaged in at the time of the fatal injury. Studies comparing death certificate entries for usual occupation and industry with information about occupation and industry at the time of death found agreement for occupation in 64% to 74% of the cases, and for industry in 60% to 76% of the cases [Karlson and Baker 1978; Baker et al. 1982; Davis 1988; Schade and Swanson 1988; Massachusetts Department of Public Health 1989]. Some studies indicate that although death certificates ask for the "usual" occupation and industry, the information recorded on the death certificates is more likely to reflect occupation or industry at time of death rather than lifetime employment [Davis 1988; Schade and Swanson 1988]. For these reasons, the possibility exists that for any surveillance system

based on death certificates, cases may be misclassified with respect to industry and occupation.

Denominator data were obtained from BLS CPS, a sample survey of the civilian noninstitutional population. These data were extracted from the BLS *Employment and Earnings* and the CPS monthly employment files [BLS 1981–1996; BLS 1992]. Fatality rates were calculated as average annual deaths per 100,000 workers. Rates were not calculated for cells with fewer than three cases because of the instability of rates based on small numbers. Frequencies and rates are presented for the civilian workforce only because denominator data are not easily obtainable for military personnel.

For further information contact
Surveillance and Field Investigations Branch
Division of Safety Research
National Institute for Occupational Safety and Health
1095 Willowdale Road, MS 1812
Morgantown, WV 26505
Telephone: 304–285–6009

Sentinel Event Notification System for Occupational Risk (SENSOR)

The original concept of SENSOR was based on a communicable disease surveillance model: providers (individual practitioners or health care facilities) reported disease cases to a health department. It had two organizational components—a network of sentinel providers (individual practitioners, laboratories, or clinics) who identified and reported the occupational cases, and a surveillance center in a State agency that analyzed reports and took action (confirmed cases and collected additional information through case followup, evaluated worksite factors, and recommended interventions). SENSOR created a cooperative, State-Federal effort to develop State capacity for recognizing, reporting, following up, and preventing selected occupational conditions. Initially, 10 States participated by focusing on one or more selected occupational conditions (acute pesticide-related illness and injury, asthma, carpal tunnel syndrome (CTS), lead poisoning, noise-induced hearing loss, and silicosis).

The SENSOR program of today is different from the original concept. Case identification has been enhanced to include not only physician reporting but information sources such as death certificates, hospital discharge data, and workers' compensation records. Intervention activities have been

Appendix A

broadened to include information dissemination, education, referral to enforcement agencies, and consultation. Currently, 13 States (California, Florida, Kentucky, Massachusetts, Michigan, Minnesota, New Jersey, New York, Ohio, Oregon, Texas, Utah, and Washington) have SENSOR programs for one or more of the following occupational conditions: acute pesticide poisoning, asthma, CTS, lead poisoning, noise-induced hearing loss, amputations, silicosis, and youth occupational injury.

For further information contact

Asthma and silicosis surveillance at NIOSH/CDC are coordinated through the:
Public Health Surveillance Team
Surveillance Branch
Division of Respiratory Disease Studies
National Institute for Occupational Safety and Health
1095 Willowdale Road, MS HG900
Morgantown, WV 26505–2888
Telephone: 304–285–6115

Pesticide-related illness and injury surveillance activities at NIOSH/CDC are coordinated through the:
Medical Section
Surveillance Branch
Division of Surveillance, Hazard Evaluations, and Field Studies
4676 Columbia Parkway, MS R21
Cincinnati, OH 45226
Telephone: 513–841–4448

CTS and noise surveillance are coordinated through the:
Surveillance Branch
Division of Surveillance, Hazard Evaluations, and Field Studies
4676 Columbia Parkway, MS R17
Cincinnati, OH 45226
Telephone: 513–841–4303

SENSOR coordination occurs through the:
Surveillance Branch
Division of Surveillance, Hazard Evaluations, and Field Studies
4676 Columbia Parkway, MS R17
Cincinnati, OH 45226
Telephone: 513–841–4303

Appendix A

Adult Blood Lead Epidemiology and Surveillance Program (ABLES)

The ABLES program is a national model for developing State-based surveillance. The surveillance of elevated blood lead levels (BLLs) provides the public health community (local, State, Federal) with essential data for monitoring adult lead poisoning and setting priorities for in-depth research, intervention, and information dissemination. The public health objective of the ABLES program is to eliminate exposures that result in workers having BLLs greater than 25 µg/dL of whole blood [DHHS 1990].

In 1998, the nationwide ABLES consisted of 27 State programs funded by NIOSH. These programs collect BLL data from local health departments, private health care providers, and private and State reporting laboratories. State ABLES use their data to (1) conduct followups with physicians, workers, and employers, (2) target onsite inspections of worksites, (3) provide referrals to cooperating agencies in the event regulatory action is necessary, and (4) conduct hazard surveillance to identify workplace exposures and control technology solutions. Findings from ABLES data have been used to identify high-risk industries, occupations, and tasks, including radiator repair shops, battery recycling operations, and construction-related jobs such as bridge repair and home remodeling. State educational materials for preventing adult and take-home lead poisoning are listed on the NIOSH Web site (www.cdc.gov/niosh/ables.html).

An essential criterion for ABLES is a State requirement that laboratories report BLL results to the State health department or designee. The lowest BLL to be reported varies from State to State. However, the reporting of *all* BLLs, elevated or not, is extremely useful for analyzing trends in these data. State ABLES programs are required to develop effective, well-defined working relationships with childhood lead poisoning prevention programs within their State. Lead may be taken home from the workplace on clothes or in cars, thus potentially exposing spouses and children. Children who come in contact with lead-exposed workers should be targeted for blood lead screening. Results are presented to the public via ABLES reports in the CDC *Morbidity and Mortality Weekly Report*.

Appendix A

For further information contact

Medical Section
Surveillance Branch
Division of Surveillance, Hazard Evaluations and Field Studies
National Institute for Occupational Safety and Health
4676 Columbia Parkway
Cincinnati, OH 45226
Telephone: 513–841–4424

Mining Injury and Employment Statistics

Data were obtained from the Mine Safety and Health Administration (MSHA) databases of reported employment and reported cases of accident/injury/illness for mine operators as well as independent contractors working on mine property as required under 30 CFR Part 50 [MSHA 1999]. The historical data (presented in Figure 6–1 for the period 1911–1995) were derived from several different sources [MSHA 1999; Adams and Wrenn 1941; Adams and Kolhos 1941; Reese et al. 1955; MSHA 1984].

According to 30 CFR Part 50, mine operators and independent contractors whose employees perform certain types of work on mine property are required to file a *Mine Accident, Injury, and Illness Report (Form 7000-1)* [30 CFR 50.20; MSHA 2000a] for reportable incidents within 10 working days after the accident or injury, or 10 working days following the illness diagnosis. The term "reportable injury" as defined by MSHA includes all incidents that require medical treatment or result in death, loss of consciousness, inability to perform all job duties, or temporary assignment or transfer to another job. Injuries involving "first-aid only" are not reportable. (First-aid only is defined as one-time treatment and subsequent observation of minor scratches, cuts, burns, splinters, etc. that do not ordinarily require medical care, even if it was provided by a physician or a registered health care professional.) Information reported on MSHA Form 7000-1 includes demographics of the injured or ill worker such as age, sex, years of total mining experience, years of experience at current mine, where the incident occurred (i.e., underground, surface, plant/mill), days away from work, days of restricted work activity, source of the injury, body part(s) injured, and a narrative description of the incident.

Also, under 30 CFR Part 50, mine operators and independent contractors whose employees perform certain types of work on mine property are required to file a *Quarterly Mine Employment and Coal Production Report (MSHA Form 7000-2)* [30 CFR 50.30; MSHA 2000b] within 15 days after the end of each calendar quarter. This information is reported in the address and employment files and includes the address and other contact

Appendix A

information, production of clean coal tonnage, average number of persons employed during the reporting period, and the corresponding number of hours worked for each type of operation (designated by MSHA as operational subunits that include underground operations, strip operations, plants or mills, etc.).

Commodity differences for type of employer (mining operators versus independent contractors).—The five commodity groups of coal (anthracite and bituminous), metal, nonmetallic minerals (nonmetal[†]), stone, and sand and gravel are based on a modification of the six 'canvass classes' designated by MSHA for mine operators. The only modification combines anthracite coal and bituminous coal into coal. Because independent contractors may work at multiple mining operations associated with a diversity of commodities, a 'canvass class' is not designated for independent contractors. Rather, independent contractors report employment under two aggregates: (1) all coal locations, and (2) all metal, nonmetal, stone, and sand and gravel locations. As a result of these reporting differences, fatality and injury rates for independent contractors can only be computed for coal and metal/nonmetal locations. However, within these two aggregates, independent contractors report employment separately for each type of operation (designated by MSHA as operational subunits that include underground operations, strip operations, plants or mills, etc.). Consequently, fatality and injury rates can be computed for both mine operators and independent contractors by type of operation.

Injury data inclusion criteria.—For the period 1988 to 1997, only cases that were coded as a degree injury 1–6 were included. This excludes reportable incidents not associated with an injury (degree injury 0), illnesses (degree injury 7), and nonoccupational injuries and illnesses that are maintained in the MSHA files because they occurred on mine property. Of those cases coded degree injury 1–6, office workers were excluded from analyses by excluding both employee hours and injuries reported for office locations (MSHA subunit code = 99).

Selection criteria for fatalities.—The number of fatalities used for the analyses varies from the number of fatalities reported in the MSHA accident/injury/illness databases as follows:

1. Seventeen fatalities attributed to and associated with a contractor code of "ZZZ" were excluded from all analyses. Although these fatalities occurred on mine property, the victims were not

---

[†]Depending on the context, the term 'nonmetal' may refer to either (1) the class of nonmetallic mineral that includes clays, salt, phosphates, etc. or (2) nonmetallic minerals, stone, and sand and gravel. In the current context, the first definition applies.

Appendix A

employees of either an independent contractor or mine operator. Rather, these victims were on mine property for other reasons (e.g., visitors, customers) when they were fatally injured.

2. Three fatalities attributed to mine operators were excluded from all analyses. Although fatally injured on mine property, the victims were nonemployees and minors (aged 5, 15, and 16).

3. Four additional fatalities were included in all analyses, two for independent contractors and two for mine operators per subsequent MSHA errata file. Two of these fatalities were originally reported in the database as nonfatal injuries.

4. One independent contractor fatality was excluded as not having occurred on mine property per subsequent MSHA errata file.

Selection criteria for lost-workday cases.—Lost-workday cases include only those cases that resulted in total or partial permanent disabilities, days away from work, or days of restricted work activity (MSHA degree of injury codes 2 through 5). The number of lost workdays were computed by summing the days away from work and days of restricted work activity, with one exception. For injuries resulting in total or partial permanent disabilities, lost workdays were the statutory days charged to the incident [MSHA 1998] whenever the statutory days exceeded the lost workdays reported or when lost workdays were unreported.

Calculation of injury rates.—Injury rates for the period 1988–1997 were computed using employment estimates derived from total hours worked. Full-time workers were calculated by dividing total hours by 2,000 hours/worker. Nonfatal injury rates were constructed per 100 full-time workers, and fatal injury rates per 100,000 full-time workers. Of note, MSHA publishes both fatal and nonfatal injury rates on the basis of 200,000 hours worked, which is equivalent to 100 full-time workers. Fatality rates for the historical data (Figure 6–1) were computed using average numbers of workers, because of the lack of exposure hours during the first few decades of this century.

Determining the type of incident associated with the injury.—MSHA's accident/injury/illness classification scheme was used to establish the type of incident associated with a fatality or nonfatal injury [MSHA 1998]. The type of incident is identical to MSHA's accident/injury/illness with two exceptions:

1. Both fatal and nonfatal cases classified as a *fall of highwall or rib* (accident/injury/illness code = 06) or as a *fall of roof or back* (accident/injury/illness code = 07) are reported under the type of incident *fall of ground*.

Appendix A

2. Nonfatal injury cases occurring underground and classified under *machinery* (accident/injury/illness = 17) were reclassified as a *fall of ground* if the source of the injury was *caving rock, ore, etc.* (MSHA source of injury code = 90). This reclassification is consistent with the way in which MSHA classifies similar incidents that resulted in a fatal injury. Typically, the victim is operating a roof bolter or continuous miner and is struck by caving rock from the mine roof or rib.

MSHA data compared with other surveillance systems.—The mining data presented in this report may differ from mining industry data for the same period using NTOF and CFOI surveillance systems. Both NTOF and CFOI use the 1987 SIC Manual [OMB 1987] to categorize fatal injuries by industry. The SIC classification scheme includes oil and gas extraction in the mining industry. MSHA excludes oil and gas extraction, as regulatory authority is delegated to OSHA. In addition, MSHA data include only incidents that occur on mine property. Therefore, an injury occurring during the course of work, but off mine property, is excluded from the MSHA file. NTOF and CFOI systems would capture this type of injury.

For more information please see *Injuries, Illnesses, and Hazardous Exposures in the Mining Industry, 1986–1995: A Surveillance Report* [NIOSH 2000]. The report summarizes available data on work-related fatal and nonfatal injuries in the mining industry for the 10-year period 1986–1995.

For further information contact

Surveillance, Statistics, and Research Support Activity
National Institute for Occupational Safety and Health
Pittsburgh Research Laboratory
626 Cochrans Mill Road
Pittsburgh, PA 15236
Telephone: 412–386–6617

or

Mining Surveillance and Statistical Support Activity
National Institute for Occupational Safety and Health
Spokane Research Laboratory
315 E. Montgomery Avenue
Spokane, WA 99207
Telephone: 509–354–8065

APPENDIX A

National Center for Health Statistics (NCHS)

NCHS is one of 12 centers, institutes, and offices of CDC. As the Nation's principal health statistics agency, NCHS provides statistical information to guide actions and policies to improve the health of the American people. NCHS surveys and data systems provide fundamental public health and health policy statistics that are used to track changes in health and health care delivery. Statistics are obtained through a broad-based program of ongoing and special studies in partnership with State government, including household interview surveys, examination surveys, surveys of health care providers, and collection of statistics on birth and death. NCHS participates with other agencies, such as NIOSH, and maintains some data systems collaboratively, such as NOMS. Further information is accessible through its Web site (www.cdc.gov/nchs/about.htm).

National Hospital Ambulatory Medical Care Survey (NHAMCS)

The National Ambulatory Medical Care Survey (NAMCS) was begun in 1973 to collect data on the use of ambulatory medical care services provided by office-based physicians. In 1992, NHAMCS was inaugurated to expand the scope of data collection to the medical services provided by hospital outpatient departments and emergency departments. Together, NAMCS and NHAMCS data provide an important tool for tracking ambulatory care use in the United States. These surveys along with a third survey, the National Survey of Ambulatory Surgery, constitute the ambulatory care component of the National Health Care Survey, which measures health care use across various types of providers.

Approximately 2,500 physicians are in the NHAMCS sample each year. The four-stage probability sample design used in the survey involves (1) primary sampling units, (2) hospitals within primary sampling units, (3) emergency departments within hospitals and/or clinics within outpatient departments, and (4) patient visits within emergency departments and/or clinics. Approximately 500 hospitals are in the sample each year. Hospitals are defined as facilities having an average patient stay of less than 30 days or those whose speciality is general (medical or surgical) or children's general. Clinic types are general medicine including internal medicine and primary care, surgery, pediatrics, obstetrics and gynecology, and other, such as neurology and psychiatry.

The Bureau of the Census is responsible for NHAMCS data collection. Information is collected on patient characteristics such as age, sex, race, expected source of payment, reason for visit, diagnoses, place of injury

Appendix A

occurrence, whether the injury was work-related, whether the injury was intentional, cause of injury, diagnostic and screening services, medications, disposition, and providers seen. Data have been collected on work-related injury visits since 1995. Beginning in 1997, verbatim text that describes the cause of injury may be analyzed. As with all probability sample surveys, the sample data may not have enough cases to produce reliable estimates for some subgroups.

For further information contact
Ambulatory Care Statistics Branch
Division of Health Care Statistics
National Center for Health Statistics
6525 Belcrest Road, Room 952
Hyattsville, MD 20782
Telephone: 301–458–4600

National Health and Nutrition Examination Survey (NHANES)

NCHS makes available public-use data from NHANES, a series of national surveys initiated in 1960. The fundamental purpose of these surveys is to characterize the health and nutritional status of the civilian noninstitutionalized population of the United States. The third National Health and Nutrition Examination Survey (NHANES III), conducted from 1988 to 1994, was a cross-sectional household interview and physical examination survey of the U.S. civilian noninstitutionalized population aged 2 months and older. NHANES III data were collected in 81 counties across the Nation from approximately 30,000 respondents among 39,696 persons selected for participation. Adults aged 17 and older constituted 20,050 respondents.

On the basis of the NHANES III adult (aged 17 and older) household interview, chronic obstructive pulmonary disease (COPD) was defined as a *yes* response to either of the following questions: (1) Has a doctor ever told you that you had chronic bronchitis? or (2) Has a doctor ever told you that you had emphysema? Asthma was defined as a *yes* response to the question, "Has a doctor ever told you that you had asthma?" Prevalence rates for COPD and for asthma were estimated for nonsmokers (using sample weights and adjustment for nonresponses) by usual industry (using the 44 industry categories as regrouped by NCHS in the NHANES III data files). Survey Data Analysis (SUDAAN) software was used to estimate variances, enabling calculation of 95% CIs for COPD and asthma prevalence rates.

APPENDIX A

For further information contact
Surveillance Branch
Division of Respiratory Disease Studies
National Institute for Occupational Safety and Health
1095 Willowdale Road
Morgantown, WV 26505–2888
Telephone: 304–285–6115

Multiple-Cause-of-Death Data

Each year since 1968, NCHS has made public-use data files available on multiple causes of death. These public-use files contain records of all U.S. deaths that are reported to State vital statistics offices (approximately 2 million annually). Each death record includes codes for up to 20 conditions listed on the death certificate, including both underlying and contributory causes of death. Other data include age, race, sex, and State and county of residence at the time of death. In addition, usual occupation and industry codes have been available for decedents from some States since 1985, and NCHS annually determines that certain quality criteria have been met by usual industry and occupation data from selected States.

Potential limitations of multiple-cause-of-death data include underreporting or overreporting of conditions on the death certificate by certifying physicians, incomplete or unclassified reporting of usual occupation and industry, and nonspecificity of codes.

For further information contact
Public Health Surveillance Team
Surveillance Branch
Division of Respiratory Disease Studies
National Institute for Occupational Safety and Health
1095 Willowdale Road
Morgantown, WV 26505–2888
Telephone: 304–285–6115

National Center for Infectious Diseases (NCID)

NCID is one of 12 centers, institutes, and offices of CDC. The mission of NCID is to prevent illness, disability, and death caused by infectious diseases in the United States and around the world. To accomplish this goal, NCID staff work in partnership with local and State public health officials, other Federal agencies, medical and public health professional associations, infectious disease experts from academic and clinical practice,

and international and public service organizations. NCID accomplishes its mission by conducting surveillance, epidemic investigations, epidemiologic and laboratory research, training, and public education programs to develop, evaluate, and promote prevention and control strategies for infectious diseases. Further information about NCID is available through its Web site (www.cdc.gov/ncidod/about.htm).

National Surveillance System for Hospital Health Care Workers (NaSH)

NaSH is a surveillance system that began in 1995 through the NCID Hospital Infections Program. NaSH focuses on surveillance of exposures and infections among hospital-based health care workers. The purpose of NaSH is to monitor national trends; identify newly emerging hazards for health care workers; assess the risk of occupational infection; and evaluate preventive measures, including engineering controls, work practices, protective equipment, and postexposure prophylaxis to prevent occupationally acquired infections. Participating hospitals are not randomly selected; they are usually hospitals that have previously participated in the National Nosocomial Infections Surveillance (NNIS) system and have also volunteered to participate in NaSH. Participating hospitals benefit by being able to conduct occupational health surveillance, analyze their data in an integrated system, and compare these data with a national database.

Initial entry of a health care worker into NaSH usually occurs during the provision of health care at the hospital's Employee Health Service for a relevant event (e.g., routine tuberculin skin test, initial assessment or followup after an exposure to blood, or initial assessment or followup after an exposure to a vaccine-preventable disease). Not all hospital employees have NaSH records.

The system collects the following data on health care workers: demographic information (identifying data is not sent to CDC), occupation, vaccination history, serologic results, immune status for vaccine-preventable diseases (including hepatitis B virus), TB exposure test and therapy status, detailed information on the nature of the exposure to blood/body fluids and bloodborne pathogens, postexposure prophylaxis treatment, information about exposures and infections from vaccine-preventable diseases such as measles, and information about exposures to infectious TB. Hospitals provide CDC with denominator data related to number of staff once a year. Every 2 years, participating hospitals distribute a survey to employees to be filled out anonymously that asks about history of needlesticks or sharps injuries; the purpose of this survey is to assess underreporting of incidents in the NaSH system.

Appendix A

For further information contact
Office of Surveillance
National Center for Infectious Diseases
1600 Clifton Road, NE, MS D59
Atlanta, GA 30333
Telephone: 800–893–0485

Sentinel Counties Study of Acute Viral Hepatitis

Although CDC conducts nationwide surveillance for acute viral hepatitis, several factors make it difficult to assess accurately changes in incidence of disease and risk factors associated with transmission: underreporting, failure to apply appropriate case definitions, and incomplete serologic testing and epidemiologic evaluation of all reported cases. To define the incidence and epidemiology of all types of viral hepatitis more accurately, a program of intensive surveillance for acute viral hepatitis was begun in several "Sentinel Counties" in 1979. Six counties currently participate in this system: Tacoma-Pierce County, WA; Pinellas County, FL; Jefferson County, AL; Denver, CO; Multnomah, OR; and San Francisco, CA.

All patients reported to the health departments participating in the Sentinel Counties Study of Acute Viral Hepatitis who meet the following clinical criteria for acute viral hepatitis are eligible for the study: (1) discrete date of onset of symptoms or jaundice and (2) liver enzymes greater than 2.5 times the upper limit of normal. For patients with non-A, non-B hepatitis (including those who test positive for hepatitis C antibodies), other possible causes of liver injury are excluded by interviewing the diagnosing physician and abstracting the medical record. All patients with acute viral hepatitis have serum drawn within 6 weeks of onset of symptoms and shipped to CDC for testing. Patients also complete a detailed epidemiologic interview.

The Sentinel Counties Study of Acute Viral Hepatitis has provided precise data on the significant sources of viral hepatitis infection in the United States and the contribution of these sources to disease incidence. In recent years, major changes have occurred in the incidence and epidemiology of the different types of viral hepatitis in the United States. Many of these changes were first recognized in the Sentinel Counties. For example, the Sentinel Counties study has been the primary source for data showing that hepatitis C is the etiologic agent of most non-A, non-B hepatitis and for describing the epidemiology and natural history of those diseases.

Appendix A

For further information contact
Hepatitis Branch
Division of Viral and Rickettsial Diseases
National Center for Infectious Diseases
1600 Clifton Road, NE, MS G37
Atlanta, GA 30333
Telephone: 404–371–5910

Viral Hepatitis Surveillance Program (VHSP)

The Hepatitis Branch of NCID operates VHSP, which obtains national surveillance data on clinical, serologic, and epidemiologic data pertaining to risk factors for viral hepatitis. Cases are submitted to the program by State governments. Limitations of the system include underreporting of cases and frequent omission of occupation on case records. [VHSP 1999]

For further information contact
Hepatitis Branch
Division of Viral and Rickettsial Diseases
National Center for Infectious Diseases
1600 Clifton Road, NE, MS G37
Atlanta, GA 30333
Telephone: 404–371–5910

National Center for HIV, STD, and TB Prevention (NCHSTP)

NCHSTP is one of 12 centers, institutes, and offices of CDC. The Center is responsible for public health surveillance, prevention research, and programs to prevent and control HIV infection and AIDS, other sexually transmitted diseases (STDs), and TB. Center staff work in collaboration with government and nongovernment partners at community, State, national, and international levels, applying well-integrated multidisciplinary programs of research, surveillance, technical assistance, and evaluation.

For further information contact
Communications Office
National Center for HIV, STD, and TB Prevention
1600 Clifton Road, NE, MS E07
Atlanta, GA 30333
Telephone: 404–639-8890

Appendix A

Surveillance of Health Care Workers with HIV/AIDS

Since 1981, all 50 States, the District of Columbia, and U.S. Trusts and Territories have reported AIDS cases, without names or other identifying information, to CDC's HIV/AIDS Reporting System (HARS). In addition to HIV risk information, the HARS case report form also requests information about past employment and occupation in the health care setting.

In 1991, CDC's HIV/AIDS Surveillance Branch developed a standardized protocol for State and local health departments to investigate in greater detail any cases of HIV infection or AIDS in health care workers without a behavioral or transfusion risk for HIV. The health departments are requested to investigate reports of health care workers who may have occupationally acquired HIV infection even if they have not yet met the criteria of the AIDS surveillance case definition and the State does not have formal requirements for HIV infection reporting. The reporting sources for potential cases of occupational transmission in health care workers include health care providers, HARS, and two sources within the CDC Hospital Infections Program: the Postexposure Prophylaxis Failure Study and the National Surveillance System for Hospital Health Care Workers.

The standardized protocol for investigating potential cases of occupational transmission in health care workers consists of a medical records review, an incident report review, discussion with the health care worker's health care provider, an interview by health department staff, and a laboratory investigation. The data are used to determine which occupations are at risk, where and how exposures to HIV commonly occur, the sources of transmission, effective prevention strategies, and HIV postexposure prophylaxis recommendations. The data also are used by manufacturers to create safer designs for medical devices and personal protective equipment.

Following an investigation by the State or local health department, cases of HIV infection in health care workers may be determined to be associated with a nonoccupational risk, or may be classified as cases of *possible* or *documented* occupational HIV transmission. Health care workers with *possible* occupational transmission are those with a history of occupational exposure to blood, other body fluids, or HIV-infected laboratory material who report no other risk factors for HIV infection but for whom no seroconversion associated with any of the occupational exposures was documented. Health care workers with *documented* occupational transmission have had documented evidence of HIV seroconversion in temporal association with an occupational exposure and have no other

Appendix A

known exposure to HIV during the same period of time; also included in this category are those persons infected with HIV strains that are closely related to the occupational exposure source by deoxyribonucleic acid (DNA) sequencing.

For further information contact

Surveillance Branch
Division of HIV/AIDS Prevention—Surveillance and Epidemiology
National Center for HIV, STD, and TB Prevention
1600 Clifton Road, NE, MS E47
Atlanta, GA 30333
Telephone: 404–639–2050

Surveillance for Tuberculosis Infection in Health Care Workers (staffTRAK–TB)

CDC recommends periodic tuberculin skin testing of health care workers with a potential for exposure to *Mycobacterium tuberculosis*. However, many health care facilities (e.g., hospitals, correctional facilities, long-term care facilities, and health departments) do not have a system for identifying and tracking workers due for tuberculin skin testing or a means of analyzing aggregate data. To facilitate the surveillance for TB infection in health care workers in health departments, CDC developed a software package called *staffTRAK–TB* to track, analyze, and report information pertaining to tuberculin skin testing surveillance in health care workers. The software allows the collection of data for each health care worker including demographic information, occupation, work location, multiple tuberculin skin test results, and results of evaluations determining if clinically active TB is present. Programmed reports include lists of workers due and overdue for skin tests, and skin test conversion rates by occupation and worksite. Standardizing types of occupations and work locations allows data from multiple facilities to be aggregated and compared. Data transfers to CDC can be performed via floppy diskettes.

In 1995, CDC implemented tuberculin skin testing demonstration projects in selected health departments and hospitals in the United States. The tuberculin skin testing demonstration project is designed to help participating sites develop model tuberculin skin test programs consistent with CDC *Guidelines for Preventing the Transmission of Mycobacterium tuberculosis in Health-Care Facilities* and to facilitate local data analysis and evaluation of tuberculin skin testing programs. Results of tuberculin skin testings are entered into the *staffTRAK–TB* software and tuberculin skin testing data **without** personal identifiers (e.g., name, social security

number, or address) and are transferred to CDC at least quarterly from each of the participating sites. This project will allow CDC to gain information about the incidence of occupationally related tuberculin skin testing conversions among health care workers in the participating sites and determine if research is needed.

For further information contact
Division of Tuberculosis Elimination
National Center for HIV, STD, and TB Prevention
1600 Clifton Road, NE, MS E10
Atlanta, GA 30333
Telephone: 404–639–8117

References Cited

Adams WW, Kolhos ME [1941]. Metal- and nonmetal- mine accidents in the United States during the calendar year 1939 (excluding coal mines). Washington, DC: U.S. Department of the Interior, Bureau of Mines, Bulletin 440.

Adams WW, Wrenn VE [1941]. Quarry accidents in the United States during the calendar year 1939. Washington, DC: U.S. Department of the Interior, Bureau of Mines, Bulletin 438.

Baker SP, Samkoff JS, Fisher RS, Van Buren CB [1982]. Fatal occupational injuries. JAMA 248(6):692–697.

BLS [1981–1996]. Employment and earnings. Washington, DC: U.S. Government Printing Office, Publication 28–43 (Issue No. 1 of each).

BLS [1992]. BLS Handbook of methods. Washington, DC: U.S. Government Printing Office, Bureau of Labor Statistics Bulletin 2414.

BLS [1997]. Occupational injuries and illnesses: counts, rates, and characteristics, 1994. Washington, DC: U.S. Department of Labor, Bureau of Labor Statistics.

Breslow NE, Day NE [1987]. Statistical methods in cancer research. Vol. II. The design and analysis of cohort studies. Lyon, France: International Agency for Research on Cancer, IARC Scientific Publications No. 82.

Bureau of the Census [1982]. 1980 census of population: alphabetical index of industries and occupations. Washington, DC: U.S. Government Printing Office, Publication CPH80-R3.

Appendix A

Bureau of the Census [1992]. 1990 census of population and housing: alphabetical index of industries and occupations. Washington, DC: U.S. Government Printing Office, Publication CPH-R-3.

CFR. Code of Federal regulations. Washington, DC: U.S. Government Printing Office, Office of the Federal Register.

Checkoway H, Pearce N, Crawford-Brown DJ [1989]. Research methods in occupational epidemiology. Monographs in epidemiology and biostatistics. Vol. 13. New York: Oxford University Press.

Davis H [1988]. The accuracy of industry data from death certificates for workplace homicide victims. Am J Public Health *78*(12):1579–1581.

DeCouflé P, Thomas TL, Pickle LW [1980]. Comparison of the proportionate mortality and standardized mortality ratio risk measures. Am J Epidemiol *111*(3):263–269.

DHHS [1990]. Healthy people 2000. National health promotion and disease prevention objectives. Washington, DC: U.S. Department of Health and Human Services, Public Health Service, DHHS (PHS) Publication No. 91-50213.

Dubrow R, Spaeth S, Adams SL, Burnett C, Petersen M, Robinson C [1993]. Proportionate mortality ratio analysis system—version V; program documentation. Cincinnati, OH: U.S. Department of Health and Human Services, Public Health Service, Centers for Disease Control and Prevention, National Institute for Occupational Safety and Health. Unpublished.

ILO [1980]. Guidelines for the use of ILO international classification of radiographs of pneumoconioses. Rev. ed. Geneva, Switzerland: International Labour Office, Occupational Safety and Health Series No. 22. (Rev.)

Karlson TA, Baker SP [1978]. Fatal occupational injuries associated with motor vehicles. In: Proceedings of the American Association for Automotive Medicine, 22nd Conference. Arlington Heights, IL. Am Assoc Automotive Med *1*:229–241.

Kim JH [1998]. Atlas of respiratory disease mortality, United States: 1982–1993. Cincinnati, OH: U.S. Department of Health and Human Services, Public Health Service, Centers for Disease Control and Prevention, National Institute for Occupational Safety and Health, DHHS (NIOSH) Publication No. 98-157. [www.cdc.gov/niosh/98-157pd.html].

Appendix A

Layne LA, Landen DD [1997]. A descriptive analysis of nonfatal occupational injuries to older workers, using a national probability sample of hospital emergency departments. J Occup Environ Med 39(9): 855–865.

Massachusetts Department of Public Health [1989]. Dying for the job: traumatic occupational deaths in Massachusetts. Boston, MA: Massachusetts Department of Public Health.

McDonald [1994]. NEISS—the National Electronic Injury Surveillance System: a tool for researchers. Washington, DC: U.S. Consumer Product Safety Commission, Division of Hazard and Injury Data Systems.

MSHA [1984]. Summary of selected injury experience and worktime for the mining industry in the United States, 1931–77 Denver, CO: U.S. Department of Labor, Mine Safety and Health Administration, IR 1132.

MSHA [1998]. Injury experience in coal mining, 1998. Denver, CO: U.S. Department of Labor, Mine Safety and Health Administration, Office of Injury and Employment Information, IR 1265, pp. 5–7.

MSHA [1999]. Quarterly employment and coal production, accidents/injuries/illnesses reported to MSHA under 30 CFR Part 50, 1986–1997. Denver, CO: U.S. Department of Labor, Mine Safety and Health Administration, Office of Injury and Employment Information.

MSHA [2000a]. MSHA form 7000-1—mine accident, injury and illness report. [www.msha.gov/forms/7000-1.pdf]. Date accessed: March 2000.

MSHA [2000b]. MSHA form 7000-2—quarterly mine employment and coal production report. [www.msha.gov/forms/7000-2.pdf]. Date accessed: March 2000.

NIOSH [1993]. Fatal injuries to workers in the United States, 1980–1989: a decade of surveillance. National and State profiles. Cincinnati, OH: U.S. Department of Health and Human Services, Public Health Service, Centers for Disease Control and Prevention, National Institute for Occupational Safety and Health, DHHS (NIOSH) Publication No. 93-108S. [www.cdc.gov/niosh/93-108s.html].

NIOSH [1997]. Mortality by occupation, industry, and cause of death, 24 reporting States, 1984–1988. Cincinnati, OH: U.S. Department of Health and Human Services, Public Health Service, Centers for Disease Control and Prevention, National Institute for Occupational Safety and

Appendix A

Health, DHHS (NIOSH) Publication No. 97-114. [www.cdc.gov/niosh/97-114.html].

NIOSH [2000]. Injuries, illnesses, and hazardous exposures in the mining industry, 1986–1995: a surveillance report. Cincinnati, OH: U.S. Department of Health and Human Services, Public Health Service, Centers for Disease Control and Prevention, National Institute for Occupational Safety and Health, DHHS (NIOSH) Publication No. 2000-117. [www.cdc.gov/niosh/mining/pubs/pdfs/iiahe.pdf].

OMB [1987]. Standard industrial classification manual. Washington, DC: Office of Management and Budget.

OSHA [2000a]. OSHA form 200—log and summary of occupational injuries and illnesses. Date accessed: March 2000.

OSHA [2000b]. OSHA form 101—supplementary record of occupational injuries and illnesses. Date accessed: March 2000.

Reese ST, Wrenn VE, Reid EJ [1955]. Injury experience in coal mining, 1952: analysis of mine safety factors, related employment, and production data. Washington, DC: U.S. Department of the Interior, Bureau of Mines, Bulletin 559.

Rosenberg HM, Burnett C, Maurer J, Spirtas R [1993]. Mortality by occupation, industry, and cause of death: 12 reporting States, 1984. Mon Vital Stat Rep 42(4), supplement.

Rothman KJ [1986]. Modern epidemiology. Boston, MA: Little, Brown and Company.

Russell J, Conroy C [1991]. Representativeness of deaths identified through the injury-at-work item on the death certificate: implications for surveillance. Am J Public Health 81(12):1613–1618.

Schade WJ, Swanson GM [1988]. Comparison of death certificate occupation and industry data with lifetime occupational histories obtained by interview: variations in the accuracy of death certificate entries. Am J Ind Med 14:121–136.

Stout N, Bell C [1991]. Effectiveness of source documents for identifying fatal occupational injuries: a synthesis of studies. Am J Public Health 81(6):725–728.

WHO [1967]. Manual of the international statistical classification of diseases, injuries, and causes of death, based on the recommendations of the Eighth Revision Conference, 1965. Geneva, Switzerland: World Health Organization.

WHO [1977]. Manual of the international statistical classification of diseases, injuries, and causes of death, based on the recommendations of the Ninth Revision Conference, 1975. Geneva, Switzerland: World Health Organization.

VHSP [1999]. Viral Hepatitis Surveillance Program. Atlanta, GA.: U.S. Department of Health and Human Services, Public Health Service, Centers for Disease Control and Prevention, National Center for Infectious Diseases. Database. [www.cdc.gov/ncidod/osr/site/surv_resources/surv_sys.htm].

Bibliography

Bureau of Labor Statistics

Austin C [1995]. An evaluation of the census of fatal occupational injuries as a system for surveillance. Compensation and Working Conditions, May, pp. 1–4.

BLS [1999]. Fatal workplace injuries: a collection of data and analysis, [www.bls.gov/iif/oshcfoi1.htm] 1992, 1993, 1994, 1995, 1996, 1997. Washington, DC: U.S. Department of Labor, Bureau of Labor Statistics.

CFOI Staff [1995]. Outdoor occupations exhibit high rates of fatal injuries. Issues in Labor Statistics, Summary 95–96, March.

Drudi D [1995]. The evolution of occupational fatality statistics in the United States. Compensation and Working Conditions, July, pp. 1–5.

Drudi D [1997]. A century-long quest for meaningful and accurate occupational injury and illness statistics. Compensation and Working Conditions, winter, pp. 19–27.

Personick M [1996]. New data highlight severity of construction falls. Compensation and Working Conditions, September, pp. 54–56.

Ruser J [1995]. A relative risk analysis of workplace fatalities. Compensation and Working Conditions, January, pp. 41–45.

Toscano G [1991]. The BLS census of fatal occupational injuries. Compensation and Working Conditions, June, pp. 1–2.

Appendix A

Toscano G [1997]. Dangerous jobs. Compensation and Working Conditions, summer, pp. 57–60.

Toscano G, Weber W [1995]. Violence in the workplace. Compensation and Working Conditions, April, pp. 1–8.

Toscano G, Windau J [1991]. Further test of a census approach to compiling data on fatal work injuries. Mon Labor Rev, October, pp. 33–36.

Toscano G, Windau J, Drudi D [1996]. Using the BLS occupational injury and illness classification system as a safety and health management tool. Compensation and Working Conditions, June, pp. 19–28.

Toscano G, Windau J, Knestaut A [1998]. Work injuries and illnesses occurring to women. Compensation and Working Conditions, summer, pp. 16–22.

Windau J, Goodrich D [1990]. Testing a census approach to compiling data on fatal work injuries. Mon Labor Rev, December, pp. 47–49.

Windau J, Jack T [1996]. Highway fatalities among the leading causes of workplace deaths. Compensation and Working Conditions, September, pp. 57–61.

Windau J, Toscano G [1994]. Workplace homicides in 1992. Compensation and Working Conditions, February, pp. 1–8.

Windau J, Jack TJ, Toscano G [1998]. State and industry fatal occupational injuries, 1992–96. Compensation and Working Conditions, summer, pp. 8–15.

National Electronic Injury Surveillance System (NEISS)

BLS (Bureau of Labor Statistics) [1997]. Employment and Earnings *44*(1).

CDC (Centers for Disease Control and Prevention) [1983]. Surveillance of occupational injuries treated in hospital emergency rooms—United States. MMWR *32*(7):89–90.

Appendix A

National Occupational Mortality Surveillance System (NOMS)

Bureau of the Census [1981]. 1980 census of population: alphabetical index of industries and occupations. 2nd ed. Washington, DC: U.S. Department of Commerce, Bureau of the Census.

Burnett CA, Dosemeci M [1994]. Using occupational mortality data for surveillance of work-related diseases of women. JOM 36(11): 1199–1203.

Chen GX, Burnett CA, Cameron LL, Alterman T, Lalich NR, Tanaka S, Althouse RB [1997]. Tuberculosis mortality and silica exposure: a case-control study based on a national mortality database for the years 1983–1992. Int J Occup Environ Health 3(3):163–170.

Guralnick L [1962]. Mortality by occupation and industry among men 20 to 64 years of age: United States, 1950. Vital Statistics—Special Reports 53(2). Washington, DC: U.S. Department of Health Education and Welfare, Public Health Service.

Guralnick L [1963a]. Mortality by occupation and cause of death among men 20 to 64 years of age: United States, 1950. Vital Statistics—Special Reports 53(3). Washington, DC: U.S. Department of Health Education and Welfare, Public Health Service.

Guralnick L [1963b]. Mortality by industry and cause of death among men 20 to 64 years of age: United States, 1950. Vital Statistics—Special Reports 53(4). Washington, DC: U.S. Department of Health Education and Welfare, Public Health Service.

Hill A [1971]. Principles of medical statistics. New York: Oxford University Press.

Kipen HM, Craner J [1992]. Sentinel pathophysiologic conditions: an adjunct to teaching occupational and environmental disease recognition and history taking. Environ Res 59:93–100.

Lalich N, Burnett C, Robinson C, Sestito J, Schuster L [1990]. A guide for the management, analysis and interpretation of occupational mortality data. Cincinnati, OH: U.S. Department of Health and Human Services, Public Health Service, Centers for Disease Control and Prevention, National Institute for Occupational Safety and Health, DHHS (NIOSH) Publication No. 90-115. [www.cdc.gov/niosh/90-115pd.html].

Mullan RJ, Murthy LI [1991]. Occupational sentinel health events: an up-dated list for physician recognition and public health surveillance. Am J Ind Med *19*:775–799

NIOSH [1999]. Work-related lung disease surveillance report 1999 (WoRLD). Cincinnati, OH: U.S. Department of Health and Human Services, Public Health Service, Centers for Disease Control and Prevention, National Institute for Occupational Safety and Health, DHHS (NIOSH) Publication No. 2000-105. [www.cdc.gov/niosh/docs/2000-105/pdfs/2000-105.pdf].

Peipins L, Burnett C, Alterman T, Lalich N [1997]. Mortality patterns among female nurses: a 27-state study, 1984 through 1990. Am J Public Health *87*(9):1539–1543.

Petralia SA, Dosemeci M, Adams EE, Zahm SH [1999]. Cancer mortality among women employed in health care occupations in 24 U.S. States, 1984–1993. Am J Ind Med *36*:159–165.

Rutstein DD, Mullan RJ, Frazier TM, Halperin WE, Melius JM, Sestito JP [1983]. Sentinel health events (occupational): a basis for physician recognition and public health surveillance. Am J Public Health *73*(9): 1054–1062.

Stockwell JR, Adess ML, Titlow TB, Zaharias GR [1991]. Use of sentinel health events (occupational) in computer assisted occupational health surveillance. Aviat Space Environ Med *62*(8):795–797.

Adult Blood Lead Epidemiology and Surveillance Program

CDC (Centers for Disease Control) [1998]. Adult blood lead epidemiology and surveillance—United States, first quarter 1998, and annual 1994–1997. MMWR *47*(42):907–11.

Multiple-Cause-of-Death Data

NCHS [1999]. Vital Statistics of the United States, 1993, Vols. I and II. Hyattsville, MD: U.S. Department of Health and Human Services, Public Health Service, National Center for Health Statistics, DHHS Publication Nos. 99-1100 and 99-1101.

Appendix B

APPENDIX B
DESCRIPTION OF INDUSTRY AND OCCUPATION CODING SYSTEMS

Overview

Appendix B presents a detailed description of the following occupational classification systems: North American Industry Classification System (NAICS), Standard Industrial Classification (SIC), Standard Occupational Classification (SOC), and Bureau of the Census Industry and Occupation Classification. Table B–1 gives an overview of these systems.

North American Industry Classification System/Standard Industrial Classification (NAICS/SIC)

Over the course of history, the United States has gone from a largely agrarian economy in its earliest period, to one based more on manufacturing, to the current more service-oriented economy. The SIC system was introduced in the 1930s to help classify the growing number of new manufacturing industries that had developed since the early 1900s. The SIC system provided a consistent framework for assigning descriptive industry codes to each establishment and for the subsequent collection, tabulation, and analysis of economic statistics by government agencies and private research firms [Murphy 1998].

The 1987 SIC includes 11 divisions and 1,004 detailed industries. Each industry is designated by a hierarchical four-digit code. For example, the industry *video tape rental* has the code 7841. The first two digits represent the major group; the third digit represents the industry group; and the fourth digit represents the detailed industry [OMB 1987].

By 1992, however, a new classification system was clearly needed to accommodate newly developed industries in such areas as information services, health care services, and high-tech manufacturing. Furthermore, the initiation of the North American Free Trade Agreement in 1994 increased

Appendix B

the need for comparable statistics from the United States, Canada, and Mexico [Levine et al. 1999].

Economic changes that have taken place in the last several decades—such as the movement toward a more service-oriented economy, the increased use of computers and other new technology, and globalization—have precipitated the need for a new system of industrial classification [Murphy 1998].

The resulting system, NAICS [OMB 1998], is a complete restructuring of the SIC. This system was organized to conform to the principle of grouping establishments by their production processes alone. Thus NAICS is a supply-based or production-oriented classification system. By contrast, the former system uses a combination of supply and demand characteristics (production and marketing activities). Supply-based categories group establishments using similar raw material inputs, capital equipment, and labor in the same industry. Demand-based categories group activities that are similar in the eyes of customers or users of the product or service. A supply-based approach creates more homogeneous categories that are better suited for economic analysis. Another advantage of NAICS is that each participating country can individualize the new system to meet its own needs as long as data can be aggregated to standard NAICS industries [Levine et al. 1999].

APPENDIX B

Table B–1. Overview of industry and occupation coding systems

Name of system	Source	Frequency	Structure of industry or occupation system	Coverage	Reference	Contact
NAICS*	Classification system of industries and definitions developed by the ECPC established by OMB	Planned to be periodically updated	20 major industry sectors, 1,170 detailed industries	All establishments in which business is conducted or services or industrial operations are performed	NAICS– United States, 1997	Executive Office of the President, OMB (www.census.gov/epcd/www/naics.html)
SIC	Classification system of industries and definitions developed by the ITCIC established by OMB	Periodically updated— this system will be replaced by NAICS and new SIC designations are not expected	11 industry divisions, 1,004 detailed industries	All establishments producing goods or services	SIC Manual, 1987	Executive Office of the President, OMB (www.osha.gov/pls/imis/sic_manual.html)
SOC	Classification system of occupations and definitions developed by the SOC Revision Policy Committee established by OMB to be used for all Federal occupational classification	Periodically updated	23 major occupational groups, 98 minor occupational groups, 452 broad occupations, 822 detailed occupations	All occupations and industries in which work is performed for pay or profit	SOC Manual, 1998	Executive Office of the President, OMB (www.bls.gov/soc/home.htm)
Bureau of the Census Occupation and Industry Classifications	Occupation and industry information from the decennial sample survey of households. Survey yields estimates of current occupational employment by industry— classifications are based on self-described occupations by respondents	Entire economy surveyed once every 10 years	6 summary occupation groups, 13 major occupation groups, 501 occupation categories 13 major industry groups, 236 industry categories	All industries except active duty military personnel	*Alphabetical Index of Industries and Occupations* *Classified Index of Industries and Occupations*	U.S. Department of Commerce, Bureau of the Census, Population Division (www.census.gov/hhes/www/ioindex/overview.html)

*Abbreviations: ECPC = Economic Classification Policy Committee; ITCIC = Interagency Technical Committee on Industrial Classification; NAICS = North American Industry Classification System; OMB = Office of Management and Budget; SIC = standard industrial classification; SOC = standard occupational classification

235

Appendix B

The following is a list of the 21 NAICS sectors:

11	Agriculture, forestry, fishing, and hunting
21	Mining
22	Utilities
23	Construction
31–33	Manufacturing, electric, gas, and sanitary services
42	Wholesale trade
44–45	Retail trade
48–49	Transportation and warehousing
51	Information
52	Finance and insurance
53	Real estate, rental, and leasing
54	Professional, scientific, and technical services
55	Management of companies and enterprises
56	Administrative and support, waste management, and remediation services
61	Educational services
62	Health care and social assistance
71	Arts, entertainment, and recreation
72	Accommodations and food services
81	Other services (except public administration)
92	Public administration
99	Unclassified establishments

Although NAICS uses a hierarchical structure much like the existing SIC, important structural differences exist between the systems. For example, NAICS uses a six-digit classification code that allows greater flexibility in the coding structure. The SIC system is limited to only four digits. Another important difference is that NAICS uses the first two digits of the six-digit code to designate the highest level of aggregation, with 21 such two-digit industry sectors under the new system. The SIC system, by contrast, has only 11 divisions. For example, the industry *software publishers* has the code 511210. The first two digits designate the highest level of aggregation, the third digit represents the subsector; the fourth digit represents the industry group, the fifth digit represents the international industry level, the sixth digit designates national detail [Murphy 1998].

Appendix B

During the transition period from SIC to NAICS, SIC codes will be assigned to create linkages between statistics classified under the two systems [Murphy 1998].

Standard Occupational Classification (SOC)

Since the early part of this century, several agencies have developed their own occupational classifications. The U.S. Employment Service needed occupational statistics for its work and developed a Convertibility List of Occupations with Conversion Tables to serve as a bridge between its statistics and information from the 1940 Census of Population. Continued revisions to the census classification scheme and publication of the third edition of the Department of Labor's *Dictionary of Occupational Titles* in 1965 encouraged the government to devise such a standard to link these different systems. This effort resulted in the 1977 Standard Occupational Classification (SOC) (revised and reissued in 1980) [U.S. Department of Commerce 1980; Levine et al. 1999].

However, the original system was not revised after 1980, and many agencies set up data collection systems with occupational classification schemes that differed from the SOC:

- *BLS*—the Occupational Employment Statistics survey classifies workers according to occupational definitions.

- *Bureau of the Census*—both the decennial Census of Population and the monthly Current Population Survey (CPS) classify workers according to the job titles given by the survey respondents.

- *Employment and Training Administration*—the *Dictionary of Occupational Titles*, which identifies and defines more than 12,000 jobs, has been replaced by the Occupational Information Network (O*NET), which adheres to the SOC.

- *Department of Education*—collects data on teachers.

- *Bureau of Health Professions*—gathers information on health occupations.

- *National Science Foundation*—surveys focus on scientists and engineers.

Observing this problem, the Bureau of Labor Statistics (BLS) hosted an International Occupational Classification Conference to establish a context for a new SOC revision process. Similarly, the Employment and Training Administration's Advisory Panel for the *Dictionary of Occupational Titles*

Appendix B

had just completed a review of the dictionary and had recommended substantial new occupations [Levine et al. 1999].

Persuaded that a reconciliation was in order, OMB subsequently invited all Federal agencies with an occupational classification system to join together to revise the SOC. The SOC Committee included representatives from BLS, the Bureau of the Census, the Employment and Training Administration, the Defense Manpower Data Center, and the Office of Personnel Management. In addition, ex-officio members included the National Science Foundation, the National Occupational Information Coordinating Committee (NOICC), and OMB. Representatives from other Federal agencies such as the U.S. Department of Education, the U.S. Department of Health and Human Services, and the Equal Employment Opportunity Commission participated in several meetings of the SOC Committee as well, or in the Federal Consultation Group [Levine et al. 1999].

The SOC Committee chose a practical approach to classification and continued the previous focus on work performed (with *skills-based considerations*) as the key classification principle for the revised (1998) SOC [Levine et al. 1999].

BLS provides information about the 1998 SOC at their Web site (www.bls.gov/soc/home.htm). This site contains links to the 1998 SOC major groups; the complete 1998 SOC hierarchical structure and detailed occupational definitions; a numerical index of detailed occupations; an SOC user's guide; and an SOC search capability, as well as SOC Federal Register notices and related documents.

The 1998 SOC is composed of four levels of aggregation: (1) major group, (2) minor group, (3) broad occupation, and (4) detailed occupation. BLS, through its establishment survey that classifies workers according to occupational definitions, is generally better able to collect data on more detailed occupations than is the Bureau of the Census, whose household surveys rely almost exclusively on job titles given by respondents to classify workers.

The following list shows the 23 major occupational groups of the 1998 SOC [Levine et al. 1999]:

Appendix B

11 Management occupations

13 Business and financial operations occupations

15 Computer and mathematical occupations

17 Architecture and engineering occupations

19 Life, physical, and social science occupations

21 Community and social services occupations

23 Legal occupations

25 Education, training, and library occupations

27 Arts, design, entertainment, sports, and media occupations

29 Healthcare practitioners and technical occupations

31 Healthcare support occupations

33 Protective service occupations

35 Food preparation and serving-related occupations

37 Building and grounds cleaning and maintenance occupations

39 Personal care and service occupations

41 Sales and related occupations

43 Office and administrative support occupations

45 Farming, fishing, and forestry occupations

47 Construction and extraction occupations

49 Installation, maintenance, and repair occupations

51 Production occupations

53 Transportation and material-moving occupations

55 Military-specific occupations

These major groups include 98 minor groups, 452 broad occupations, and 822 detailed occupations in the SOC [Levine et al. 1999].

The 1980 SOC included 22 divisions (comparable to major groups in the 1998 SOC), 60 major groups (comparable to minor groups in the 1998 SOC), 223 minor groups (comparable to broad occupations in the 1998 SOC), and 664 unit groups (comparable to detailed occupations in the 1998 SOC) [Levine et al. 1999]. Each occupation is designated by a six-digit code. For example, the occupation *printing machine operator* has the code 51-5023. The hyphen between the second and third digit is

Appendix B

used only for presentation clarity. The first two digits represent the major group; the third digit represents the minor group; the fourth and fifth digits represent the broad occupation; and the sixth digit represents the detailed occupation.

To facilitate consistent classification by data collection agencies across surveys, the 1998 SOC associates some 30,000 job titles with detailed occupations. Because many of these job titles are industry-specific, the industries are also listed for many titles. To further facilitate consistent classification, each detailed occupation has a definition that uniquely defines the workers included. Most historical comparisons with older classification systems are still possible.

The SOC Committee proposed that a permanent review committee be established to keep the SOC current, and OMB is considering the proposal. In addition, it was proposed that the review committee provide timely advice to the Bureau of the Census during its 2000 Census occupations coding operation, particularly with respect to the proper classification of unfamiliar job descriptions and job titles. The next major revision of the SOC is expected to begin in 2005, in preparation for the 2010 Census of Population [Levine et al. 1999].

Bureau of the Census

The census occupational data result from the decennial Census of Population and Housing, the monthly CPS, and other demographic surveys conducted by the Bureau of the Census. The most currently available decennial census occupational data are from the 1990 census, which collected data from about 17% of households. The job titles given by the survey respondents were classified into 501 occupations in 236 industries. The monthly CPS collects occupational data from about 50,000 of the approximately 118 million households in the United States. The CPS data provide national occupational trend information. The CPS uses the decennial census classification structure of occupational coding [Thompson 1981; Bureau of the Census 1992a].

The alphabetical and classified indexes used by the Bureau of the Census in its coding operation presents a listing of some 21,000 industry and 30,000 occupation titles that have appeared on schedule returns, together with the census code for each, but includes no descriptive material [Bureau of the Census 1992a, 1992b; Miller et al. 1980].

The 1990 census occupational classification structure is arranged into 6 summary and 13 major occupational groups and contains 501 occupational

Appendix B

categories, each of which is assigned a 3-digit numeric code. For example, the occupation *dental hygienist* has the code 204. However, the major group can only be determined by referring to the classification manual; in this case it is within the first major group—technical, sales, and administrative support occupations.

The 1990 census industry classification structure is arranged into 13 major industry groups and contains 236 industry categories, each of which is assigned a 3-digit numeric code. For example, the industry *drafting service* has the code 741. The major group is determined by referring to the classification manual; in this case it is within the major industry category business and repair services.

NAICS United States has been adopted for statistical use by all Federal agencies, including the Bureau of the Census. Government-wide implementation is underway and will continue at least through the year 2004. Planning is underway for implementing NAICS in the current programs of the Bureau of the Census, however, the Bureau's plans for implementing NAICS United States in current survey programs, including the 2000 decennial census, are not yet final [U.S. Economic Classification Policy Committee 1998].

The Bureau of the Census uses the SOC (last revised in 1980) to classify responses to its household surveys. The SOC is currently undergoing revision, and the proposed 1998 SOC was released in the Federal Register in August 1998. The revised SOC will be used to classify responses to the 2000 decennial census and will be adapted for use with household surveys shortly thereafter [Bureau of the Census 1999].

NOICC is a committee with representation from 10 Federal agencies. It maintains the NOICC Master Crosswalk, a computerized database that shows relationships among the major occupational and educational classification systems used by the Federal government. A formal crosswalk is available electronically at (www.state.ia.us/ncdc/xw_xwalk.html). Since the major occupation coding systems are being revised, a new crosswalk will be developed to reflect the changes.

Appendix B

References Cited

Bureau of the Census [1992a]. 1990 census of population and housing: alphabetical index of industries and occupations. Washington, DC: U.S. Government Printing Office, Publication CPH-R-3.

Bureau of the Census [1992b]. 1990 census of population and housing: classified index of industries and occupations. Washington, DC: U.S. Government Printing Office, Publication CPH-R-4.

Bureau of the Census [1999]. Occupation. [www.census.gov/hhes/www/occupation.html]. Date accessed: August 30, 1999.

Levine C, Salmon L, Weinberg DH [1999]. Revising the standard occupational classification system. Mon Labor Rev 122(5):36–45.

Miller AR, Treiman DJ, Cain PS, Roos PA, eds. [1980]. Work, jobs, and occupations. A critical review of the Dictionary of occupational titles. Washington, DC: National Academy Press. [www.nap.edu/books/0309030935/html].

Murphy JB [1998]. Introducing the North American Industry Classification System (NAICS). Mon Labor Rev 121(7):43–47.

OMB [1987]. Standard industrial classification manual, 1987. Washington, DC: Executive Office of the President, Office of Management and Budget.

OMB [1998]. North American Industry Classification System—United States, 1997. Washington, DC: Executive Office of the President, Office of Management and Budget.

Thompson J [1981]. BLS job cross-classification system relates information from six sources. Mon Labor Rev 104:40–44.

U.S. Department of Commerce [1980]. Standard occupational classification manual. Washington, DC: U.S. Department of Commerce, Office of Federal Statistics Policy and Standards.

U.S. Economic Classification Policy Committee [1998]. New data for a new economy. Washington, DC: U.S. Government Printing Office.

Glossary

Glossary

Adult Blood Lead Epidemiology and Surveillance Program (ABLES): A surveillance system for identifying and preventing cases of elevated blood levels (BLLs) among workers in the United States. Twenty-eight States participated in ABLES in 1999 by collecting laboratory-reported blood lead results and by targeting high-risk industries and occupations, physicians, workers, and worksites for outreach, intervention, and research.

American Association of Poison Control Centers: A nationwide resource that provides information about all aspects of poisoning and refers patients to treatment centers.

Causality: The relating of causes to the effects they produce. Most of epidemiology concerns causality, and several types of causes can be distinguished. It must be emphasized, however, that epidemiologic evidence by itself is insufficient to establish causality, although it can provide powerful circumstantial evidence. A cause is termed "necessary" when it must always precede an effect. This effect need not be the sole result of the one cause. A cause is termed "sufficient" when it inevitably initiates or produces an effect. Any given cause may be necessary, sufficient, neither, or both.

CDC National HIV/AIDS Reporting System (HARS): This Centers for Disease Control and Prevention (CDC) reporting system contains information about U.S. AIDS and HIV case reports, including data by State, metropolitan statistical area, mode of exposure to HIV, sex, race/ethnicity, age group, vital status, and case definition category.

Census of Fatal Occupational Injuries (CFOI): A national census of occupational injury fatalities, including self-employed workers, agricultural workers, and government workers. CFOI uses multiple sources of information such as death certificates, OSHA reports, workers' compensation data, police reports, and newspaper clippings. CFOI program data are collected in cooperation with BLS to ensure that data are comparable among States. States provide data to BLS for inclusion in a national database and maintain their own State databases. Data are currently available for the years 1992–1997.

Glossary

Coal Workers' X-Ray Surveillance Program (CWXSP): An ongoing, congressionally mandated program to provide periodic chest X-rays to working underground coal miners for identifying early pneumoconiosis and facilitating the transfer of affected workers to a job with lower dust concentrations. The program has been in effect since 1970.

Etiology: Literally, the science of causes; causality; in common usage, cause. See also *causality* and *pathogenesis*.

Fatality Assessment and Control Evaluation (FACE): A National Institute for Occupational Safety and Health (NIOSH) field investigation program with two arms: (1) a NIOSH/State cooperative program in which 15 States conduct State censuses of fatal occupational injuries and investigate specific types of these; and (2) a NIOSH intramural program that investigates specific types of fatalities at the request of 5 States. Fatalities specifically investigated by FACE include falls, machinery-related events, and logging fatalities.

Incidence rate: The rate at which new events occur in a population. The numerator is the number of new events that occur in a defined period; the denominator is the population at risk of experiencing the event during this period, sometimes expressed as person-time.

Long latency period: (Synonym: latency.) Delay between exposure to a disease-causing agent and manifestation of the disease. For example, after exposure to ionizing radiation, the average latency period is 5 years before the development of leukemia and more than 20 years before the development of certain other malignant conditions. The term *latent period* or *latency* is often used synonymously with *induction period* (the period between exposure to a disease-causing agent and manifestation of the disease). Latency has also been defined as the period from disease initiation to disease detection. In infectious disease epidemiology, this period corresponds with the period between exposure and onset of infectiousness (which may be shorter or longer than the incubation period).

Median: A measure of central tendency. The simplest division of a set of measurements is into two parts—the lower and the upper half. The point on the scale that divides the group in this way is called the "median."

National Center for Health Statistics (NCHS): A center within CDC that is responsible for the collection, analyses, and dissemination of health statistics. NCHS has two major types of data systems: systems based on population data collected through personal interviews or examinations;

systems based on individual records, with data collected from State and local vital and medical records.

National Center for Infectious Diseases (NCID): A center within CDC whose mission is to prevent illness, disability, and death caused by infectious diseases in the United States and around the world. NCID accomplishes its mission by conducting surveillance, epidemic investigations, epidemiologic and laboratory research, training, and public education programs to develop, evaluate, and promote prevention and control strategies for infectious diseases.

National Electronic Injury Surveillance System (NEISS): A data system maintained by the Consumer Product Safety Commission (CPSC) to monitor consumer-product-related injuries representing a national sample of U.S. emergency departments. In an interagency agreement with NIOSH, NEISS also collects and codes data on all work-related injuries from emergency departments, regardless of consumer product involvement.

National Health and Nutrition Examination Survey (NHANES): An ongoing Federal survey administered by the National Center for Health Statistics (NCHS) to provide researchers with information about the health and nutrition status of the U.S. population, prevalence of selected diseases, and associated risk factors.

National Hospital Ambulatory Medical Care Survey (NHAMCS): A national survey designed to collect data on the utilization and provision of ambulatory care services in hospital emergency and outpatient departments. Findings are based on a national sample of visits to the emergency departments and outpatient departments of approximately 500 noninstitutional general and short-stay hospitals. Annual surveys were begun in 1992.

National Occupational Mortality Surveillance System (NOMS): A mortality statistics database derived from public-use vital statistics data disseminated by the National Center for Health Statistics (NCHS). Since the early 1980s, NIOSH, NCHS, and the National Cancer Institute have supported the collection and coding of decedents' usual occupation and industry information for State vital statistics programs. NOMS uses data from these cooperating States and States that received cooperative agreements through early NIOSH State-based surveillance programs. Usual occupation and industry of the decedent are coded according to the Bureau of the Census classification system. Cause of death is coded according to the World Health Organization's *Manual of the International*

Glossary

Statistical Classification of Diseases, Injuries, and Causes of Death, Based on Recommendations of the Ninth Revision Conference, 1975.

National Surveillance System for Hospital Health Care Workers (NaSH): A surveillance system that focuses on surveillance of exposures and infections among hospital-based health care workers. The purpose of NaSH is to monitor national trends; identify newly emerging hazards for HCWs; assess the risk of occupational infection; and evaluate preventive measures, including engineering controls, work practices, protective equipment, and postexposure prophylaxis to prevent occupationally acquired infections.

National Surveillance System for Pneumoconiosis Mortality (NSSPM): An annually updated pneumoconiosis surveillance system developed by NIOSH. The NSSPM includes information about all U.S. decedents with death-certificate mention of pneumoconiosis since 1968. The system is based on death certificate data files made available annually by the National Center for Health Statistics (NCHS). Records are currently available for more than 100,000 pneumoconiosis decedents; they include information about demographic characteristics, year of death, underlying and contributing causes of death, and (since 1985 for deaths occurring in about half of the States) usual industry and occupation.

National Traumatic Occupational Fatalities Surveillance System (NTOF): A nationwide surveillance system for occupational injury deaths. NTOF is based on death certificates as a sole source of case identification. The system has been estimated to include an average of 81% of all occupational injury deaths nationwide. NTOF data are currently available for 1980 through 1995. NTOF is the most comprehensive source of data on occupational injury deaths before 1992.

Noise-induced hearing loss: A sensorineural hearing loss caused by repeated exposure to high-intensity sound levels. Noise-induced hearing loss is characterized by irreversible damage to the sensory hair cells located within the inner ear. The condition is usually preventable by limiting noise exposures or by using personal hearing protective devices.

Pathogenesis: The postulated mechanisms by which the etiologic agent produces disease. The difference between etiology and pathogenesis should be noted: The etiology of a disease or disability consists of the postulated causes that initiate the pathogenetic mechanisms. Control of these causes might lead to prevention of the disease.

GLOSSARY

Prevalence rate (ratio): The total number of all persons who have an attribute or disease at a particular time (or during a particular period) divided by the population at risk of having the attribute or disease at this point in time or midway through the period. A problem may arise with calculating period prevalence rates because of the difficulty of defining the most appropriate denominator. This is a proportion, not a rate.

Proportionate mortality ratio (PMR): Ratio of the proportion of deaths from a specific cause in an exposed population compared with the comparable ratio in the nonexposed population. For example, the proportion of deaths from disease X in the exposed population could be compared with the proportion of deaths from disease X in the nonexposed population.

Sentinel Event Notification System for Occupational Risk (SENSOR): A NIOSH cooperative agreement with State health departments or other State agencies that develops generalizable, condition-specific strategies for State-based surveillance of occupational diseases and injuries. Efforts have focused on standardization of variables collected by the State programs, creation of software to facilitate adoption of the surveillance systems by additional States, comparison of SENSOR findings to other surveillance data sources, collaboration with the Council of State and Territorial Epidemiologists (CSTE) on building infrastructure for State-based surveillance, further development of State-based hazard surveillance, and publication and dissemination of SENSOR reports.

Surveillance for Tuberculosis Infection in Health Care Workers (*staffTRAK–TB*): CDC recommends periodic tuberculosis (TB) skin testing of health care workers with potential for exposure to *Mycobacterium tuberculosis*. *staffTRAK–TB* was developed to track, analyze, and report demographic, occupation, work location, and multiple TB skin-testing results to determine whether clinically active TB is present.

Surveillance: The systematic, ongoing collection and/or acquisition of information for occupational diseases, injuries, and hazards. Surveillance includes the analysis and interpretation of surveillance data, the dissemination of data or information derived from surveillance to appropriate audiences for prevention and control, and the development of surveillance methodology.

Survey of Occupational Injuries and Illnesses (SOII): An annual survey of a large sample of U.S. employers (approximately 250,000) maintained by the Bureau of Labor Statistics (BLS). The sample is drawn to provide

Glossary

national and State estimates for those States that participate in this Federal/State cooperative program (about 40). The annual survey excludes government workers, the self-employed, and employees of small farms. Employers report information from their injury and illness logs. For employers not required to keep logs, recordkeeping forms are provided at the beginning of the study period.

Toxic Exposure Surveillance System (TESS): A State-based surveillance system for identifying, investigating, and preventing pesticide-related illnesses and injuries. TESS is maintained by the American Association of Poison Control Centers.

Viral Hepatitis Surveillance Program (VHSP): The Hepatitis Branch of the National Center for Infectious Diseases (NCID) operates the Viral Hepatitis Surveillance Program (VHSP), which obtains national surveillance data on clinical, serologic, and epidemiologic data pertaining to risk factors for viral hepatitis.

www.ingramcontent.com/pod-product-compliance
Lightning Source LLC
Chambersburg PA
CBHW081720170526
45167CB00009B/3650